In Vitro Percutaneous Absorption: Principles, Fundamentals, and Applications

Editors

Robert L. Bronaugh, Ph.D.
Supervisory Research Pharmacologist
Division of Toxicological Studies
U.S. Food and Drug Administration
Washington, D.C.

Howard I. Maibach, M.D.
Professor of Dermatology
School of Medicine
University of California
San Francisco, California

CRC Press
Boca Raton Ann Arbor Boston London

Library of Congress Cataloging-in-Publication Data

In vitro percutaneous absorption : principles, fundamentals, and
   applications / editors, Robert L. Bronaugh, Howard I. Maibach.
      p. cm
      Includes bibliographical references.
      Includes index.
      ISBN 0-8493-4748-3
      1. Skin absorption.   2. Skin absorption--Research--Methodology.
   3. Skin--Cultures and culture media.   I. Bronaugh, Robert L., 1942-
      II. Maibach, Howard I.
      [DNLM: 1. Administration, Cutaneous.   2. Models, Biological.
   3. Skin--metabolism.   4. Skin Absorption--physiology.   WR 102 135]
   QP88.5.I46   1991
   612.7'91--dc20
   DNLM/DLC
   for Library of Congress   90-15183
                                                                                CIP

Direct all inquiries to CRC Press, Inc., 2000 Corporate Blvd., N.W., Boca Raton, Florida 33431.

© 1991 by CRC Press, Inc.

International Standard Book Number 0-8493-4748-3

Library of Congress Card Number 90-15183
Printed in the United States

PREFACE

Increasing numbers of investigators are entering the field of percutaneous absorption as interest in dermal exposure to toxic chemicals and transdermal delivery of drugs continues to grow. Frequently *in vitro* methods are selected for permeation measurements because of considerations of cost, desire for human data, and relative ease in obtaining absorption rates. It has also become clear that skin metabolism should be considered in evaluating the pharmacological or toxicological effects of a topically applied chemical. Skin metabolism can be measured accurately only by *in vitro* means since systemic metabolism interferes with *in vivo* biotransformation data.

This book is unique in that it addresses strictly *in vitro* percutaneous absorption issues. Many more variables must be considered in an *in vitro* experiment including: diffusion cell design, receptor fluid, preparation of skin, and temperature. These topics and many others are discussed in detail by recognized experts in the field. Not only is experimental methodology described, but underlying principles are discussed. It is our hope that both new and experienced investigators will benefit from this approach and will be able to more adequately design experiments to suit their own particular needs.

Robert L. Bronaugh
Howard I. Maibach

THE EDITORS

Robert L. Bronaugh, Ph.D. is a Supervisory Research Pharmacologist at the Food and Drug Administration (FDA) in Washington, D.C. He has a B.S. in pharmacy (University of New Mexico) and a Ph.D. in pharmacology (University of Colorado). After postdoctoral work at the University of Colorado Medical School and New York University Medical School, he joined INTERx Research Corporation and became involved in the development of pro-drugs for dermal delivery. Dr. Bronaugh joined the Dermal and Ocular Toxicology Branch of FDA in 1978. His major research interests involve the study of percutaneous absorption and metabolism of cosmetics, drugs, and environmental chemicals. He is a member of the Society of Toxicology, the American Association of Pharmaceutical Scientists, and the Society of Cosmetic Chemists. He serves as a reviewer for journals in the fields of toxicology, pharmacy, and cosmetic science.

Dr. Bronaugh has presented approximately 50 guest lectures at universities and institutes and has over 100 articles published in scientific publications.

Howard I. Maibach, M.D. is Professor of Dermatology at the School of Medicine, University of California, San Francisco.

Dr. Maibach graduated from Tulane University, New Orleans, Louisiana (A.B. and M.D.) and received his research and clinical training at the University of Pennsylvania, Philadelphia. He received an honorary doctorate from the University of Paris Sud in 1988.

Dr. Maibach is a member of the International Contact Dermatitis Research Group, the North American Contact Dermatitis Group, and the European Environmental Contact Dermatitis Group. He has published more than 800 papers and 35 textbooks.

CONTRIBUTORS

Brian W. Barry, Ph.D., D.Sc.
School of Pharmacy
University of Bradford
Bradford, England

Robert L. Bronaugh, Ph.D.
Division of Toxicological Studies
Food and Drug Administration
Washington, D.C.

Daniel A. W. Bucks, Ph.D.
Department of Dermatology
University of California
San Francisco, California

Steven W. Collier, M.S.
Division of Toxicological Studies
Food and Drug Administration
Washington, D.C.

Christopher L. Gummer, D. Phil.
Procter and Gamble
Health and Beauty Care
Egham, England

Richard H. Guy, Ph.D.
Department of Dermatology
University of California
San Francisco, California

Barbara W. Kemppainen, Ph.D.
Department of Physiology and
  Pharmacology
College of Veterinary Medicine
Auburn University
Auburn, Alabama

Kiyoshi Kubota, M.D.
Department of Dermatology
University of California
School of Medicine
San Francisco, California

Howard I. Maibach, M.D.
Department of Dermatology
University of California Hospital
San Francisco, California

William G. Reifenrath, Ph.D.
Division of Ocular Hazards
Letterman Army Institute of Research
Presidio of San Francisco, California

Jim Edmond Riviere, D.V.M., Ph.D.
Cutaneous Pharmacology and Toxicology
  Center
College of Veterinary Medicine
North Carolina State University
Raleigh, North Carolina

Alain Rolland, Pharm. D., Ph.D.
Centre International de Recherches
  Dermatologiques (CIRD) GALDERMA
Sophia Antipolis
Valbonne, France

R. C. Scott, Ph.D.
Department of Metabolism and
  Pharmacokinetics
ICI Central Toxicology Laboratory
Cheshire, England

Jan E. Storm, Ph.D.
Division of Toxicological Studies
Food and Drug Administration
Washington, D.C.

Christian Surber, Ph.D.
Department of Dermatology
University of Basel
Basel, Switzerland

Ronald C. Wester, Ph.D.
Department of Dermatology
University of California
School of Medicine
San Francisco, California

Klaus-P. Wilhelm, M.D.
Department of Dermatology
Medizinische Universität zu Lübeck
Lübeck, Germany

Joel L. Zatz, Ph.D.
Department of Pharmaceutics
Rutgers University
College of Pharmacy
Piscataway, New Jersey

TABLE OF CONTENTS

Chapter 1

PREPARATION OF HUMAN AND ANIMAL SKIN

Robert L. Bronaugh and Steven W. Collier

TABLE OF CONTENTS

I. LIVING OR NONLIVING SKIN

For a number of years *in vitro* absorption studies have been performed using nonliving skin in diffusion cells often with normal saline as the receptor fluid. The term diffusion cell accurately indicated the type of process being measured—the passive diffusion of substrate from one side of the barrier to the other. For many compounds, accurate rates of absorption can be determined in this way. For some compounds that are metabolized by enzymes at a high specific activity, a more accurate picture of absorption may be obtained by maintaining the viability of skin in the diffusion cell as discussed later in this chapter and in this book. The knowledge of the formation of therapeutic or toxic metabolites in skin can be of great value in assessing the effect of the absorption of a compound in contact with skin.

II. HUMAN OR ANIMAL SKIN

Human skin, of course, would be preferable for all permeation studies; but its use is not always possible because of limitations in availability. If one is willing to use only freshly obtained viable human skin (surgical specimens) this further limits supply since the use of cadaver skin would be excluded. Therefore, the use of animal skin is usually necessary at some point in an *in vitro* permeability study. This can best be accomplished by "calibrating" an animal membrane by comparing the absorption of the test compound through the skin of the animal model and through human skin. The skin of a hairless animal is most satisfactory for the preparation of a membrane for an absorption study. As discussed below, hair interferes greatly in the preparation of split-thickness preparations of skin. The skin of the hairless guinea pig and rat are preferable to the skin of the hairless mouse because hairless mouse skin, like all mouse skin, is very thin and therefore much more permeable than human skin.

III. THE BARRIER LAYER

A membrane used in an *in vitro* study should simulate as close as possible the barrier layer in skin. The barrier layer refers to the thickness of skin a compound must diffuse through *in vivo* before being taken up by blood vessels in the upper papillary dermis, and then entering the systemic circulation. This includes the whole epidermis and a small portion of the dermal tissue. This distance certainly varies based on the type of skin used but is typically in the range of 100 to 200 μm.

If additional dermal tissue is present on the skin membrane used in a diffusion experiment, the effects of this tissue on absorption depend on the solubility properties of the chemical. A water soluble substance will diffuse readily through the aqueous dermal tissue and its absorption will only be effected minimally by the presence of additional tissue. However, hydrophobic compounds will diffuse through this tissue very slowly and will therefore appear to be absorbed much more slowly than observed in *in vivo* studies.

IV. FULL- OR SPLIT-THICKNESS SKIN

The use of full-thickness skin is really only justifiable when using animal skin that is already very thin, such as occurs in the mouse[1] (400 μm) or rabbit. With the skin of other animals, such as the rat[2] (800 to 870 μm), guinea pig, monkey, and pig, full-thickness skin is almost 1 mm in thickness; and with human skin it can be several mm thick.[3] Therefore, some means should be used to prepare a membrane that more accurately reflects the barrier layer in thickness. As mentioned above, this is particularly important when examining a hydrophobic compound.

V. PREPARATION OF SPLIT-THICKNESS SKIN

A. DERMATOME

The use of the dermatome is the best way of preparing biological membranes for percutaneous absorption studies. Unlike other methods which will be described, a dermatome can be used with hairless or haired skin and it can be used without adversely effecting the viability of the membrane. For the last eight years we have prepared membranes almost exclusively with a dermatome[4,5] — the Padgett Electodermatome (Padgett Instruments, Kansas City, MO).

Full-thickness skin from a human subject or an animal is pinned (epidermis side up) to the surface of a block for cutting. Styrofoam is convenient because it can be readily shaped to the desired size by cutting with a knife and because pins for anchoring the skin can be easily punched into this material. The width of the styrofoam block must be less than the cutting edge of the dermatome blade so that the blade can rest on the surface of the skin. The piece of skin should overlap the edges of the block so that it can be attached with the pins to the side of the block (so that the pins are out of the way of the dermatome blade).

The depth of the cut is controlled with a lever on the side of the dermatome head with the indicated calibrations being in thousandths of an inch. The actual thickness of the membrane obtained is a result of the pressure applied and the angle of holding the dermatome as it is pushed across the skin. We have found it helpful to check the thickness of each membrane prepared with the dermatome by using a micrometer (Mitutoyo micrometer, 0.01 to 9 mm, L. A. Benson Inc., Baltimore, MD). With a little practice it becomes possible to make skin sections of reproducible thickness.

B. SEPARATION AT THE DERMAL/EPIDERMAL JUNCTION
1. Heat Separation

Elevation of temperature has been used for a number of years to loosen the bond between the epidermis and dermis for the preparation of epidermal sheets. Baumberger et al.[6] placed full-thickness human skin on a hot plate for 2 min at a temperature of $50°C$ and found that the epidermis could easily be removed from the dermis using blunt dissection. Heated water is used more commonly for this separation because of better temperature control which leads to more reproducible results.[7,8] Water is heated in a beaker to $60°C$. Full-thickness human skin (underlying muscle and fascia removed) is suspended in the water with forceps for 30 s. The epidermis is then removed from the dermis using forceps to gently pull the epidermis away. The ease of separation varies somewhat from specimen to specimen. The dissection can be facilitated by pinning the skin to a block so that it is stretched tight.

Heat separation of skin is useless, however, for preparing a membrane for absorption studies with haired skin. The hair shafts remain in the dermis causing holes to be created in the epidermal membrane as it is pulled from the dermis. Therefore, only the skin from hairless animals can be separated using this procedure. A different length of exposure to heat may be required for separation of skin from different species.

The effect of heat on the viability of skin would be of concern in absorption/metabolism studies. The enzymatic hydrolysis of diisopropyl fluorophosphate in human skin appeared to be inactivated during the heat separation procedure.[3] However, separation of epidermis from dermis using $60°C$ heat for 30 s resulted in a reduction of aryl hydrocarbon hydroxylase activity of only 10 to 15%.[9]

2. Chemical Separation

The separation of epidermis from dermis has been achieved after soaking full-thickness human and animal skin in different chemical solutions. Exposure times are usually in hours and so viability of skin is likely lost.

The effects of $2M$ solutions of various salt anions and cations on separation of the epidermis was investigated with human skin.[10] The separation of the epidermis was achieved by acids and bases at such pHs that cause swelling of the collagen. The most potent anions were the bromide, thiocyanate, and iodide ions; while the acetate, sulfate, and citrate ions, were found to be ineffective in causing separation.

The advantage of chemical separation is that it appears to be effective in separating the epidermis from dermis of haired animals under certain conditions. The barrier properties remain unaltered because the hair shaft comes out of the dermis during separation and stays in place in the epidermis. Scott and co-workers[11] report that after soaking 28-day-old Wistar rat skin in $2M$ sodium bromide for 24 h, epidermal membranes could be separated from the dermis. However, this procedure was ineffective with skin of older rats (7 to 8 weeks of age).

3. Enzyme Separation

A few studies are reported in the literature concerning the separation of the epidermis and dermis by incubation of skin in enzyme preparations. The protease dispase was shown to produce an epidermal sheet that could be easily peeled from the dermis of human skin after a 24-h incubation at 4°C.[12] A crude bacterial collagenase was found effective in the preparation of epidermal sheets at a concentration of 0.1 and 0.2% after a 3-h incubation at 37°C.[13] Unlike dispase separation, most of the cells were found to be nonviable following separation, but differences in temperature of the incubations may be responsible. The use of the enzymes pancreatin and trypsin for epidermal-dermal separation has also been described.[14] The referenced articles should be consulted for more details.

VI. BARRIER INTEGRITY OF MEMBRANES

In studies, involving the use of human cadaver skin for diffusion measurements, we examined the use of a standard compound (tritiated water) for assessing the integrity of the barrier layer.[15] A barrier check should be done even with newly obtained cadaver skin and it should also be repeated following storage. The ^{3}H-water permeability of human cadaver skin was measured initially and then skin was stored in an airtight plastic freezer bag at -20°C. After various time periods, the skin was thawed and the barrier properties rechecked with ^{3}H-water (Table 1). Significant increases in water permeability were seen with the skin of two donors (no. 39 and 40) after 2 months of storage. In most cases of storage ranging from 2 to 12 months, water permeability did not change, indicating a maintenance of the barrier properties.

^{3}H-Water permeation has often been measured by applying an excess of compound to the surface of the skin and measuring the steady state rate of permeation; such as by taking at least 4 to 5 measurements at hourly intervals following application of ^{3}H-water. A permeability constant can be determined by dividing this rate by the initial concentration of applied material.

We observed that a much more rapid evaluation of water permeability could be made in what we called the 20 min test. Using diffusion cells with an exposed area of skin of 0.32 cm^2, 100 ul of ^{3}H-water (approximately 0.3 uCi) was applied to the surface of the skin so that it was completely covered. The tops of the cells were occluded with Parafilm. After 20 min, the unabsorbed material was blotted from the surface of the skin with a cotton-tipped applicator and then the surface of the skin was rinsed once with distilled water. Effluent from the flow cell was collected for an additional 60 min, at which time radioactivity in the effluent had returned to baseline. ^{3}H-water absorption was expressed as the percent of the applied dose absorbed. A good correlation ($r = .98; p < 0.01$) was obtained between values measured with the 20-min test and with permeability constant values, both obtained from the same skin samples (Figure 1).

TABLE 1
Effect of Length of Frozen Storage on Water Permeation[a]

Skin donor no.	Initial Kp × 10³ (cm/h)	Final Kp × 10³ (cm/h)	Storage length (months)
34	0.98 ± 0.13	0.84 ± 0.13	2
39	1.98 ± 0.08	3.48 ± 0.58[b]	2
40	1.22 ± 0.28	2.02 ± 0.05[b]	2
41	1.43 ± 0.15	1.57 ± 0.06	2
36	1.05 ± 0.19	1.36 ± 0.22	5
26	0.93 ± 0.03	1.35 ± 0.23	12
29	2.13 ± 0.09	1.89 ± 0.10	12
30	1.81 ± 0.10	2.24 ± 0.26	12
32	0.97 ± 0.08	1.06 ± 0.07	12

[a] Values are the mean ± SE of 3 to 4 determinations.
[b] Statistically different from initial Kp value by Student t test, $p < 0.05$.

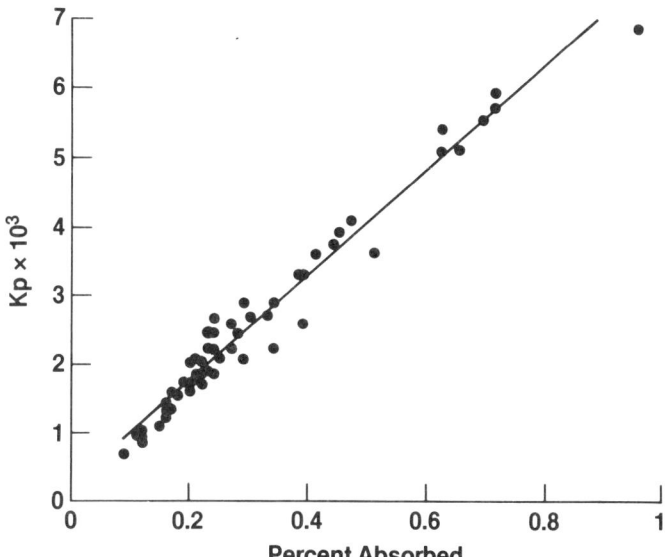

FIGURE 1. Correlation of results from two methods of measuring water permeation through skin: permeability constant (K_p) values and the percent dose absorbed (20 min test).

REFERENCES

1. **Behl, C. R., Flynn, G. L., Kurihara, T., Smith, W. M., Bellantone, N. H., Gataitan, O., and Higuchi, W. I.,** Age and anatomical site influences on alkanol permeation of skin of the male hairless mouse, *J. Soc. Cosmet. Chem.*, 35, 237, 1984.
2. **Yang, J. J., Roy, T. A., Mackerer, C. R.,** Percutaneous absorption of benzo(a)pyrene in the rat: comparison of *in vivo* and *in vitro* results, *Toxicol. Ind. Health,* 2, 409, 1986.
3. **Loden, M.,** The *in vitro* hydrolysis of diisopropyl fluorophosphate during penetration through human full-thickness skin and isolated epidermis, *J. Invset. Dermatol.*, 85, 335, 1985.

4. **Bronaugh, R. L. and Stewart, R. F.,** Methods for *in vitro* percutaneous absorption studies. III. Hydrophobic compounds, *J. Pharm. Sci.,* 73, 1255, 1984.

5. **Bronaugh, R. L. and Stewart, R. F.,** Methods for *in vitro* percutaneous absorption studies. VI. Preparation of the barrier layer, *J. Pharm. Sci.,* 75, 487, 1986.

6. **Baumberger, P. J., Suntzeff, V., and Cowdry, E. V.,** Methods for the separation of epidermis from dermis and some physiologic and chemical properties of isolated epidermis, *J. Natl. Cancer Inst.,* 2, 413, 1942.

7. **Scheuplein, R. J.,** Mechanism of percutaneous absorption. I. Routes of penetration and influence of solubility, *J. Invest. Dermatol.,* 45, 334, 1965.

8. **Bronaugh, R. L., Congdon, E. R., and Scheuplein, R. J.,** The effect of cosmetic vehicles on the penetration of *N*-nitrosodiethanolamine through excised human skin, *J. Invest. Dermatol.,* 76, 94, 1981.

9. **Mukhtar, H. and Bickers, D. R.,** Drug metabolism in skin, *Drug. Metab. Dispos.,* 9, 311, 1981.

10. **Felsher, Z.,** Studies on the adherence of the epidermis to the corium, *J. Invest. Dermatol.,* 8, 35, 1947.

11. **Scott, R. C., Walker, M., and Dugard, P. H.,** *In vitro* percutaneous absorption experiments: a technique for the production of intact epidermal membranes from rat skin, *J. Soc. Cosmet. Chem.,* 37, 35, 1986.

12. **Kitano, Y. and Okada, N.,** Separation of the epidermal sheet by dispase, *Brit. J. Dermatol.,* 108, 555, 1983.

13. **Hentzer, B. and Kobayasi, T.,** Enzymatic liberation of viable cells of human skin, *Acta Dermatovener (Stockholm),* 58, 197, 1978.

14. **Omar, A. and Krebs, A.,** An analysis of pancreatic enzymes used in epidermal separation, *Arch. Derm. Res.,* 253, 203, 1975.

15. **Bronaugh, R. L., Stewart, R. F., and Simon, M.,** Methods for *in vitro* percutaneous absorption studies. VII. Use of excised human skin, *J. Pharm. Sci.,* 75, 1094, 1986.

Chapter 2

DIFFUSION CELL DESIGN

Christopher L. Gummer and Howard I. Maibach

TABLE OF CONTENTS

I. INTRODUCTION TO DIFFUSION CELL DESIGN

The design of an experiment must be totally dependent on the overall aims of the study. This is standard dogma. However, many skin penetration experiments start with ''Let's see how much penetrates....'' and there ends the design. Often, little consideration is given to the prevailing *in vivo* conditions which the experiment may be used to mimic, yet the data may be used to predict the penetration of powerful drugs and toxic substances or even safety-in-use profiles. It is surprising, therefore, that when *in vitro* skin penetration experiments are conducted the investigator doesn't give just a little more thought to the experimental design. It is only in recent years that a concerted effort has been made to understand those parameters of diffusion cell design that affect the penetration of the molecule under study.[2]

The aims of *in vitro* skin penetration experiments are as follows:

1. To predict reliably and quantitatively the penetration of a molecule through the skin under typical *in vivo* conditions, e.g., from creams, lotions, drug delivery devices, or for toxicology and safety-in-use assessments.
2. To study the penetration of molecules through the skin under carefully controlled standard conditions for the investigation of penetration kinetics vs. molecular structure.

In this chapter I will not debate which cell, i.e., horizontal (Figure 1) or vertical cell (Figure 2 and 3), flow-through (Figures 2 and 3A and B) or static (Figure 3C and D) cell should be used. Unfortunately this is often a matter of personal preference. Instead I will highlight those areas of cell design that must be considered when starting skin penetration studies.

Ideally, the diffusion cell experiment will predict exactly, skin penetration. However, one finds that the results give only a good indication of the expected penetration. This does not usually result from poor cell design but rather the failure to control the large number of variables and from expecting too much from a simple cell design.

II. MATERIALS SUITABLE FOR CELL CONSTRUCTION

Glass is usually the material of choice. It is easy to shape, easily cleaned, and resistant to almost all chemicals. Few molecules show a high affinity for glass. It also has the major benefit of being completely transparent so you can see what's going on inside. Many experiments have been conducted with large bubbles in either the donor or receptor compartments. Other materials such as perspex (plexiglass), Teflon, and metal tend to show high levels of adsorption of the penetrating molecule and are often very difficult to clean and observe.

It should be standard procedure to check whether the molecule under investigation sticks to the cell body or its component pieces.

III. THE DONOR COMPARTMENT

This is the first of three compartments in the diffusion cell and shows the greatest difference between vertical and horizontal diffusion cell design. The aim of this compartment is to deliver the molecule in a way that is representative of the proposed *in vivo* conditions. Areas such as the vagina, rectum, eyelid, etc. may all be potential delivery sites and each has its own special environment that can affect penetration kinetics. Even variations between skin sites should be considered.[1] This compartment must allow ease of application of the penetrant in an appropriate form, e.g., liquid, cream, controlled delivery patch, etc. However, it must also provide for control of the ambient conditions if comparisons between

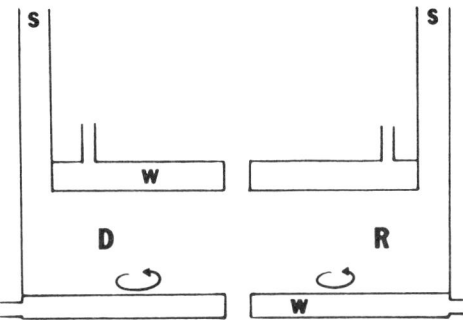

FIGURE 1. Elevation of a horizontal diffusion cell. D, donor compartment; R, receptor compartment; S, sample ports, W, water jacket. The skin sample is placed in a vertical plane relative to the cell body. (From Gummer, C. L., in *Transdermal Drug Delivery, Developmental Issues and Research Initiatives*, Hadgraft, J. and Guy, R. H., Marcel Dekker, New York, 1989, 117. With permission.)

vehicle formulations, or if structure/function relationships are to be studied. For example, experiments run through the night will be subject to large variations in ambient temperatures compared to those during the day. This is particularly important where volatile ingredients, including water, are involved. Reifenrath et al.[8] have given considerably more attention to the donor compartment owing to the volatile nature of the penetrants under study. They have shown in a series of experiments that both increased temperature (20 to 32°C) and increased relative humidity in the donor compartment will promote penetration. Elevation of the pH was confined to increasing the penetration of hydrophilic compounds. Similarly Gummer and Maibach[4] showed that simple occlusion of the donor compartment resulted in increased penetration of both ethanol and methanol and at the same time appeared to promote back diffusion through the skin of liquid from the receptor compartment.

Horizontal cells lend themselves well to the control of ambient conditions as they represent a closed environment using heated water jackets. They also allow the penetrant formulation to be stirred. Therefore, the cells are very suitable for studies such as the penetration of homologous molecular series under standard conditions. However, it is well established that increasing skin hydration increases penetration. One must question their suitability if the ultimate aim is to deliver the penetrant from a small volume of thinly spread cream or from a volatile lotion. In contrast, there is probably little difference in skin hydration and the resultant penetration between a horizontal cell and a well-established drug delivery patch or occlusive dressing.

Few donor compartments are designed with the facility to wash the skin. Yet, *in vivo*, many drug formulations may be applied directly after washing with a surfactant which is known to change the barrier properties of the skin. Although not normally a problem, ignoring this area may underestimate the penetration of potentially toxic compounds.

IV. THE BARRIER COMPARTMENT

Typically, the barrier is skin although synthetic materials may be used.[6] The main concerns are that the skin is held tightly, thus preventing leaks, and is not damaged when the cell is assembled. The author has found increased penetration in some vertical cells when the donor formulation has contacted the area of skin compressed between the two cell halves. This must be of even greater concern in horizontal cells where the whole of the barrier surface, including the area compressed between the two cell halves, is in constant contact with the donor phase. The actual method of clamping the cell and preventing leaks needs

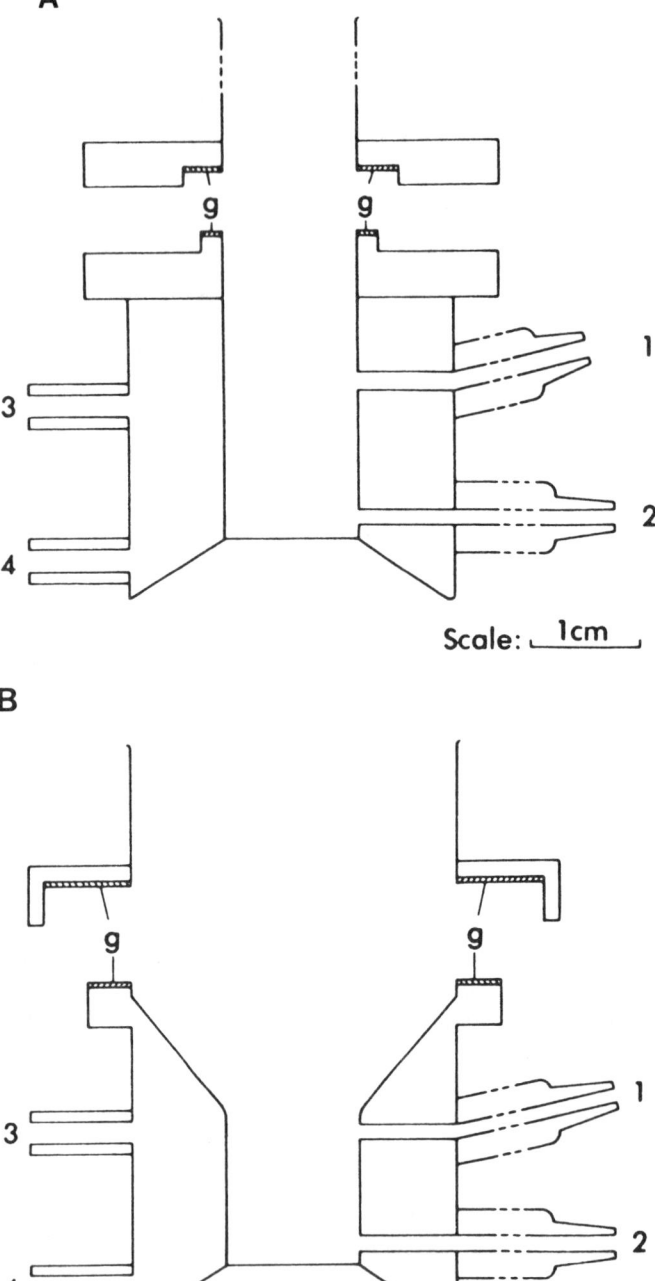

FIGURE 2.   Elevation of vertical diffusion cells. Cell A is designed to hold mouse skin; cell B, to hold guinea pig skin. Both cells have flow-through receptor compartments. The barrier is retained between the ground glass junction (g); 1 and 2, receptor compartment ports; 3 and 4, water jacket ports. (From Gummer, C. L., in *Transdermal Drug Delivery, Developmental Issues and Research Initiatives*, Hadgraft, J. and Guy, R. H., Marcel Dekker, New York, 1989, 117. With permission.)

careful thought. Uniform pressure around the cell should be the aim although most simple pinch clamps rarely achieve this. Depending on the skin type, e.g., hairless mouse, additional O-rings may be used to help prevent leaks. However, the author has found that these are unnecessary if the joint between the two cell halves is suitably designed for the skin type, i.e., with opposing surfaces perfectly flat and a wide area of compression. Rebates built into the two adjoining surfaces (Figure 2) have been found useful but they must not cut through the skin.[3] In addition, O-rings need cleaning and may show an affinity for the molecule under study. Although unproven, it is also reasonable to assume that the skin should be stretched to a similar degree as that found *in vivo*. The author has found that this is easily achieved by using cells with an overlapping top (Figure 2b).[3] As the top is pressed into place it gently pulls down on the edge of the skin giving reasonably uniform tension in all directions. This appears to be an improvement on cells with directly opposing edges where it is difficult to maintain a uniform skin tension as the cell is assembled. Consideration must also be given to the great range of skin thickness found between animal species. A cell designed for hairless mouse skin will never properly accommodate guinea pig skin.[3] Of equal importance, though rarely considered, is the maintainance of the barrier properties.[7] This is discussed more fully in Chapter 5. Typically the main emphasis has been on minimizing the area of skin used while maximizing ease of use. It must be remembered that the skin is a living organ, capable of metabolizing many of the molecules under investigation. It also has a very complex biochemistry yet little attempt is made to keep it alive. This may not be so unreasonable when it is well established that the principal barrier is the stratum corneum. However, molecules such as 5-fluorouracil are readily metabolized and despite penetrating rapidly into skin show little evidence of penetration in a standard experiment[8] simply because they don't penetrate through the skin into the receptor compartment. There is therefore, a need to consider the quality of the barrier, particularly for compounds that are targeted at the skin for dermatoses or malignant diseases or for molecules that can be expected to be metablized by the skin. Also, for toxicological studies, 'dead skin' may produce erroneous results.

Temperature control of the barrier is not normally considered to be a problem due to its close proximity to heated receptor compartments. However, based on the experiments of Reifenrath et al.[9] which show that temperature above the skin does indeed have an effect, it may need more consideration than one may expect if *in vivo* predictions are to be made.

One must also remember that the barrier has two sides, each with its own special needs that should, where possible, be maintained. The skin surface is covered in a lipophilic film of sebum and epidermal lipids[7] whereas the underside is bathed in aqueous interstitial fluids. Molecules may find preferential penetration in skin samples carefully prepared by cleaning the surface with alcohol, swabs thus removing the lipid barrier just prior to assembling the diffusion cell.

V. THE RECEPTOR COMPARTMENT

This is the third compartment of the diffusion cell and has achieved the bulk of attention in diffusion cell design.[3] In both vertical and horizontal cells it is easily temperature controlled and stirred. It is here that glass has the advantage over other materials in cell design. It is essential that the investigator can have a clear view of the underside of the barrier in order to remove air bubbles and to observe signs of barrier breakdown, common in long duration experiments.

The overiding aim of most investigators has been to make this compartment as small as possible to reduce the volumes of receptor phase handled and in turn increase the relative concentration of the penetrating molecule. This minimizes the number of samples and increases the accuracy of assay techniques. However, it is vital that infinite sink conditions

A

B

FIGURE 3.   Typical diffusion cells with tops removed and filled with water containing a crystal of potassium permanganate. Boundary layers have formed in C and D. The improved receptor compartments (A and B) show complete stirring. (From Gummer, C. L., in *Transdermal Drug Delivery, Developmental Issues and Research Initiatives*, Hadgraft, J. and Guy, R. H., Marcel Dekker, New York, 1989, 117. With permission.)

are maintained so that diffusion into this compartment does not become the rate limiting step. It is always worthwhile calculating the solubility of the molecule under study, particularly when many show very low solubility in physiological saline, the most commonly used receptor fluid.

Surprisingly, not all receptor phases are well stirred. Gummer et al.[3] clearly showed that boundary layers were established in standard cell receptor compartments. Normally these would be considered to adversely affect the diffusion of the molecule into the receptor fluid as they are known to have quantifiable effects in standard partitioning experiments. However, the author has analyzed data from each of the diffusion cells shown and reached the same prediction of *in vivo* penetration. The only real difference in the data is that better stirring, combined with better overall cell design results in less variable data and hence reduces the number of samples and increases the confidence in the results. As a method of

C

D

FIGURE 3 (continued).

qualifying adequate stirring in new diffusion cells the author adds a crystal of potassium permanganate to the receptor fluid. Stirring is deemed to be adequate if the whole of the fluid turns mauve within 30 s. For the majority of cells, if it doesn't happen immediately, it won't happen at all. In general, stirring efficiency is defined by a ratio of receptor phase height to diameter of the compartment with better stirring in short, squat cells compared to a tall compartment of equal volume. It should not be assumed that a small volume equals efficient stirring.

A typical problem of the receptor compartment is that in an effort to reduce the volume the available area of barrier is also reduced. We have overcome this by designing a funnel shaped compartment that maximizes available skin area and minimizes the receptor volume.[3]

Temperature control of the receptor compartment is essential in order to avoid the effects of changes in ambient conditions that will affect the kinetics of penetration through the skin, and also the solubility of the penetrant in the receptor fluid. The whole of the compartment should be temperature controlled. This is most easily achieved with flow through water jackets. The practice of placing cells in heated blocks makes access difficult. In addition, heating blocks rarely provide uniform heating. While it is predicted that temperature will directly affect penetration it is difficult to find the definitive publication that addresses the issue of "at what temperature should such experiments be conducted". One would predict that temperatures of 32 to 37°C would be the nearest to the *in vivo* situation. However, an attempt must be made to record the skin temperature if experiments are to be repeated and compared.

A. SAMPLING FROM THE RECEPTOR COMPARTMENT

Both horizontal and vertical diffusion cells lend themselves to easy sampling of the penetrant from the receptor phase. However, the sampling apparatus is a direct extension of the receptor compartment and must therefore, receive serious consideration, particularly in the choice of materials used and how the sample is handled. For example, there is little point building the cell from glass, then collecting the penetrant through highly absorptive polyethylene tubing and then dripping the collected solution into open vials on a fraction collector to allow evaporation. The author knows to his cost that up to 25% of the collected dose may be lost by evaporation from a fraction collector overnight.[4] It is, therefore, well worth conducting a simple preexperiment of spiking a sample of receptor fluid and leaving it uncovered overnight to establish whether any loss will occur during the experiment.

B. TWO SAMPLING METHODS AVAILABLE TO THE INVESTIGATOR

1. Aliquot Sampling

Aliquot sampling is the method which is typical of static diffusion cells. The main problems are high labor input, irregular sampling intervals, and disturbing the experiment. Ideally, aliquot size is kept to a minimum, e.g., 100 µl and the receptor compartment replenished after sampling. However, the trend towards smaller receptor compartments means that the aliquot represents an increasing percentage of the total volume. Then, depending on the interval between samples, the concentration and hence diffusion kinetics into the receptor may change. All of these changes should, where possible, be factored into the final kinetic calculations. Of greater importance, particularly in vertical cells, is the chance introduction of air bubbles under the barrier which will alter the area of skin in contact with the receptor phase. Similarly, in cells that show poor stirring, the aliquot may not be representative of the whole receptor phase.

2. Continuous Sampling

This sampling is typical of flow-through cell designs. The main attraction of this method is that once established, the experiment can be left unattended to run its course. Disadvantages are the comparatively large overall receptor phase volume which includes the total volume pumped through the cell, long runs of sampling tubing, establishing suitable flow rates to maintain infinite sink conditions, and to keep the experiment running. It has been shown[5] that varying the flow rate between 2 to 7.5 ml/h has no significant effect on the penetration of benzo(a)pyrene. It should not, however, be accepted carte blanche that the flow rate in the experimenter's own system will not affect penetration. It should be understood that flow-through cells do not result in a complete exchange of receptor fluid at each time interval. They may, however, be more indicative of the *in vivo* situation, where the penetrating molecule is continually removed from the skin by the microvasculature.

VI. RECOMMENDATIONS FOR DIFFUSION CELL DESIGN

1. All materials used in the construction of the experiment should be assessed for their ability to adsorb the molecule undertest.
2. The cell must be easy to handle, easy to use, and robust enough for routine work.
3. Donor compartment should reflect the prevailing *in vivo* conditions.
   A. Easy access to deliver the penetrant to the barrier and to allow washing or replenishment
   B. Stirred where possible (if applicable)
   C. Temperature- and humidity-controlled
   D. Control of evaporation for volatile vehicles and penetrants
4. Barrier compartment
   A. Suitable to hold the barrier in place
   B. Prevent leaks
   C. Prevent excessive barrier damage during assembly
   D. Designed for the skin type under study
5. Receptor compartment
   A. Either flow-through or static
   B. Temperature-controlled
   C. Sufficient volume to maintain infinite sink conditions
   D. Stirred without obvious formation of boundary layers
   E. Permit observation of the underside of the barrier
   F. Allow collection and assay of the penetrant

In conclusion, the *in vitro* experiment and its results are easily manipulated; hence, the investigator must pay close attention to the design of the diffusion apparatus in order to produce the highest quality and most applicable data. The operating conditions must be rigorously controlled and be at least representative of the conditions that exist at the *in vivo* site for which the data is to be applied.

There are a number of high quality, commercially available diffusion cells which, when operated correctly will give first class results. New ones should only be developed if they improve the ease of operation, improve the quality of data and its relevance to the *in vivo* situation, or it can show the quantifiable control of variables known to affect the penetration of molecules through the skin.

REFERENCES

1. **Feldman, R. J. and Maibach, H. I.,** Regional variation in percutaneous penetration of [^{14}C] cortisol in man, *J. Invest Dermatol.,* 48, 181, 1967.
2. **Gummer, C. L.,** *In vitro* evaluation of transdermal delivery, in *Transdermal Drug Delivery, Developmental Issues and Research Initiatives,* Hadgraft, J. and Guy, R. H., Eds., Marcel Dekker, New York, 1989, 117.
3. **Gummer, C. L., Hinz, R. S., and Maibach, H. I.,** The skin penetration cell: a design update. *Int. J. Pharmaceut.,* 40, 104, 1987.
4. **Gummer, C. L. and Maibach, H. I.,** The penetration of ^{14}C methanol through excised guinea pig skin *in vitro, Food and Chem. Toxicol.,* 24, 4, 305, 1986.
5. **Holland, J. M., Kao, J. Y., and Whitaker, M. J.,** A multi sample apparatus for kinetic evaluation of skin penetration *in vitro, Toxicol. Appl. Pharmacol.,* 72, 272
6. **Nacht, S. and Yeung, D.,** Artificial membranes and skin permeability, in *Percutaneous Penetration,* Bronaugh, R. L. and Maibach, H. I., Eds., Marcel Dekker, New York, 1985, 373.
7. **Wertz, W. W. and Downing D. T.,** Stratum corneum: biological and biochemical considerations, in *Transdermal Drug Delivery, Developmental Issues and Research Initiatives,* Hadgraft, J. and Guy, R. R., Eds., Marcel Dekker, New York, 1989.
8. **Gummer, C. L. and Maibach, H. I.,** Unpublished data.

Chapter 3

A FLOW-THROUGH DIFFUSION CELL

Robert L. Bronaugh

TABLE OF CONTENTS

I. BACKGROUND

Although many different designs have been utilized for diffusion cell studies, there are really only two basic types: the one-chambered and the two-chambered cell. Each type has its own utility in percutaneous absorption studies.

Variations of the two-chambered cell have been used for years to create conditions in which the diffusion of a compound in solution can be measured from one side of the membrane to the other.[1] An infinite dose (one that is large enough to maintain constant concentration during the course of an experiment) is added to one side of the membrane and its rate of diffusion across a concentration gradient into a solution on the opposite side is determined. Usually the solutions on both sides of the membrane are stirred to ensure uniform concentrations. Studies comparing permeation through skin to Fickian diffusion through a membrane are performed in this fashion. The two-chambered cell is useful for studying mechanisms of diffusion through skin. It also is applicable to the measurement of absorption from drug delivery devices where compounds are applied to skin at an infinite dose and a steady-state rate of delivery is desired.

The exposure of skin to permeating substances usually occurs under conditions that are different from those created in the two-chambered cell. Some substances are intentionally applied to skin in creams or lotions during the use of drug and cosmetic products. Other chemicals, often of toxicological interest, come in contact with skin in a wide variety of vehicles in our environment. Often, the amount of penetrating substance on the surface of the skin is relatively small (finite dosage), and as permeation proceeds, a steady-state rate of absorption is not attained. In these examples, absorption of the chemicals through skin can only be studied in a one-chambered cell. The surface of the skin in this type of cell is open to the environment, so that thin layers of material can be applied in vehicles relevant to *in vivo* exposure. The skin is not excessively hydrated by continued exposure to an aqueous solution as in the two-chambered cell. The chamber beneath the skin serves as a container for the receptor fluid that is continually stirred; samples are taken through a side-arm for subsequent determination of rates of absorption. If desired, infinite doses can also be applied to the skin in the one-chambered cell for determination of steady-state absorption kinetics. Finite dose techniques and the design of a static diffusion cell were described by Franz.[2]

A flow-through cell system[3] was introduced to automate sample collection from a one-chambered cell. It also facilitates the maintenance of viability of skin since the physiological receptor fluid is continually replaced. Receptor fluid eluting from the diffusion cells is automatically collected in a fraction collector.

Special attention may be necessary in measuring the permeability of highly volatile compounds when the skin is not occluded to prevent evaporation. The short walls on the tops of some diffusion cells can protect the skin surface from air currents and it has been suggested that this protection may be responsible for differences between *in vivo* and *in vitro* results.[4,5] Diffusion cells have been designed to collect evaporating material above the surface of the skin.[6,7] These cells have proven particularly useful in studies of the effectiveness of mosquito repellents and in studies of volatile compounds that require mass balance determinations.

II. FLOW CELL DESIGN

Important features in the design of a flow-through cell are discussed under the following headings: receptor volume, construction material, maintenance of physiological temperature, ease of assembly of skin, and mixing of receptor contents.

A. RECEPTOR VOLUME

Maybe the most important single feature of flow cell design is the volume of the diffusion cell receptor. The volume of the receptor must be small (<0.5 ml) so that it can be completely flushed out during sample collection intervals with a manageable volume of receptor fluid. As a general rule, the flow rate should be about 5 to 10 times that of the volume of the receptor, i.e., a 0.5 ml receptor requires a flow rate of at least 2.5 ml/h. Too often investigators modify existing one-cell chambers such as the Franz cell that have receptor volumes as large as 5 ml, and then pump receptor fluid through the cell at 1 or 2 ml/h. Clearly the receptor contents are not rapidly removed from the cell following absorption through the skin. The time course of absorption will be skewed to the right with absorption of a chemical appearing to occur at a later time than it actually happened. The importance of a small receptor was also stressed by Barry in his discussion of flow cells.[8] In our laboratory, the receptor fluid is pumped beneath the skin through a chamber with a volume of only 0.13 to 0.26 ml (depending on skin surface area of the diffusion cell). This small volume allows the receptor contents to be rapidly and completely flushed out with flow rates of 1.5 ml/h or greater.

B. CONSTRUCTION MATERIAL

The main goal here is to have cells made of a material that does not bind or retain test compounds during the course of a study. Some initial flow cells prepared by us were made of plexiglass because of its transparent properties. We discovered that radiolabeled compounds diffused into this porous material and could not be washed off. The compounds would leach out into the receptor fluid of subsequent experiments confounding the results. Glass and Teflon are preferred for construction of cells because of their inert properties. The flow-through cells used in our laboratory are machined from Teflon and are fitted with a glass window in the bottom to facilitate viewing of the receptor contents and thereby verifying the absence of air bubbles. Some very lipophilic molecules have been found to adhere to the surface of the cells but can easily be removed by soaking overnight in water.

C. MAINTENANCE OF PHYSIOLOGICAL TEMPERATURE

Skin surface temperature in a diffusion cell during an absorption/metabolism study should be maintained at a physiological temperature (about 32°C). This can be accomplished in a number of ways including: (1) placing cells in a heated holder; (2) jacketing the cells and running heated water through the compartment; (3) placing cells in an environmental chamber heated to the correct temperature.

D. EASE OF ASSEMBLY OF SKIN

Skin samples must be easy to assemble in a suitable flow cell. Most commonly, diffusion cell halves have been clamped together in some way to assemble the cell prior to use. In the cell developed by Bronaugh,[3] the skin membrane is placed in the lower portion of the cell and the cap is screwed into place. The inside portion of the cap swivels so that the skin is not twisted during the tightening process. This design was chosen because of the small size of the receptor.

E. MIXING OF RECEPTOR CONTENTS

Stirring bars are commonly used to mix the receptor contents of many types of diffusion cells. However, flow cells with small receptor volumes are unique. The receptor fluid flowing through the cell may be sufficient to provide adequate mixing. If so, additional mixing may not be performed since the small receptor volume makes mixing with stirring bars more difficult to achieve. We have found that for water-soluble compounds, adequate mixing in our flow cell is achieved without stirring.[3] For water-insoluble compounds, additional mixing

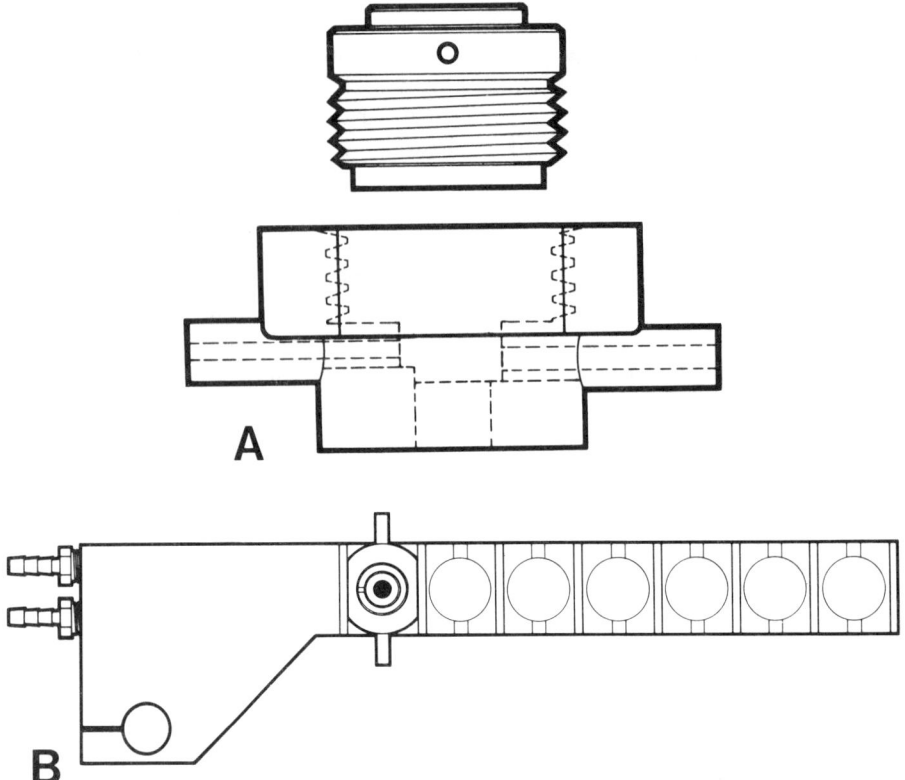

FIGURE 1.   Flow-through cell and holding block. (A) Cross section of diffusion cell. (B) Aluminum holding block used to position cells over vials in fraction collector and to maintain the cells at a physiological temperature.

promotes further partitioning of material from the skin into the receptor fluid. This is discussed further under Section IV.

III. VALIDATION OF FLOW-THROUGH CELL

The validation of the flow-through cell (Figure 1) used in our laboratories has been reported.[3] A minimum flow rate is required for good mixing and for removing absorbed compound in a rapid manner. But flow rates above this minimum (in our case about 1.5 ml/h) resulted in no change in absorption rate. An exception to this might be with the use of compounds that have limited solubility in the receptor fluid. Crutcher and Maibach[9] found that absorption of testosterone and testosterone propionate varied with the flow rates utilized in their system, probably because they had not yet reached the critical rate of perfusing their relatively large surface area of exposed skin (4.5 cm[2]). The two compounds have very limited water solubility, another possible explanation for the increase in absorption with increase in flow rate.

The accuracy of data from the flow-through cell was initially determined by comparing the results to those obtained in the standard static diffusion cell system. Good agreement was found between both types of cells for both absorption profiles (Figures 2 and 3) and numerical comparisons (Table 1) for the absorption of water, cortisone, and benzoic acid. This is evidence of good mixing and rapid removal of absorbed material from the flow-through cell receptor.

Diffusion cell absorption values were then compared with results from *in vivo* studies

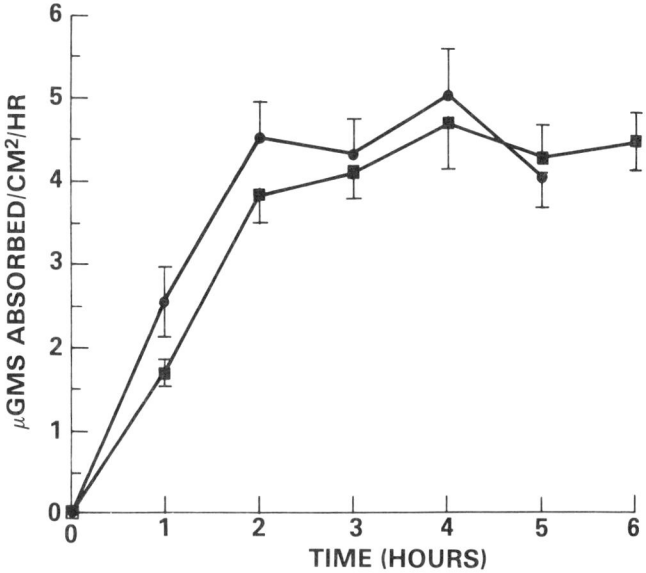

FIGURE 2. Comparison of tritiated water absorption in the flow-through
(■) and static (●) cells.

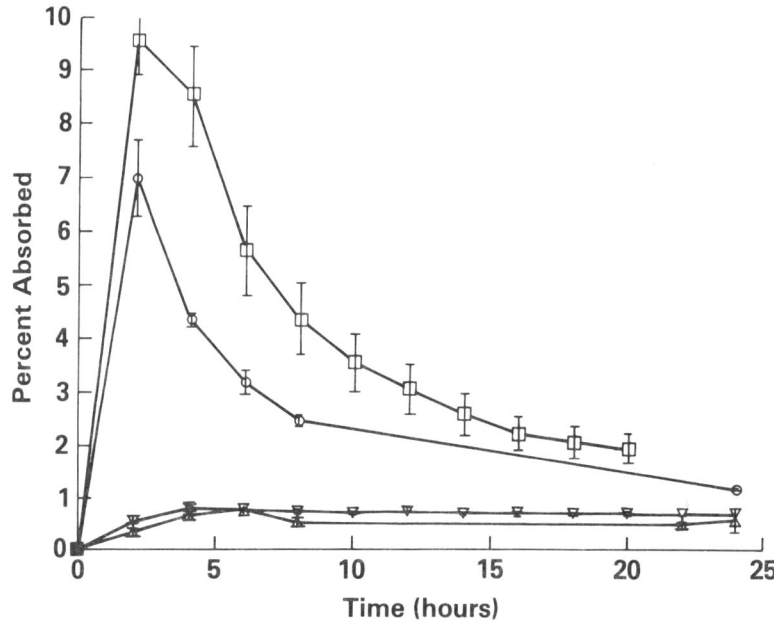

FIGURE 3. Comparison of cortisone and benzoic acid absorption in flow-through
and static cells. Flow-through cell: ▽ = cortisone, □ = benzoic acid; static cell: △
= cortisone, ○ = benzoic acid.

in rats. The *in vivo* absorption of cortisone and benzoic acid in a petrolatum vehicle was
similar to that obtained in the two types of diffusion cells (Table 2).

IV. MODIFICATIONS

The absorption of water-insoluble compounds is difficult to measure in diffusion cells
because of their lack of free partitioning from skin into the diffusion cell receptor fluid.

TABLE 1
Comparison of Flow-Through and Static
Diffusion Cells[a]

| Compound | Absorption[b] | |
	Flow-through	Static
Water	4.3 ± 0.4 (5)	4.4 ± 0.2 (5)
Cortisone	8.5 ± 0.9 (5)	6.3 ± 0.8 (8)
Benzoic acid	45.9 ± 7.6 (5)	48.6 ± 3.8 (6)

[a] Values are the mean ± SE of the number of determinations in parentheses. [^{3}H]Cortisone and [^{14}C]benzoic acid were applied in an acetone vehicle, [^{3}H]water in a water vehicle. The values obtained for each compound in the two types of cells were not significantly different from each other when compared by the two-tailed Student's t test, $p < 0.05$.

[b] For cortisone and benzoic acid, absorption is expressed as the percentage of the applied dose absorbed in 24 h. For water, the steady-state rate of absorption of the radiolabeled molecule is given ($\mu g/cm^2/h$).

TABLE 2
Comparison of *In Vivo* and *In Vitro* Absorption (Percentage
Applied Dose Absorbed)[a]

| Compound | *In vivo* | *In vitro* | |
		Flow cell	Static cell
Cortisone	19.6 ± 1.3 (4)	20.1 ± 1.1 (6)	22.8 ± 2.7 (5)
Benzoic acid	37.0 ± 2.8 (8)	28.3 ± 3.0 (6)	35.5 ± 5.2 (5)

[a] Values are the Mean ± SE of the number of determinations in parentheses. Compounds were applied in a petrolatum vehicle. The values obtained for each compound by the three methods were not significantly different from each other when compared by the two-tailed Student's t test, $p < 0.05$.

Increased mixing of the receptor contents beneath skin can facilitate this partitioning process.[10] Small magnetic stirring bars can be added to the receptor prior to assembly of cells. These bars can then be rotated with motor-driven magnets placed underneath the heating blocks (Crown Glass Company, Somerville, NJ).

The effluent from diffusion cells should be collected in a fraction collector immediately after leaving the cell so that the time of sample collection closely corresponds to the time of penetration through skin. We have found this easy to achieve by allowing the effluent from cells to drip into 20-ml scintillation vials in a fraction collector. To accommodate magnetic stirring devices, the cells must be moved several inches higher above the fraction collector. Effluent side-arms for cells have been made with a right-angle positioned downward to more accurately direct the drops from the cells into the collecting vials.

The ideal area of exposed skin in a flow-through cell can be determined by several factors. A small area of skin ($< 1 \ cm^2$) allows a small receptor beneath the skin. This allows a minimum amount of receptor fluid to flush out the absorbed material into the collecting vials. A large area of skin is more useful for attaching drug delivery devices and increasing the sensitivity of measurements of absorbed material and enzyme reactions. We have selected cells with an exposed surface area of skin = 0.64 cm^2. For slow penetrating compounds and for enzymic reactions at low levels, radioisotopic or other sensitive techniques may be required for quantitation.

V. CONCLUSIONS

For many percutaneous absorption studies, the use of static or flow-through cells would yield the same results. The ease of use of a flow cell combined with the ability to obtain round-the-clock sampling are attractive to many investigators. The viability of skin can also be maintained in a flow-through cell.[11] It would be difficult to provide the nutrients required for skin viability if the skin was assembled in a static cell.

REFERENCES

1. **Scheuplein, R. J.,** Mechanism of percutaneous absorption. I. Routes of penetration and the influence of solubility, *J. Invest. Dermatol.,* 45, 334, 1965.
2. **Franz, T. J.,** On the relevance of *in vitro* data, *J. Invest. Dermatol.,* 64, 190, 1975.
3. **Bronaugh, R. L. and Stewart, R. F.,** Methods for *in vitro* percutaneous absorption studies. IV. The flow-through diffusion cell, *J. Pharm. Scio.,* 74, 64, 1985.
4. **Bronaugh, R. L. and Maibach, H. I.,** Percutaneous absorption of nitroaromatic compounds: *in vivo* and *in vitro* studies in the human and monkey, *J. Invest. Cermatol.,* 84, 180, 1985.
5. **Bronaugh, R. L., Stewart, R. F., Wester, R. C., Bucks, D., Maibach, H. I., and J. Anderson, J.,** Comparison of percutaneous absorption of fragrances by humans and monkeys, *Fd. Chem. Toxicol.,* 23, 111, 1985.
6. **Spencer, T. S., Hill, J. A., Feldmann, R. J., and Maibach, H. I.,** Evaporation of diethyltoluamide from human skin *in vivo* and *in vitro, J. Invest. Dermatol.,* 72, 317, 1979.
7. **Reifenrath, W. G. and Robinson, P. B.,** *In vitro* skin evaporation and penetration characteristics of mosquito repellents, *J. Pharm. Sci.,* 71, 1014, 1982.
8. **Barry, B. W.,** Methods for studying percutaneous absorption, in *Dermatological Formulations: Percutaneous Absorption,* Marcel Dekker, New York, 1983, 234.
9. **Crutcher, W. and Maibach, H. I.,** The effect of perfusion rate on *in vitro* percutaneous penetration, *J. Invest. Dermatol.,* 53, 264, 1979.
10. **Bronaugh, R. L., Collier, S. W., and Stewart, R. F.,** *In vitro* percutaneous absorption of a hydrophobic compound, *Toxicologist,* 9, 241, 1989.
11. **Collier, S. W., Sheikh, N. M., Sakr., A., Lichtin, J. L., Stewart, R. F., and Bronaugh, R. L.,** Maintenance of skin viability during *in vitro* percutaneous absorption/metabolism studies, *Toxicol. Appl. Pharmacol.,* 99, 522, 1989.

Chapter 4

INDIVIDUAL AND REGIONAL VARIATION WITH *IN VITRO* PERCUTANEOUS ABSORPTION

Ronald C. Wester and Howard I. Maibach

TABLE OF CONTENTS

I. INTRODUCTION

Percutaneous absorption in man and animals *in vivo* shows individual differences and regional variation. Feldmann and Maibach[1,2] were the first to show this in human volunteers, and the concept has been shown to be true for animals.[3,4] *In vivo* percutaneous absorption variation has been ascribed to a multiplicity of events such as skin thickness, blood flow, lipid content, number of hair follicles, etc. Some of these factors are changed when skin is removed from the body and placed in a diffusion cell. Therefore the existence of individual variation and regional variations *in vitro* remain questionable.

II. HUMAN SKIN INDIVIDUALISM

Table 1 presents the results on *in vitro* percutaneous absorption of a test article in three different human skin sources. The data were summarized over five different formulations (each run in all three skin sources). What is of interest is the receptor fluid accumulation in human skin source #2. This is illustrated in Figure 1, and what it shows is that the barrier properties of human skin source #2 were such that no skin absorption occurred. A formulation comparison with only that human skin source would have provided completely negative data.

Table 2 gives the *in vitro* percutaneous absorption of pentadeconic acid from two formulations (A and B) in two human skin sources. Skin content is illustrated in Figure 2, and it shows that formulations were only distinguishable in one of the human skin samples. This suggests that decisions based on only one human skin source may be misleading.

Table 3 gives the *in vitro* percutaneous absorption of taurocholic acid. The data are summarized over five formulations and individualized for each human skin source and repeated skin absorption for each of the skin sources. What is of interest is the receptor fluid content (24-h accumulation) and the totally different results in skin source #3 compared to the other two skin sources (#1 and #2). Again, individual variations can be significant, but this only becomes apparent when other skin sources are also used.

Table 4 compares the *in vitro* percutaneous absorption of DDT and benzo(a)pyrene in two different human skin sources. Skin content differences should be noted for both individual skin source and for the chemical penetrating into human skin.

III. HUMAN SKIN REGIONAL VARIATION

Table 5 compares the *in vitro* permeability of coumarin, griseofulvin, and propranolol across human abdominal skin and scalp skin.[5] For coumarin and propranolol, the scalp showed higher permeability. *In vivo* in man, Feldmann and Maibach[1,2] identified the scalp as a higher absorbing area; so these *in vitro* results of Ritschel et al.[5] agree with the *in vivo* results.

IV. ANIMAL SKIN REGIONAL VARIATION

Bronaugh[6] reported the effect of body site (back vs. abdomen) on male rat skin permeability. Abdominal rat skin was more permeable to water, urea, and cortisone. Skin thickness (stratum corneum, whole epidermis, whole skin) is less for the abdomen than for the back (Table 6). With the hairless mouse, Behl et al.[7] showed dorsal skin to be more permeable than abdominal skin (reverse that of the male rat). Hairless mouse abdominal skin is thicker than dorsal skin (also reverse that of the male rat).

TABLE 1
Individual Variation in *In Vitro* Percutaneous Absorption

Human skin source	Skin	Percent dose (Mean ± SD)		Total recovery
		Surface wash	Receptor fluid	
#1	5.0 ± 2.4	85.7 ± 7.8	2.5 ± 4.5	93.2 ± 6.0
#2	3.2 ± 2.8	83.7 ± 9.5	0.3 ± 0.2	87.7 ± 8.1
#3	2.6 ± 1.0	72.1 ± 12.3	4.4 ± 5.0	79.0 ± 12.5

Note: Skin sources are from (#1) 68-year-old white male abdominal, (#2) 69-year-old white male thigh, and (#3) 33-year-old white male abdominal.

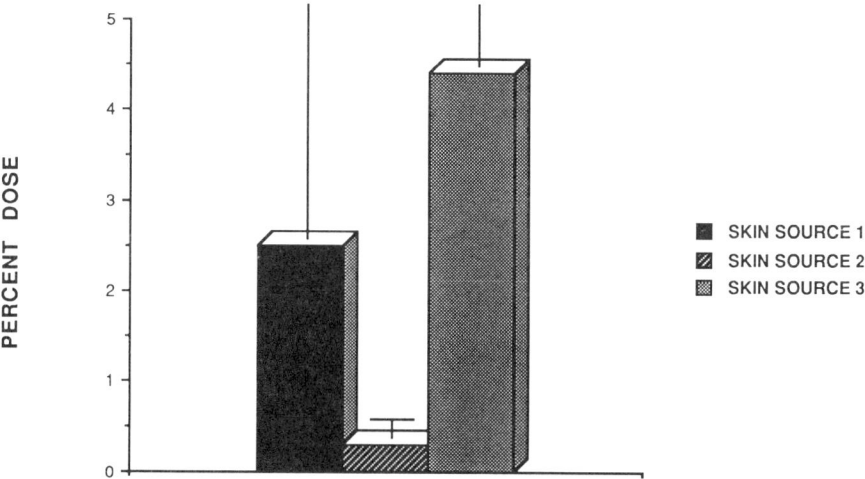

FIGURE 1. Receptor fluid accumulation of a test article for 24-h *in vitro* diffusion through three different human skin sources. Please note that skin source #2 yielded no skin permeability.

TABLE 2
In Vitro Percutaneous Absorption of Pentadeconic Acid

Skin source	Formulation	Skin	Percent dose (Mean ± SD)		Total recovery
			Surface wash	Receptor fluid	
#1	A	5.4 ± 0.8	80.4 ± 8.9	0.05 ± 0.04	85.9 ± 9.5
	B	6.4 ± 1.8	85.6 ± 5.2	0.15 ± 0.03	92.1 ± 4.5
#2	A	5.8 ± 1.5[a]	86.2 ± 6.6	0.03 ± 0.03	92.1 ± 5.2
	B	14.2 ± 23.0[a]	80.0 ± 5.3	0.03 ± 0.01	94.2 ± 4.3

Note: Skin sources are from (#1) 43-year-old white female thigh and (#2) 7-year-old Hispanic female thigh.

[a] Statistically different ($p = 0.006$).

V. CONCLUSION

Individual variation and regional variation exists for *in vivo* percutaneous absorption, and as shown in this chapter, the same variability exists for *in vitro* percutaneous absorption. For human skin this can be critical because of human skin availability, and the tendency to perhaps only use one human skin source to conserve supply. We recommend, if possible, the use of multiple human skin sources (Table 7 *in vitro* study design).

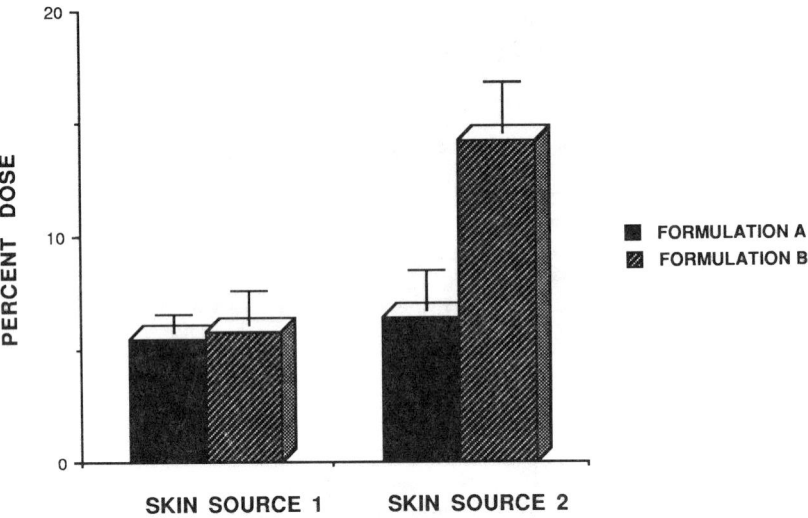

FIGURE 2.   Pentadeconic acid human skin content following *in vitro* diffusion from two formulations in two human skin sources. Note that formulations were distinguished in only one human skin source.

TABLE 3
In Vitro Percutaneous Absorption of Taurocholic Acid

Skin source	Skin content[a]	Receptor fluid content[a]
#1	0.5 ± 0.3	8.9 ± 12.9
#1 Repeat	1.3 ± 1.3	8.2 ± 6.0[b]
#2	1.4 ± 1.8	16.5 ± 18.0
#3 Repeat	3.4 ± 6.3	11.7 ± 14.3
#3	1.4 ± 1.5	0.2 ± 0.3
#3 Repeat	0.8 ± 0.8	0.3 ± 0.7[b]

Note: Skin sources are from (#1) 41-year-old white male thigh; (#2) 32-year-old white male thigh; and (#3) 53-year-old white male thigh.

[a]   Mean percent dose ± SD; N = 5.
[b]   Significant difference ($p < 0.02$).

TABLE 4
Individual Human Skin Source Variation in *In Vitro* Percutaneous Absorption

Human skin source	Skin content[a]	
	DDT	Benzo(a)pyrene
#1	6.7 ± 1.5[b]	27.4 ± 10.9
#2	29.6 ± 7.1[b]	20.0 ± 8.6

Note: Skin sources are from (#1) 41-year-old white male thigh and (#2) 21-year-old white female thigh.

[a]   Mean percent dose ± SD (N = 3) of skin content following 24-h *in vitro* percutaneous absorption study.
[b]   Significant difference ($p = 0.005$).

TABLE 5
In Vitro Permeability Across Human Abdominal and Scalp Skin

Drug	Steady-state flux (mg/cm²/h)	
	Abdominal skin	Scalp skin
Coumarin	130 ± 78[a]	172 ± 64[2]
Griseofulvin	10 ± 6	16 ± 8
Propranolol	30 ± 6[a]	42 ± 11[2]

[a] Significant difference ($p < 0.05$) abdominal vs. scalp skin.

From Ritschel, W. A., Sabouni, A., and Hussain, A. S., *Meth. Findings Exp. Clin. Pharmacol.*, 11, 643, 1989. With permission.

TABLE 6
Effect of Body Site on Rat Skin Permeability

Compound	Permeability constant (cm/h $\times 10^4$)
Water	
Back	4.9 ± 0.4
Abdomen	13.1 ± 2.1
Urea	
Back	1.6 ± 0.5
Abdomen	18.8 ± 5.5
Cortisone	
Back	1.7 ± 0.4
Abdomen	12.2 ± 0.6

From Bronaugh, R. L., in *Percutaneous Absorption,* Bronaugh, R. and Maibach, H., Eds., Marcel Dekker, New York, 267, 1985. With permission.

TABLE 7
In Vitro Study Design

Human skin source	Number of replicates	
	Formulation #1	Formulation #2
A	4	4
B	4	4
C	4	4
Total	12	12

Note: Assay: reservoir fluid; skin; skin surface wash; apparatus wash; material balance.

From Bronaugh, R. L., in *Percutaneous Absorption,* 2nd ed., Bronaugh, R. and Maibach, H., Eds., Marcel Dekker, New York, 653, 1989. With permission.

REFERENCES

1. **Feldmann, R. J. and Maibach, H. I.,** Regional variation in percutaneous penetration of [^{14}C]cortisol in man, *J. Invest. Dermatol.,* 48, 181, 1967.
2. **Maibach, H. I., Feldmann, R. J., Milby, T. H., and Serat, W. F.** , Regional variation in percutaneous penetration in man, *Arch. Environ. Health,* 23, 208, 1971.
3. **Wester, R. C., Noonan, P. K., and Maibach, H. I.,** Variation on percutaneous absorption of testosterone in the rhesus monkey due to anatomic site of application and frequency of applications, *Arch. Dermatol. Res.,* 267, 229, 1980.
4. **Moody, R. P. and Franklin, C. A.,** Percutaneous absorption of the insecticides fenitrothion and aminocarb, *J. Toxicol. Environ. Health,* 20, 209, 1987.
5. **Ritschel, W. A., Sabouni, A., and Hussain, A. S.,** Percutaneous absorption of coumarin, griseofulvin and propranolol across human scalp and abdominal skin, *Meth. Findings Exp. Clin. Pharmacol.,* 11, 643, 1989.
6. **Bronaugh, R. L.,** Determination of percutaneous absorption by *in vitro* techniques, in *Percutaneous Absorption,* Bronaugh, R. and Maibach., H., Eds., Marcel Dekker, New York, 267, 1985.
7. **Behl, C. R., Bellantone, H. H., and Flynn, G. L.,** Influence of age on percutaneous absorption of drug substances, in *Percutaneous Absorption,* Bronaugh, R. and Maibach, H., Eds., Marcel Dekker, New York, 183, 1985.

Chapter 5

RECEPTOR FLUIDS

Steven W. Collier and Robert L. Bronaugh

TABLE OF CONTENTS

I. INTRODUCTION

The investigators of *in vitro* percutaneous absorption are confronted with a variety of models and techniques at their disposal. The ability to choose the most suitable model or technique to answer a question and properly interpret the results of an experiment comes from an understanding of the process of percutaneous absorption and the properties of the *in vitro* experiment. A common criticism of *in vitro* percutaneous absorption data is the lack of correlation of some compounds to *in vivo* data. This is often caused by inadequate methodology and/or improper assessment of results. The selection of an appropriate receptor fluid used in an *in vitro* percutaneous absorption study is crucial to obtaining meaningful data and properly interpreting them. This section explains the process of *in vitro* percutaneous penetration in terms of partitioning between skin and the receptor fluid and how the selection and evaluation of the receptor fluid is dependent on the goals and constraints of the study.

II. DIFFUSION AND PARTITIONING

The investigation of skin as a diffusional barrier is often conducted *in vitro*. These studies are typically conducted with infinite dosing techniques in two-chambered diffusion cells[1] with identical solvents in both chambers. The chambers are generally of large enough volume so that the donor compartment concentration is not measurably depleted during the study and the compound diffusing into the receptor compartment is not appreciably concentrated (sink conditions). With this arrangement, a steady-state flux can be obtained and the diffusivity (D) and permeability constant (k_p) for the compound calculated from Fick's equation:

$$J_s = \frac{D_m \Delta C_m}{L} = k_p C_o \qquad (1)$$

where J_s is the steady-state flux; D_m, the membrane diffusivity; ΔC_m, the concentration difference across the membrane; L, the length of the membrane; and C_o, the donor compartment concentration. If the donor and receptor chambers are well stirred, the boundary diffusion layers between vehicle and skin and between receptor and skin are minimized. The penetrating compound must partition into the membrane from the vehicle, and later, out of the membrane and into the receptor fluid. The membrane concentrations cannot be assumed to be equal to the concentrations of the vehicle and receptor phases.[2] The membrane concentration on the donor side (C_{vm}) is dependent on the vehicle concentration and the vehicle/membrane partition coefficient (K_{vm}):

$$C_{vm} = K_{vm} C_o \qquad (2)$$

On the receptor side, the membrane concentration (C_{mr}) is related to the concentration in the receptor phase (C_r) by the membrane/receptor partition coefficient (K_{mr}):

$$C_{mr} = K_{mr} C_r \qquad (3)$$

The concentration difference across the membrane in Equation 1, C_m, is equal to $C_{vm} - C_{mr}$. Substituting Equations 2 and 3 into 1 gives the partitioning dependent form of Fick's equation:

$$J_s = \frac{(K_{vm} C_o - K_{mr} C_r) D_m}{L} \qquad (4)$$

At steady state with identical vehicle and receptor solutions used ($K_{vm} = K_{mr} = K$) and sink conditions ($C_r = 0$), the equation simplifies to the form:

$$J_s = \frac{KC_oD_m}{L} = k_pC_o \qquad (5)$$

While Equation 5 is useful for determining the physicochemical parameters of a membrane, the experimental conditions employed are not strictly relevant to exposure and absorption conditions of environmental contaminants, cosmetic compounds, and topical drug products. When exposures are low and absorption rapid, a steady state may not be reached. It is also unlikely that K_{vm} will equal K_{mr}.

The test compound in *in vitro* and *in vivo* studies is often applied to skin in low volumes of a volatile vehicle such as acetone (approximatley 10 μl/cm^2). The volatile solvent vehicle delivers the test compound into solution in the surface stratum corneum lipids and then rapidly evaporates without appreciably altering the barrier properties of the skin membrane. Dosage by this technique is reported as surface density (μg/cm^2). In this manner, partitioning from vehicle to membrane is eliminated. Equation 4 is reduced to:

$$J_s = \frac{(C_o - K_{mr}C_r)D_m}{L} \qquad (6)$$

For the *in vivo* case, the partitioning denoted by K_{mr} is the mass transfer from the epidermis, through the basement membrane, into the capillary plexus of the papillary dermis where it is removed by the *in vivo* receptor fluid, blood. The *in vivo* mass transfer is given by $l\phi$, where l is the diffusional distance, and ϕ is the basal blood flow. Most compounds are rapidly removed by the capillary circulation and the term $l\phi$ is small in comparison to the term D_m.[3,4] Under these conditions, the percutaneous absorption of a compound is solely determined by its exposure concentration and its diffusivity in the skin membrane. *In vitro* studies which simulate these conditions utilize one-chamber static or flow-through diffusion cells as discussed in Chapters 3 and 4.

III. SKIN STRUCTURE AND THE *IN VITRO* RECEPTOR

Skin is a laminated structure whose layers impart its structural and protective functions. The outer layer, stratum corneum, is a highly compacted layer of terminally differentiated keratinocytes cabled together at desmosomal, tight, and gap junctions and suspended in an excreted lipoidal matrix interspersed with membrane-coating granules. Adnexal appendages erupt through the surface and provide thermoregulatory and sensory functions. The complex structure of the stratum corneum imparts with it some complex diffusional properties. The physicochemical properties of the penetrating compound such as lipophilicity, polarity, size, and solubility determine diffusional paths through the stratum corneum. Much evidence exists to classify the routes of absorption into polar pathways, nonpolar pathways, and shunt diffusion. Electrolyte diffusion through the skin is very small. The lipoidal nature of the stratum corneum gives diffusional preference to compounds which are most soluble in it.

Below the stratum corneum lies the viable layer of the epidermis. It is a continuum of cells replicating at the lowest layer, stratum germinativum, and differentiating as they migrate outward to renew the stratum corneum. As these daughter cells of the stratum germinativum differentiate, they extrude their cytoplasm and eventually die as they become stratum corneum. While these cells are still viable, they contain sufficient cytoplasm and are surrounded by the interstitial fluid necessary to support their metabolic needs. In contrast to the lipoidal nature of the stratum corneum, the viable epidermis is of much greater aqueous character.

Compounds which are absorbed through the stratum corneum encounter this different environment as they diffuse toward the papillary dermis. For very lipophilic compounds, the greatest diffusional resistance may be encountered in the viable epidermal tissue. For less lipophilic compounds, it is the stratum corneum which limits the rate of diffusion.

Connective tissue comprises the bulk of the dermis. This lies below the epidermis to which it is attached by a basement membrane. The dermis structurally supports the epidermis and, through its capillary circulation, provides oxygen and nutrients and removes waste products (and percutaneously absorbed material) by diffusion through the basement membrane. The dermal/epidermal junction is not a smooth interface. Rather, it is marked by rete pegs, projections of epidermis into the dermis, and hair follicles, which are deeper invaginations of epithelium into the dermis.

The results of *in vitro* studies are heavily dependent on the source of skin and the manner in which it is prepared (Chapter 1). Full-thickness skin is removed from the donor and used as is prepared by scraping away subcutaneous tissue or preparing a dermatome section. The thickness of the skin preparation becomes the diffusional distance the compound must travel to the receptor phase and not the distance from the surface to the capillary plexus since the capillary circulation is lost. Exceptions to this are *in vitro* models which perfuse receptor solutions through a skin flap with an isolated vasculature. Skin flap perfusion as a technique for the study of dermatotoxicology, physiology, percutaneous absorption, and metabolism has been demonstrated with cat,[5] dog,[6] human,[7] and pig skin.[8]

In studies conducted with excised skin, the residual dermal and subcutaneous tissue can be extremely important. This is a relatively lipophilic domain of considerable mass on most skin preparations. The thickness of the epidermis on humans may average 100 μm and be thinner on most other model species. A 200 μm thick dermatome section is usually only obtainable on sparsely haired skin and then only through practiced technique. Half or more of this comparatively thin section still consists of dermis. The absorbed compound must partition between each membrane layer before finally partitioning into the receptor fluid.

The partitioning of two very lipophilic compounds — acetyl ethyl tetramethyl tetralin (AETT) and benzo[a]pyrene (B[a]P) — and a more hydrophilic compound — estradiol — between tissue and receptor fluids was studied. Dermatomed skin was enzymatically separated and the dermal and epidermal portions were dosed with the test compounds and equilibrated with HEPES-buffered Hanks' balanced salt solution (HHBSS); HHBSS + 4% BSA; and 50:50 ethanol/water receptor solutions. The volume of receptor solution chosen is equivalent to that perfused beneath a skin section in a flow-through diffusion cell for 24 h at a rate of 1.5 ml/h (36 ml). The dosed tissue disks were incubated in the receptor solution at 37°C for 24 h and the receptor/tissue partition coefficient determined. In general, partitioning between epidermis — receptor, and dermis — receptor was very similar. The addition of 4% BSA to HHBSS improved partitioning an average of 3- to 13-fold for all compounds. The use of 50:50 EtOH/H$_2$O increased partitioning 13- to 200-fold depending on tissue and compound. Figure 1 shows the effect of the receptor fluids on the partitioning of AETT, B[a]P, and estradiol. The partitioning of the two most lipophilic compounds, AETT and B[a]P, favored the epidermis and dermis when HHBSS was used as the receptor. This is reflected as a negative log P (the receptor/tissue partition coefficient). The addition of 4% BSA to HHBSS shifts the partitioning to one favoring the receptor fluid. The use of 50% ethanol (EtOH) in water increased the partitioning of AETT and B[a]P; however, it dehydrates the epidermal and dermal tissue, a condition incompatible with viability of the membrane and functioning of most cutaneous metabolic functioning. Estradiol partitioning favored the receptor fluid in all three cases. The addition of 4% BSA increased partitioning into HHBSS but the use of 50% ethanol in water did not significantly increase partitioning. Ionizable salts are used in receptor fluids, cell and tissue culture media, and other physiological buffers to provide pH control, physiologic osmolarity, and appropriate extracellular ionic concen-

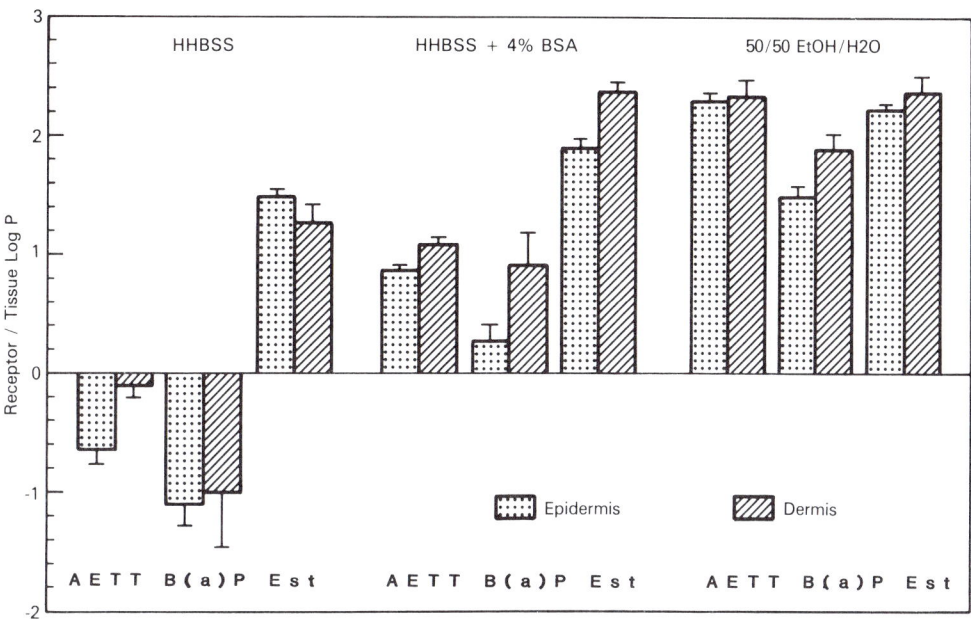

FIGURE 1. Log tissue/receptor fluid partition coefficients for partitioning of AETT, B[a]P, and estradiol (Est) between epidermal and dermal tissue sections and HHBSS, HHBSS + 4% BSA, and 50:50 ethanol/water receptor solutions. Dermatome sections (200 μm) of hairless guinea pig skin were obtained and enzymatically separated by floating (dermis side down) on the surface of a collagenase/dispase containing balanced salt solution for 70 min and an 80-min balanced salt solution rinse. Disks (8 mm diameter) were produced from the epidermal and dermal sheets with a steel punch, patted dry and allowed to air dry for 20 min. Approximately 1 to 3 μg of AETT, B[a]P, or estradiol in 10 μl MeOH were pipetted onto the dermal disks and the stratum corneum side of the epidermal disks. The MeOH vehicle evaporated rapidly and the disks were incubated in 36 ml of the receptor solutions. Values are the Means ± SEM of three determinations.

trations for maintaining cellular membrane potentials. The presence of electrolytes in a receptor fluid decreases the solubility (and therefore the partitioning) of penetrating none-lectrolyte compound by the "salting-out" effect. Though partitioning is typically greater from the dermal disks, the dermis remains an appreciable reservoir for these compounds.

The dermal tissue on skin sections used in *in vitro* percutaneous absorption studies can serve as a sink for lipophilic compounds which do not partition readily from the dermis into the receptor fluid. In general, the thicker the skin section used, the smaller the quantity of lipophilic test compound which will appear in the receptor fluid. Failure to consider this effect has led to some investigators' inability to correlate *in vitro* with *in vivo* studies. If only cumulative absorption data are needed, the addition of the material recovered in the receptor fluid with that remaining in the skin at the end of the experiment gives the total absorption during the experiment.[9] The skin absorption of the two water-insoluble compounds, AETT and 1,1'-(2,2,2-trichloroethylidene)*bis*[4-chlorobenzene] (DDT), was measured by using normal saline and PEG 20 oleyl ether (0.5%) receptor fluids (Table 1). Both compounds partitioned poorly into the saline receptor fluid and much more readily into the surfactant solution, which resulted in a more accurate value for absorption. For both compounds, however, values obtained by combining the receptor fluid and skin contents (a measure of the total compound absorbed) for the two receptor fluids were in much better agreement. No significant difference was found in the AETT values. This technique has been applied to the measurement of the percutaneous absorption of B[a]P.[10] Following perfusion of skin sections with Eagle's modified minimal essential media (MEM) with 10% fetal bovine serum (FBS) in flow-through diffusion cells for 24 h, the unabsorbed surface

TABLE 1
Percutaneous Absorption of AETT and DDT in the Fuzzy Rat
(% Applied Dose Absorbed)[a]

Compound and receptor fluid	Receptor fluid contents	Skin contents	Total absorbed
AETT			
Normal saline (5)	1.2 ± 0.1	15.3 ± 1.4	16.5 ± 1.4
PEG 20 oleyl ether (5)	15.1 ± 0.9	5.9 ± 0.3	22.8 ± 3.0[b]
DDT			
Normal saline (5)	2.1 ± 0.1	54.8 ± 2.6	56.9 ± 2.6
PEG 20 oleyl ether (4)	66.0 ± 5.2	4.6 ± 0.9	70.6 ± 4.9

[a] Values are the Mean ± SEM of the number of determinations in parentheses. AETT was applied in a petrolatum vehicle (5 mg/cm²), and DDT was applied in acetone (15 μl/cm²).

[b] No significant difference from corresponding normal saline value (*t* test, *p*<0.05).

Adapted from Bronaugh, R. L. and Stewart, R. F., *J. Pharm. Sci.*, 75, 487, 1986.

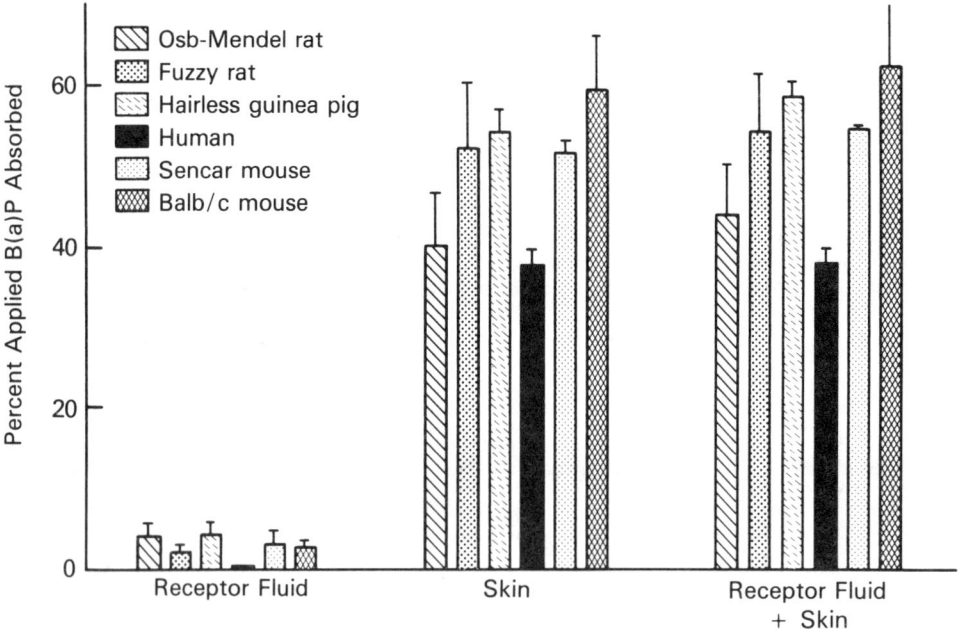

FIGURE 2.   Percentage of applied B[a]P absorbed in 24 h by skin in flow-through diffusion cells. Applied dose was 3 μg B[a]P/cm² in 10 μl acetone. Values from two to five skin samples from each experimental subject were averaged and constituted one experiment. Values are Means ± SEM of three to four experiments. Rat, guinea pig, and human skin were 200 μm dermatome sections; mouse skin was full thickness. The asterisk indicates a value significantly different from those for hairless guinea pig, fuzzy rat, and Balb/c mouse (LSD test, *p*<0.05) and different from Sencar mouse (LSD test, *p*<0.10).[9]

material was removed by washing with a 1% commercial detergent solution. The receptor fluid and skin levels were determined, summed, and compared for species differences (Figure 2). This method gives good agreement with *in vivo* results.[11,12] Failure to account for the B[a]P remaining in the skin at the end of the experiment results in a serious underestimation of the percutaneous absorption of the compound.

IV. MEASURING PENETRATION RATES *IN VITRO*

Ideally, *in vitro* studies should provide the investigator with kinetic data on rates of penetration and the absorption profile. With these data, the effects of different exposure conditions on pharmacokinetic parameters can be studied. For lipophilic compounds which form appreciable reservoirs in the viable epidermis and dermis, two approaches can be used to obtain kinetic data. If several diffusion experiments are run simultaneously using similar skin membranes from the same source, the experiments can be concluded at different time-points and the penetration determined by summing the material in the receptor fluid and that remaining in the skin. Alternatively, the partitioning of the test compound from the skin into the receptor fluid can be increased so that the diffusivity of the material in the skin becomes the rate-limiting determinant of the appearance of the test compound in the receptor fluid. This is the case when K_{mr} in Equation 6 is very large and C_r is low so that the concentration difference across the skin approaches C_o. As discussed earlier, the addition of materials such as ethanol to the receptor fluid will increase partitioning of lipophilic compounds. One concern over the use of these additives is that they may alter the membrane properties of the skin and by themselves increase the diffusivity of the test compound through the skin, giving an artificially high estimate of percutaneous absorption. Another concern is that they are nonphysiological and disrupt the natural metabolic status of the skin.

Brown and Ulsamer[13] increased hexachlorophene partitioning through the inclusion of 3% BSA in the receptor fluid. Hoelgaard and Møllgaard[14] incorporated 0.5% poloxamer 188 in a phosphate buffer for the study of linoleic acid absorption. Bronaugh[15] systematically approached the problem of improving partitioning without altering skin permeability in the study of the *in vitro* percutaneous absorption of AETT and another fragrance ingredient, cinnamyl anthranilate. These compounds were applied in a petrolatum vehicle to rat skin dermatome sections mounted on static diffusion cells. The permeability of the more hydrophilic compound, cortisone, was concurrently determined with the percutaneous absorption of the test materials (Table 2). Cortisone is freely soluble in all receptor fluids tested and so its appearance in the receptor fluid is limited by its permeability through the skin. An increase in cortisone absorption with a test receptor fluid is an indication of membrane alteration. The addition of surfactants, BSA, methanol, or ethanol or the use of rabbit serum increased the appearance of AETT and cinnamyl anthranilate in the receptor fluid. The addition of the surfactants, (octoxynol 9, poloxamer 188,) or use of 50:50 methanol to water, or 40:60 ethanol to water significantly increased the permeability of cortisone, indicating damage to the membrane. The use of PEG-20 oleyl ether at up to 6% concentration did not affect cortisone permeability while enhancing the appearance of AETT and cinnamyl anthranilate in the receptor fluid. The use of PEG-20 oleyl ether in the receptor fluid abolished glucose utilization in the skin. More recently, Bronaugh et al.[16] investigated the effects of skin preparation, receptor fluid, and stirring of the flow-through diffusion cell on the *in vitro* percutaneous absorption of AETT. It was found that the elimination of subcutaneous and dermal connective tissue, the addition of 4% BSA, and stirring of the receptor fluid in the diffusion cells increased the appearance of AETT in the receptor fluid (Table 3). The total absorption (skin + receptor fluid) remained relatively constant between 50 and 60%. The formation of a diffusion boundary layer in any diffusion cell diminishes the concentration gradient across the membrane and therefore reduces the flux of the test compound through the membrane. The use of a receptor fluid additive such as BSA increases the viscosity of the receptor fluid and so increases the likelihood of laminar flow to occur in the diffusion cell. The hydrodynamic boundary layer beneath the skin is increased and diffusion across the skin decreases. Stirring the contents of the diffusion cell greatly diminishes or eliminates the boundary layer to provide a maximal concentration gradient across the skin and a flux limited by the inherent permeability of the test compound in the skin. Also examined in this

TABLE 2
Effect of Receptor Fluid and Skin Thickness on Lipophilic Compound Absorption

3-Phenyl-2-propenyl 2-aminobenzoate (cinnamyl anthranilate)

Receptor fluid	Percent applied dose absorbed in 5 d[a]	Cortisone permeability constant $\times$ 10^{5a}
Normal saline (4)[b]	5.0 ± 0.3	3.8 ± 0.7
1.5% PEG-20 oleyl ether (4)[b]	5.4 ± 0.9	—
Normal saline (4)	5.8 ± 0.4	7.1 ± 0.5
1.5 % PEG-20 oleyl ether (10)	15.5 ± 1.2[c]	6.1 ± 0.5
6% PEG-20 oleyl ether (8)	27.9 ± 1.8[c,d]	7.0 ± 0.9
20% PEG-20 oleyl ether (8)	18.3 ± 1.8[c]	9.3 ± 0.9
Rabbit serum (4)	8.8 ± 0.6[c]	6.8 ± 0.8
3% Bovine serum albumin (4)	12.1 ± 1.2[c]	5.4 ± 0.2
50:50 Methanol-water (4)	27.1 ± 2.0[c]	17.2 ± 0.2[c]
1.5% Octoxynol 9 (4)	17.9 ± 1.1[c]	10.8 ± 0.5[c]
6% Octoxynol 9 (4)	38.4 ± 2.9[c]	14.5 ± 1.3[c]
6% Poloxamer 188 (4)	7.3 ± 1.3	9.8 ± 0.6[c]

Acetyl ethyl tetramethyltetralin (AETT)

Normal saline (6)[b]	0.08 ± 0.01	3.8 ± 0.7[e]
1.5% PEG-20 oleyl ether (6)[b]	0.24 ± 0.07	—
Normal saline (4)	0.20 ± 0.06	6.3 ± 0.3
1.5% PEG-20 oleyl ether (4)	2.3 ± 0.4[c]	4.9 ± 0.2
6% PEG-20 oleyl ether (4)	6.0 ± 0.9[c,f]	7.0 ± 0.9[e]
50:50 Glycerol-water (3)	0.14 ± 0.03	4.7 ± 0.9
40:60 Ethanol-water (4)	6.1 ± 1.2[c,g]	21.7 ± 3.3[c]

[a]   Values are the Mean ± SEM; the number of determinations is in parentheses. For most experiments, a 350 μm section from the surface of whole rat skin was prepared with a dermatome. Compounds were applied to the skin in a petrolatum vehicle. *In vivo* absorption of cinnamyl anthranilate was 45.6% *in vivo* absorption of AETT was 18.9%

[b]   Whole skin.

[c]   Significant increase when compared with results from the appropriate saline control (dermatome section) by one-tailed Student's *t* test, $p < 0.05$.

[d]   Significant increase when compared with results from all receptor fluid for cinnamyl anthranilate except methanol-water and 6% octoxynol 9, $p < 0.05$.

[e]   Value determined in experiments with cinnamyl anthranilate.

[f]   Significant increase when compared with results from all receptor fluids for AETT except ethanol-water, $p < 0.05$.

[g]   Value determined at 4 d.

Adapted from Bronaugh, R. L. and Stewart, R. F., *J. Pharm. Sci.*, 73, 1255, 1984.

study was a reperfusion fluid based on the fluorocarbon liquid, FC-43 (Fluorinert® FC-43, 3M Company, St. Paul, MN). Results equivalent to that of 4% BSA in HHBSS were obtained with the FC-43 emulsion as the receptor fluid. Skin sections perfused with FC-43 emulsion or with the aqueous phase containing 3% poloxamer 188 surfactant, 2% carboxymethyl-cellulose, and a balanced salt solution maintained aerobic glucose utilization equivalent to HHBSS perfused 200 μm hairless guinea pig skin sections (2.4 to 3.3 nmoles glucose utilized/hour). While this suggests that certain surfactant solutions or emulsions may acceptably increase lipophilic compound partitioning while maintaining skin viability, further study on their properties and effects on tissue is warranted before routine use. Additionally,

TABLE 3
**Effect of Changes in Skin Thickness, Receptor Fluid, and
Stirring on AETT Absorption in the Hairless Guinea Pig[a]**

	Percent applied dose absorbed		
Skin preparation	Receptor fluid	Skin	Total
Full thickness			
unstirred	0.3 ± 0.1	60.5 ± 1.7	60.8 ± 1.8
Dermatomed (200 μm)			
unstirred	3.3 ± 0.4	49.2 ± 2.5	52.5 ± 1.9
4% BSA unstirred	13.6 ± 1.5	46.2 ± 3.9	59.8 ± 2.6
4% BSA stirred	26.6 ± 1.6	29.5 ± 1.6	56.1 ± 3.5
FC-43 emulsion			
stirred	29.7 ± 3.8	31.6 ± 4.4	61.3 ± 6.4
Epidermis			
4% BSA unstirred	14.6 ± 0.8	35.0 ± 1.5	49.7 ± 2.2
4% BSA stirred	33.1 ± 1.1	16.0 ± 1.5	49.2 ± 1.0

[a] Values are the Mean ± SEM of 3 to 11 determinations. Values were calculated from radioactivity measured in the skin and receptor fluid. Skin was
washed with soap and water at 24 h. Unless otherwise indicated, the study
was terminated immediately after the wash and the receptor fluid was HHBSS
+ 4% bovine serum albumin.

extraction of a test material and metabolites from an emulsion during a study in which skin
metabolism of a compound is assessed is more difficult than extraction from a simple solution.
At this time it appears that alternate conditions using 4% BSA in a physiological buffer as
the receptor fluid and thin dermatome sections with rigorous mixing in a flow-through
diffusion cell give results equivalent to those obtained with surfactant solutions and do so
in a more physiological environment.

V. NONVIABLE SKIN IN PERCUTANEOUS ABSORPTION STUDIES

Many of the conditions described above are deleterious to viable tissue. The use of
viable skin is conceptually appealing in *in vitro* studies; however, it may not be necessary
in all cases. For human studies, cadaver skin is often easier to obtain than viable skin and
so practicality limits some investigators to the level of complexity in their *in vitro* experiments. Receptor fluids used with nonviable skin should mimic physiological conditions as
closely as possible. Osmolarity and pH are two properties easily adjusted and maintained.
A buffered saline solution at pH 7.4 and osmolarity of approximately 283 mosmole sufficiently mimics these properties. Any additives to a buffered saline solution may alter pH
and osmolarity, and pH adjustments or reagent concentration adjustments may be required.

Lengthy diffusion cell experiments require microbiological preservation of the receptor
fluid. Antibiotics such as the penicillins and aminoglycosides are used in cell culture media
and are applicable to aqueous receptor fluids. Sodium azide has been used as an antimicrobial
in buffer solutions where nonviable tissue is used.

VI. SKIN VIABILITY MAINTENANCE AND ITS DETERMINATION

A. CELL SURVIVAL AND GROWTH

The requirements of cells for their growth and replication include proper pH, osmolarity,
the proper ratio of potassium and sodium, phosphate, calcium, magnesium, and chloride.[17]

These requirements are met through the use of an appropriate BSS. In addition to BSS, the proper temperature must be maintained and dissolved oxygen sufficient to meet the metabolic requirements of the cultured cells must be available. For protein synthesis, which is necessary for growth and replication, the required amino acid precursors must be present. Other growth requirements include vitamins, lipids (essential fatty acids, phospholipids, lecithin, and cholesterol), hormones, attachment and spreading factors, and carbohydrate energy sources.[18-21] As required for protein synthesis, the lipid, attachment and spreading factors, and hormonal requirements are typically met by a serum supplement.[22]

The requirements for growth of cells are much more stringent than those required for cell survival. Ham defines survival as "the maintenance of viability" and states that the requirements for cell survival in the absence of growth may be much simpler than previously believed.[17]

B. pH AND OSMOLARITY

Of primary importance for organ or tissue culture is the maintenance of pH in a physiological range in the media. For most mammalian cells, the tolerable range of pH is quite narrow though the optimum pH varies slightly from cell type to cell type.[17] Generally, the pH range of 7.2 to 7.4 is accepted as physiological. Although the bicarbonate buffering system is most commonly used in tissue culture applications, there are other inorganic (phosphate) and organic buffers (reviewed by Waymouth[24]) which may be used. Among the organic buffers is N-2-hydroxyethylpiperazine-N'-2-ethanesulfonic acid (HEPES) (reviewed by Eagle[23]). Bicarbonate buffer systems require an atmosphere containing 5% CO_2 to maintain a pH of 7.4. Phosphate buffer systems utilize a phosphoric acid buffer system having a second pK_a of 7.2. The pK_a of HEPES is 7.6. Both the pKa's of HEPES and phosphoric acid are closer to the physiological pH range than the bicarbonate/carbonic acid buffer system ($pK_a = 6.3$) and thus have a higher buffer capacity at pH 7.4, for a given concentration, than does a carbonic acid system. Another advantage to the phosphate and HEPES buffers is they both lack CO_2 as a requirement for maintenance of the buffer system, and, therefore, their pH is easier to regulate. The adjustment of the pH of these buffer systems may be made with $0.3N$ HCl or $0.3N$ NaOH. As Hanks has noted, the combination of these solutions yields isotonic NaCl and thus can be used for BSS pH adjustment without changing the osmotic pressure of the adjusted solution.[25]

The pH and osmolarity of the internal milieu of plant and animal cells shows remarkable similarities regardless of species. Both organic molecules and inorganic ions contribute to the osmotic pressure of the solution. The major contributors to the osmotic pressure of culture medium are sodium and chloride ions. Cells may have different media osmotic pressure requirements but generally require a narrow range of osmolarities for optimal growth. Osmotic requirements of cells in culture have been reviewed by Waymouth.[26]

C. SKIN ORGAN CULTURE AND KERATINOCYTE CELL CULTURE

The growth of skin explants in culture (organ culture) is desirable from scientific and medical perspectives. For studies in physiology and toxicology, viable differentiated skin organ maintained *in vitro* is an excellent isolated study system. From a reconstructive surgical or burn treatment perspective, a repository of healthy human skin for grafting is a favorable clinical option. The medical requirements for skin has resulted in the establishment of skin banks. Generally, allografts are stored frozen[27,28] for later thawing and transplantation, although the freezing of skin tissue for later culturing[29] and the use of combined tissue culture and cryopreservation techniques has been investigated.[30]

Successful culture of human skin was achieved over 70 years ago by Ljunggren (cited in Paul[31]) although not widely studied until the second half of this century. Blank et al. were successful in maintaining viability of human skin on agar and utilized this *in vitro* system for the study of fungal infections.[32] Reaven and Cox described a skin organ culture system for the study of environmental effects on skin growth.[33] Among their observations was the demonstration of increased epidermal cell mitosis following stratum corneum tape-stripping *in vitro* and a decreased but prolonged mitotic burst in stripped skin maintained at 32°C. By examining mitotic indices, they were also able to develop and optimize a serum-free medium capable of maintaining skin in culture for several weeks. As a continuation of the work of Reaven and Cox, Yasuno et al. studied the effects of lower incubation temperatures on skin organ culture.[34] Using MEM supplemented with 35% calf serum, the authors found lower initial, but higher prolonged mitotic rates in cultures maintained at 32°C instead of the customary 37°C. This phenomenon may reflect an evolutionary adaptation in skin growth due to the skin's lower surface temperature. Chapman et al. were able to culture split thickness porcine skin for at least 3 weeks without the use of fetal bovine serum (FBS) in Dulbecco minimal essential media.[35]

Though not fully perfected as a complete substitute for donated skin, skin substitutes fabricated from cultured keratinocytes hold promise as an eventual proxy in many skin studies. Much of the interest in keratinocyte culture has been in the control of differentiation of the epidermal cells. The importance of a defined medium for adequate control of the growth and differentiation of keratinocytes has been recognized.[36-38] Through the reduction or elimination of the serum variable in cell culture, more predictable results are obtained.[39] Construction of the cultured skin equivalent requires a suitable substrate (often collagen) for support and attachment. The details of such substrates, natural[40] and artificial[41] have been described.

Following successful culturing on the substrate and attainment of cell layer confluency, differentiation of the keratinocyte culture is induced by raising the culture to the air-liquid interface.[42] Studies conducted in this manner with murine keratinocytes have produced viable, morphologically correct skin tissue that has been utilized to provide evidence that lamellar granule extrusion in the upper granular layer is partly responsible for the hydrophobic barrier of the stratum corneum.[43] While the cultivation of structurally correct, viable, differentiated skin tissue is now routine, these methods have yet to provide a suitable substitute for donor-produced skin in percutaneous absorption/metabolism studies. Defects in the barrier properties in cultured skin equivalents due to lack of continuity and uniformity of the stratum corneum currently limit their use for *in vitro* percutaneous penetration studies.

On the other end of the spectrum from the use of cultured skin equivalents are the use of skin-flap techniques. Skin flaps are harvested pieces of split-thickness with an intact isolated vasculature. The nutrient requirements of the flap are provided by circulating serum or culture fluid through the isolated vasculature. Waste products and, in the case of percutaneous penetration metabolism studies, penetrating parent compound(s) and metabolite(s) are collected from the skin flap's venous flow.

D. VIABILITY ASSESSMENT

The purpose of biological experimentation is to gain information on the functioning of a specimen by manipulating environmental variables and making observations of the response. Because *in vitro* studies inevitably invite comparison to the situation *in vivo*, the normal functioning of the cell or tissue of study is essential for valid conclusions. Logically, the first step in validating an *in vitro* system must be a determination of the integrity and viability of the tissue of interest. The techniques for assessment of viability are applicable to tissues of most types and origins. From a medical standpoint, a tissue specific viability assay for a skin section might be its ability to be successfully grafted.[44] Perry[45] and Malinin and Perry[46] have compiled a lengthy list of *in vitro* skin viability tests:

1. Mitotic figures
2. Cellular activity

3. Transformation *in vitro*
4. Vital dye exclusion
5. 'Take' and survival time
6. Autoradiography
7. pH changes *in vitro*
8. Hair growth
9. Pigment changes
10. Ultrastructural changes
11. Changes in histological tinctorial properties
12. Respiration quotients
13. Enzyme activity

The scope of many of these methods overlap. Taylor suggested that a battery of tests be used to gather as much data as possible to make an assessment of tissue viability.[47] These types of tests can be broken down into three broad categories: (1) operational, (2) biochemical, and (3) histological; these being concerned with, respectively, overall functioning of the skin as an organ, the specific biochemical processes in the cell, and the histopathological appearance of the skin specimen as compared to a fresh specimen.

1. Protein Synthesis — Amino Acid Incorporation

Amino acids are the basic building blocks of proteins. For a cell to grow, protein must be synthesized for structural, enzymatic, and genetic material. A sensitive method for detection of protein synthesis is that of introducing a radiolabeled amino acid into the local environment of the tissue to be tested and after a suitable period, measuring the level of isotopic enrichment in the tissue protein. For greatest sensitivity, an essential amino acid should be used for the incorporation assay. Radiolabeled leucine,[48] thymidine,[34] and alanine[49] accumulation have been described for cell growth and viability measurements.

2. Histological Appearance

The most direct evidence of the condition of a tissue specimen is its appearance. On the tissue level, light microscopy reveals the overall structure and integrity of the specimen. Living tissue is naturally transparent. Fixation, sectioning, and staining are required for visualization of the inner structure of cells.[50] Staining with hematoxylin and eosin (H&E) is the most commonly used method for cellular visualization. Hematoxylin stains deoxyribonucleic acid (DNA) blue or black. Eosin stains ribonucleic acid (RNA) pink or red. The H & E staining produces pink cytoplasmic and intercellular structures and dark purple or blue nuclear structures. While not much fine detail is observable by this method, the character of the nucleus and the extent of the cytoplasm are evident. Morphological changes which are recognized as necrosis are the result of degradation of the contents of the dead cell.[51,52] The term necrosis is used to describe cell death and degradative changes. Conceptually, cell death is synonymous with irreversible changes, but, along the continuum, it is not possible to recognize a precise point where cell death occurs. Signs of reversible cell injury recognizable by light microscopy include the following:

1. Cellular swelling
2. Changes in staining characteristics
3. Accumulation of intracellular lipid
4. Alterations of nuclear morphology (such as chromatin clumping)

Signs of cell injury recognizable by electron microscopy include the following:

1. Loss of surface specialization

2. Changes in cell outline (blebs)
3. Mitochondrial alterations
4. Dilation of endoplasmic reticulum
5. Disaggregation of ribosomes
6. Membrane whorls (myelin figures)
7. Lysosomal swelling and rupture (late change)

Signs of irreversible cell injury and necrosis include all of those already mentioned and include changes associated with denaturation and dissolution of proteins and other cellular constituents. These changes may be manifested as follows:

1. Increased eosinophilia of cytoplasm due to loss of ribosomal RNA and enhanced binding of eosin to altered proteins
2. Hyalinized or vacuolated cell cytoplasm
3. Calcification of the cell
4. Pyknosis of the nucleus
5. Karyorrhexis — nuclear fragmentation
6. Karyolysis — nuclear dissolution

Signs of cell death such as pyknotic cell nuclei, cell membrane disruption, and structural disorganization at the tissue level are readily observable when compared to sections prepared from fresh tissue controls. More subtle changes in cell ultrastructure, signaling the onset of cell death, are observable by electron microscopy.

3. Biochemical Indicators of Viability

The measurement of biochemical events *in vitro* which are present *in vivo* give evidence of normal functioning in explanted tissues or cells in culture. Evidence of intact cell membranes is provided by the cell's ability to exclude the vital dye, trypan blue. Dye inclusion (or exclusion) by large numbers of cells can be evaluated by flow cytometry as described by Morasca and Erba.[53] These techniques are not, however, readily applicable to aggregated cells such as tissue explants. The continued functioning of an enzymatic system *in vitro* has been used as an assessment of viability. Enzyme assays such as the succinate dehydrogenase technique have been used in split-thickness human skin sections.[54] The appearance of an active enzyme in skin flap perfusion fluid (lactate dehydrogenase) has been ascribed to leakage from dead cells and, therefore, inversely correlated to tissue viability.[8] The presence (or absence) of an active enzyme in the tissue or perfusion fluid addresses one aspect of cell function. The sole reliance of dye exclusion tests, tissue culture techniques, or single enzyme assays as skin viability indicators has been discouraged.[55]

Oxygen consumption (respiration) by living tissue has been a widely described sensitive technique for measuring biochemical activity. The use of respiration as an assessment of viability is an operational definition of the entire organ, not just single cells or those isolated in a histological sample. Respirometric techniques including Warburg manometry have been used for the study of respiration by skin *in vitro*.[56-58] The Clark membrane-covered surface electrode for oxygen-tension measurements, as described by Penneys,[59] has been proposed as a screening procedure in skin-bank allograft procurements.[60] Regardless of the technique used, the measurement of oxygen consumption by tissue is a measurement of respiration. Respiration is a chemical energy conversion process which takes place in cell mitochondria. Glucose enters the cell in response to the body's energy needs as regulated by the hormones, glucagon and insulin. Within the cell, the final product of glycolysis (pyruvate) is utilized in the citric acid cycle for the production of NADH, which is in turn utilized by the mitochondria for the formation of adenosine triphosphate (ATP) via oxidative phosphorylation. Mitochondrial oxidation of a molecule of NADH by an atom of oxygen results in the net formation of three molecules of ATP. It is this consumption of oxygen that the above methods measure as respiration. Alternately, the utilization of glucose can be measured by other means as an assessment of respiration for an indication of tissue viability.

4. Glucose Utilization

The use of the ability of a tissue to utilize glucose as an indicator of viability is based on the need for a number of enzymes to function repeatedly and in concert for such conversions to take place. Therefore, the utilization of glucose in skin surpasses the continued functioning of a single enzyme for viability assessment.[55] Glucose utilization is performed sequentially in different compartments of the cell. Glycolysis, the conversion of glucose to pyruvate, is performed in the cytosol. The utilization of acetyl-CoA in the citric acid cycle is performed in cell mitochondria. The oxidation of pyruvate to lactate in the cytosol produces ATP, though it is less efficient than the citric acid cycle and is considered to be a metabolic dead end. The earliest demonstration of glucose utilization by skin was the measurement of anaerobic conversion of glucose to lactic acid (Wohlgemuth; 1927, cited in Freinkel[61]). Later studies of carbohydrate metabolism of skin were carried out by Cruickshank et al. using a differential capillary respirometer.[62] The determination of respiratory quotients in human epidermis and dermis by use of Warburg manometry was performed by Leibsohn et al.[63] Freinkel utilized [14C]glucose to determine the contribution of different pathways to the utilization of glucose in human skin.[64] Freinkel's study revealed the predominance of the anaerobic pathway in glucose utilization by skin, with the aerobic utilization of glucose by the citric acid cycle accounting for only about 2% of the total amount catabolized.

The predominance of anaerobic glucose utilization in skin has not been explained. Hypotheses for this condition include the ability of anaerobic glycolysis to resist oxygen and temperature changes which result from the skin's exterior body location, or perhaps the acidic conditions produced by anaerobic respiration may be required for some other skin function.[61] Anaerobic glycolysis has been found to increase in skin during wound healing and ischemia.[65-67] Reviews of epidermal energy metabolism have been prepared by Decker[68] and Freinkel.[61] The utilization of glucose in skin, though following a significantly different pattern than other tissues, is a requirement for the maintenance of cellular energetics and its measurement is an effective indicator of viability.

5. Viable Skin in Flow-Through Diffusion Cells

Organ culture of intact skin has been used in flow-through diffusion cells for percutaneous absorption studies.[9,10,69-73] Conditions for maintaining viable rat skin sections for 24 h in flow-through diffusion cells have been described.[74] Dermatome skin sections (200 μm) were mounted in Teflon diffusion cells and perfused with various receptor fluids at a flow rate of 1.5 ml/h. Distilled water, normal saline (NS), phosphate-buffered saline with glucose (PBSG), Dulbecco modified phosphate-buffered saline (DMPBS), Eagle's modified minimal essential media (MEM), and HEPES-buffered Hanks' balanced salt solution (HHBSS) were used as receptor fluids and evaluated for their ability to maintain skin section viability. The rates of aerobic and anaerobic glucose utilization, rates of estrogen and testosterone metabolism, and histological appearances of skin sections perfused for different durations were used as indicators of tissue viability. MEM, DMPBS, and HHBSS were equivalent in their ability to maintain rates of glucose utilization (Figures 3 and 4). Perfusion with PBSG resulted in declining rates of glucose utilization with no respiration after 12 h of perfusion. Distilled water and NS were unable to maintain glucose utilization. MEM, DMPBS, and MEM receptor fluids were able to maintain steroid metabolism. Histopathological evaluation of MEM, DMPBS, and MEM perfused sections showed normal tissue similar to unperfused fresh control skin. Studies with mouse, hairless guinea pig, and human skin sections have shown similar results.

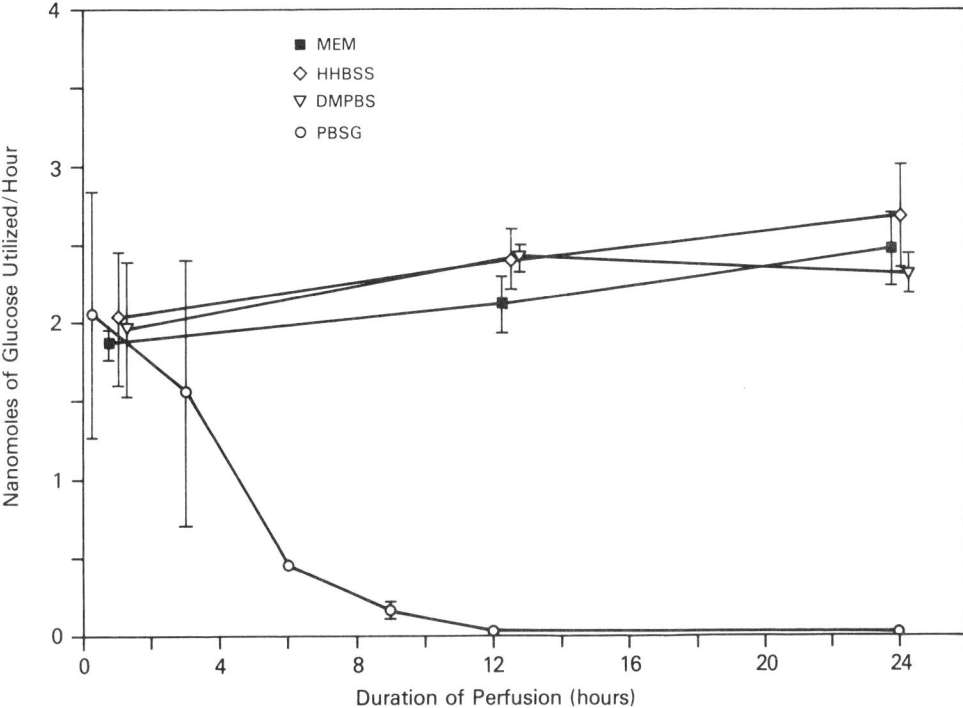

FIGURE 3.   Rates of aerobic glucose utilization in skin sections perfused over a 24 h period. Skin sections from fuzzy rats were perfused in flow-through diffusion cells with Eagle's minimal essential media (MEM), HEPES-buffered Hanks' balanced salt solution (HHBSS), Dulbecco modified phosphate-buffered saline (DMPBS), or phosphate-buffered saline with 0.1% (w/v) glucose (PBSG). Results are expressed as Means ± SEM (N = 3).[74]

The results of this study indicate that the metabolic viability of 200 μm dermatome skin sections can be maintained for 24 h in flow-through diffusion cells. The indicators of tissue viability — aerobic and anaerobic glucose utilization, histopathological appearance of the perfused skin, and the maintenance of steroid metabolism — suggest that MEM, with or without FBS, HHBSS, or DMPBS, will adequately sustain viability of the perfused skin sections. The requirements for metabolic maintenance of skin sections are such that normal saline, distilled water, or phosphate buffered saline do not lend themselves to percutaneous absorption experiments where metabolism of the penetrating compound is to be assessed.

Cell survival requirements *in vitro* differ from those necessary for cellular growth.[17] Percutaneous absorption/metabolism studies are generally conducted for finite durations and so active multiplication of the perfused tissue is not necessary for most investigations. Serum supplements are complex mixtures containing growth factors, hormones, vitamins, and other proteins. The inclusion of a serum supplement in a skin perfusion fluid is not necessary for the survival of the tissue, can interfere with metabolite extraction, affects metabolic functioning, and is of variable composition.[75,76] Serum use might be justified in the study of lipophilic compound penetration as a means of providing a more lipophilic receptor fluid; however, the use of bovine serum albumin for this purpose instead of whole serum may represent a less costly and better-defined alternative. Amino acids, included in MEM, are nutrient requirements for cells to synthesize protein and replicate, but are not survival requirements in this system. The concentrations of amino acids in MEM can also be a significant analytical problem in terms of sample cleanup prior to penetrant and metabolite analysis. Vitamin supplements, as provided by MEM, are cellular growth requirements and can have beneficial effects as antioxidants during skin flap reperfusion.[77] Their inclusion,

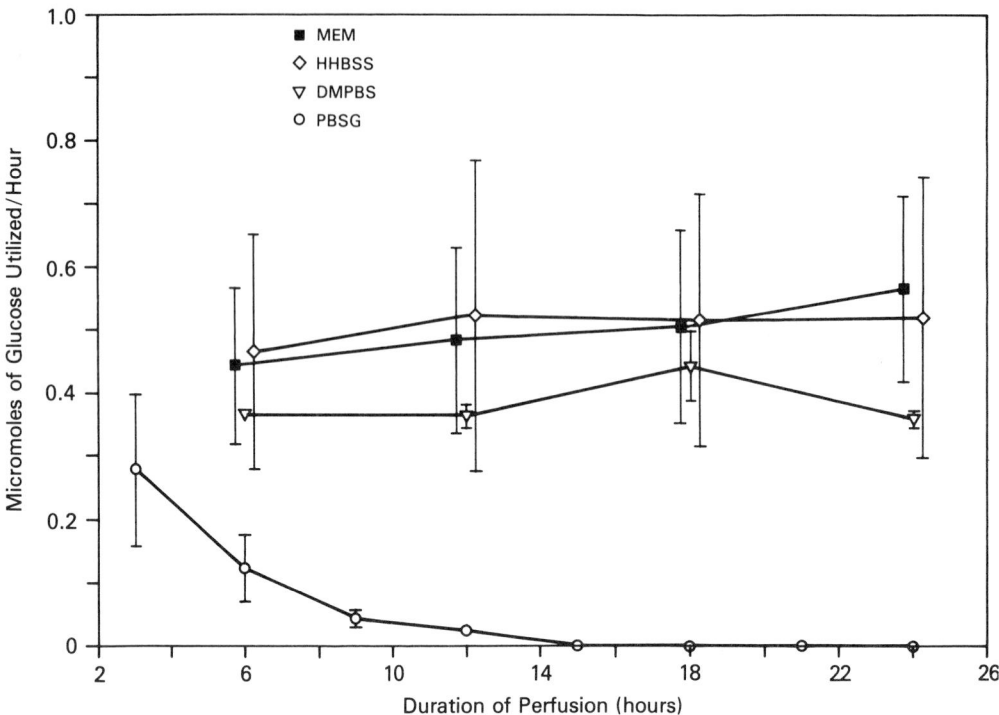

FIGURE 4.   Rates of anaerobic glucose utilization in skin sections perfused over a 24-h period. Skin sections from fuzzy rats were perfused in flow-through diffusion cells with Eagle's minimal essential media (MEM), HEPES-buffered Hanks' balanced salt solution (HHBSS), Dulbecco modified phosphate-buffered saline (DMPBS), or phosphate-buffered saline with 0.1% (w/v) glucose (PBSG). Results are expressed as Means ± SEM (N = 3).[74]

however, is not required for maintenance of tissue viability for the 24-h perfusion period. MEM utilizes a bicarbonate buffer system requiring an atmosphere containing 5% CO_2 to maintain a pH of 7.4. DMPBS utilizes a phosphate buffer system. The second pK_a of phosphoric acid is 7.2, for HEPES, used as the buffer in HHBSS, the pK_a is 7.6. Both the pK_a's of HEPES and phosphoric acid are closer to the physiological pH of 7.4 than the carbonic acid buffer system of MEM (pK_a = 6.3) and thus have a higher buffer capacity at pH 7.4, for a given concentration, than does a carbonic acid system. Neither DMPBS nor HHBSS require CO_2 for maintenance of the buffer system, and, therefore, their pH is easier to regulate. HEPES has been shown to be cytotoxic to some cells at high concentrations[78] but was not observed in this study at the concentration used (25 mM).

A carbohydrate source, such as glucose, is necessary for the continued metabolic functioning of the tissue explant.[17,79] Analysis of lactate in receptor fluid fractions may thereby represent a method of checking for metabolic functioning at the start, during, or at the conclusion of a percutaneous absorption study.

The viability of perfused skin sections can be maintained in flow-through diffusion cells with MEM, DMPBS, or HHBSS. The simplicity of HHBSS and DMPBS and the stability of the HEPES and phosphate buffer systems make them good choices for percutaneous absorption studies utilizing viable skin. Other studies have used physiological receptor fluids such as medium 1640[80] or Tyrode's solution.[81]

VII. CONCLUSIONS

The selection of the receptor fluid used in an *in vitro* percutaneous absorption study should be made after careful consideration of the requirements of the study. For the measurement of lipophilic compounds, a skin section preparation with minimal dermis and a receptor fluid into which the test compound freely partitions must be used. For preservation of the tissue viability during studies of duration more than a few hours, a physiological receptor fluid with a carbohydrate energy source must be used and either constantly perfused or regularly replaced. Surfactant additives or fluorocarbon emulsions may be effective in increasing partitioning of lipophilic compounds while maintaining skin viability; however, a validation of the system used (including permeability effects) must be undertaken. Inclusion of 4% BSA into a physiological buffer is more similar to the *in vivo* receptor fluid (blood) and is adequate for many studies. Organic solvents or solvent solutions destroy viable tissue and in long studies or in high concentrations alter skin permeability. Percutaneous absorption studies with nonviable skin should use a receptor fluid with physiological pH (7.4) and osmolarity (approximately 282 mosmoles). Adequate preservation of all receptor fluids is necessary to prevent microbial catabolism of the test compound and tissue.

REFERENCES

1. **Blank, I. H. and Scheuplein, R. J.,** Transport into and within the skin, *Br. J. Dermatol.,* 81, (Suppl. 4), 4, 1969.
2. **Higuchi, T.,** Physical chemical analysis of percutaneous absorption process from creams and ointments, *J. Soc. Cosmet. Chem. II,* 2, 85, 1960.
3. **Kety, S. S.,** Measurement of regional circulation by the clearance or radioactive sodium, *Am. Heart J.,* 38, 321, 1949.
4. **Scheuplein, R. J. and Blank, I. H.,** Permeability of the skin, *Physiol. Rev.,* 51, 702, 1971.
5. **Feldberg, W. and Patton, W. D. M.,** Release of histamine from skin and muscle in the cat by opium alkaloids and other histamine liberators, *J. Physiol.,* 114, 490, 1951.
6. **Kjaersgaard, A. R.,** Perfusion of isolated dog skin, *J. Invest. Dermatol.,* 22, 135, 1954.
7. **Hiernickel, H.,** An improved method for *in vitro* perfusion of human skin, *Brit. J. Dermatol.,* 112, 299, 1985.
8. **Riviere, J. E., Bowman, K. F., Monteiro-Riviere, N. A., Dix, L. P., and Carver, M. P.,** The isolated perfused porcine skin flap (IPPSF). I. A novel *in vitro* model for percutaneous absorption and cutaneous toxicology studies, *Fund. Appl. Toxicol.,* 7, 444, 1986.
9. **Bronaugh, R. L., Stewart, R. F., and Storm, J. E.,** Extent of cutaneous metabolism during percutaneous absorption of xenobiotics, *Toxicol. Appl. Pharmacol.,* 99, 534, 1989.
10. **Storm, J. E., Collier, S. W., Stewart, R. F., and Bronaugh, R. L.,** Metabolism of xenobiotics during percutaneous penetration: role of absorption rate and cutaneous enzyme activity, *Fund. Appl. Toxicol.,* 15 132, 1990.
11. **Bronaugh, R. L. and Stewart, R. F.,** Methods for *in vitro* percutaneous absorption. VI. Preparation of the barrier layer, *J. Pharm. Sci.,* 75, 487, 1986.
12. **Yang, J. J., Roy, T. A., and Mackerer, C. R.,** Percutaneous absorption of benzo[a]pyrene in the rat: comparison of *in vivo* and *in vitro* results, *Toxicol. Ind. Health,* 2, 409, 1986.
13. **Brown, D. W. C. and Ulsamer, A. G.,** Percutaneous penetration of hexachlorophene as related to receptor solutions, *Fd. Cosmet. Toxicol.,* 13, 81, 1973.
14. **Hoelgaard, A. and Møllgaard, B.,** Permeation of linoleic through skin *in vitro, J. Pharm. Pharmacol.,* 34, 610, 1982.
15. **Bronaugh, R. L. and Stewart, R. F.,** Methods for *in vitro* percutaneous absorption studies. III. Hydrophobic compounds, *J. Pharm. Sci.,* 73, 1255, 1984.
16. **Bronaugh, R. L., Collier, S. W., and Stewart, R. F.** *In vitro* percutaneous absorption of a hydrophobic compound through viable hairless guinea pig skin, *Toxicologist,* 9, 241, 1989.
17. **Ham, R. G.,** Survival and growth requirements of nontransformed cells, in *Handbook of Experimental Pharmacology; Tissue Growth Factors,* 57, Baserga, R., Ed., Springer-Verlag, New York, 1981, 14.
18. **Levintow, L. and Eagle, H.,** Biochemistry of cultured mammalian cells, *Annu. Rev. Biochem.,* 30, 605, 1961.
19. **Waymouth, C.,** Construction and use of synthetic media, in *Cells and Tissues in Culture,* Willmer, E. N., Ed., Academic Press, New York, 1965, 99.
20. **Ham, R. G. and McKeehan, W. L.,** Media and growth requirements, *Methods Enzymol.,* 58, 44, 1979.

21. **Rizzino, A., Rizzino, H., and Sato, G.,** Defined media and the determination of nutritional and hormonal requirements of mammalian cells in culture, *Nutr. Rev.,* 37, 369, 1979.
22. **Barnes, D. and Sato, G.,** Serum-free cell culture: a unifying approach, *Cell,* 22, 649, 1981.
23. **Eagle, H.,** Some effects of environmental pH on cellular metabolism and function, in *Control of Proliferation in Animal Cells,* Clarkson, B. and Baserga, R., Eds., Cold Spring Harbor, New York, 1, 1974.
24. **Waymouth, C.,** Major ions, buffer systems, pH, osmolality, and water quality, in *The Growth Requirements of Vertebrate Cells in vitro,* Waymouth, C., Ham, R. G., and Chapple, P. J., Eds., Cambridge University, New York, 1981, 107.
25. **Hanks, J. H.,** Balanced salt solutions, inorganic requirements and pH control, in *An Introduction to Cell and Tissue Culture,* Scherer, W. F., Ed., Burgess, Minneapolis, 1955, 5.
26. **Waymouth, C.,** Osmolarity of mammalian blood and of media for culture of mammalian cells, *In Vitro,* 6, 109, 1970.
27. **May, S. R., Still, J. M., and Atkinson, W. B.,** Recent developments in skin banking and the clinical uses of cryopreserved skin, *J. Med. Assoc. Ga.,* 73, 233, 1984.
28. **Baxter, C., Aggarwal, S., and Diller, K. R.,** Cryopreservation of skin: a review, *Transplant Proc.,* 17, 112, 1985.
29. **Taylor, H. A., Riley, S. E., Parks, S. E., and Stevenson, R. E.,** Long-term storage of tissue samples for cell culture, *In Vitro,* 14, 476, 1978.
30. **Nathan, P., Robb, E. C., Harper, A. D., and Ballantyne, D. L.,** Long-term skin preservation by combined use of tissue culture and freezing techniques, *Cryobiology,* 15, 133, 1978.
31. **Paul, J., Ed.,** in *Cell and Tissue Culture,* Williams and Wilkins, Baltimore, 1961, 1.
32. **Blank, H., Sagami, S., Boyd, C., and Roth, F. J., Jr.,** The pathogenesis of superficial fungus infections in cultured human skin, *Arch. Derm.,* 79, 524, 1959.
33. **Reaven, E. P. and Cox, A. J.,** Organ culture and human skin, *J. Invest. Dermatol.,* 44, 151, 1965.
34. **Yasuno, H., Sotomatsu, S., Maeda, M., Sato, M., Nishimura, A., and Matsubara, M.,** Organ culture of adult human skin; effect of culture temperature, *J. Dermatol. (Tokyo),* 8, 267, 1981.
35. **Chapman, S. J., Walsh, A., Beckett, E., and Vickers, C. F. H.,** A fully differentiating epidermal model with extended viability: development and partial characterization. *J. Invest. Dermatol.,* 93, 762, 1989.
36. **Eisinger, M., Lee, J. S., Hefton, J. M., Darzynkiewicz, Z., Chiao, J. W., and De Harven, E.,** Human epidermal cell cultures: growth and differentiation in the absence of dermal components or medium supplements, *Proc. Natl. Acad. Sci.,* 76, 5340, 1979.
37. **Peehl, D. M. and Ham, R. G.,** Growth and differentiation of human keratinocytes with small amounts of dialyzed serum, *In Vitro,* 16, 516, 1980.
38. **Peehl, D. M. and Ham, R. G.,** Clonal growth of human keratinocytes with small amounts of dialyzed serum, *In Vitro,* 16, 526, 1980.
39. **Dils, R. R.,** Explants and disaggregated tissue preparations as model systems in nutritional research: advantages and pitfalls, *Proc. Nutrition. Soc.,* 43, 133, 1984.
40. **Asselineau, D. and Prunieras, M.,** Reconstruction of 'simplified' skin: control of fabrication, *Brit. J. Dermatol.,* 111, 219, 1984.
41. **Vaughan, F. L., Gray, R. H., and Bernstein, I. A.,** Growth and differentiation of primary rat keratinocytes on synthetic membranes, *In Vitro Cell. Dev. Biol.,* 22, 141, 1986.
42. **Bernstam, L. I., Vaughan, F. L., and Bernstein, I. A.,** Keratinocytes grown at the air-liquid interface, *In Vitro Cell. Dev. Biol.,* 22, 695, 1986.
43. **Madison, K. C., Swartzendruber, D. C., Wertz, P. W., and Downing, D. T.,** Lamellar granule extrusion and stratum corneum intercellular lamellae in murine keratinocyte cultures, *J. Invest. Dermatol.,* 90, 110, 1988.
44. **Jensen, H. S.,** Skin viability studies *in vitro, Scand. J. Plast.,* 31 *Reconstr. Surg.,* 18, 55, 1984.
45. **Perry, V. P.,** A review of skin preservation, *Cryobiology,* 3, 109, 1966.
46. **Malinin, T. I. and Perry, V . P.,** A review of tissue and organ viability assay, *Cryobiology,* 4, 104, 1967.
47. **Taylor, A. C.,** Cryopreservation of skin: discussion and comments, *Cryobiology,* 3, 192, 1966.
48. **Moore, K. G., Schofield, B. H., Higuchi, K., Kajiki, A., Au, K., Pula, P. J., Bassett, D. P., and Dannenberg, A. M., Jr.,** Two sensitive *in vitro* monitors of chemical toxicity to human and animal skin (in short term organ culture). I. Paranuclear vacuolization in glycol methacrylate tissue sections. II. Interference with [^{14}C]leucine incorporation, *J. Toxicol. Cutaneous Ocul. Toxicol.,* 5, 285, 1986.
49. **Siekevitz, P.,** Uptake of radioactive alanine *in vitro* into the proteins of rat liver fractions, *J. Biol. Chem.,* 195, 549, 1952.
50. **Bloom, W. and Fawcett, D. W.,** Methods of histology and cytology, in *A Textbook of Histology,* W. B. Saunders, Philadelphia, 1968, 10.
51. **Cheville, N. F.,** Cell Degeneration, in *Cell Pathology,* Iowa State University, Ames, 1983, 75.
52. **Cotran, R. S., Kumar, V., and Robbins, S. L.,** Cellular injury and adaptation, in *Robbins Pathological Basis of Disease,* 4th ed., W. B. Saunders, Philadelphia, 1989.

53. **Morasca, L. and Erba, E.,** Flow cytometry, in *Animal Cell Culture—A Practical Approach,* Freshney, R. I., Ed., IRL, Washington, D.C., 1986, 125.

54. **Brown, R. F. R. and Groves, A. R.,** Succinate dehydrogenase as an index of thermal injury to skin, *Br. J. Exp. Path.,* 54, 117, 1973.

55. **May, S. R. and DeClement, F. A.,** Skin banking. III. Cadaveric allograft skin viability, *J. Burn Care Rehab.,* 2, 128, 1981.

56. **Cruickshank, C. N. D.,** Continuous observation of the respiration of skin *in vitro, Exp. Cell Res.,* 7, 374, 1954.

57. **Umbriet, W. W.,** The Warburg constant volume respirometer, in *Manometric Techniques,* Umbreit, W. W., Burns, R. H., and Stauffer, J. R., Eds., 4th ed., Burgess, Minneapolis, 1964, 1.

58. **Døssing, M. and Sørensen, B.,** Freeze-dried and nonfreeze-dried split-skin allografts on excised burns, *Burns,* 2, 36, 1975.

59. **Penneys, R.,** Some experiments on the use of the miniaturized Clark32 electrode, and the open-tip electrode, for the measurements of tissue oxygen tension, *Polarography,* Proc. 3rd Int. Congr., 2, 951, 1964.

60. **Jensen, H. S. and Alsbjörn, B. F.,** Viability in split-skin biopsies measured by a surface oxygen electrode, *Scand. J. Clin. Invest.,* 44, 423, 1984.

61. **Freinkel, R. K.,** Carbohydrate metabolism of epidermis, in *Biochemistry and Physiology of the Skin,* 1, Goldsmith, Ed., Oxford University, New York, 1983, 328.

62. **Cruickshank, C. N. D., Trotter, M. D., and Cooper, J. R.,** Studies on the carbohydrate metabolism of skin, *Biochemistry,* 66, 285, 1957.

63. **Leibsoh, E., Appel, B., Ullrick, W. C., and Tye, M. J.,** Respiration of human skin, normal values, pathological changes and effect of certain agents on oxygen uptake, *J. Invest. Dermatol.,* 30, 1, 1958.

64. **Freinkel, R. K.,** Metabolism of glucose-C-14 by human skin *in vitro, J. Invest. Dermatol.,* 34, 37, 1960.

65. **Halprin, K. M. and Ohkawara, A.,** Glucose and glycogen metabolism in the human epidermis, *J. Invest. Dermatol.,* 46, 43, 1966.

66. **Im, M. J. C. and Hoopes, J. E.,** Energy metabolism in healing skin wounds, *J. Surg. Res.,* 10, 459, 1970.

67. **Harmon, C. S., Masser, M. R., and Phizackerley, P. J. R.,** Effect of ischemia and reperfusion of pig skin flaps on epidermal glycogen metabolism, *J. Invest. Dermatol.,* 86, 69, 1986.

68. **Decker, R. H.,** Nature and regulation of energy metabolism in the epidermis, *J. Invest. Dermatol.,* 57, 351, 1971.

69. **Holland, J. M., Kao, J. Y., and Whitaker, M. J.,** A multisample apparatus for kinetic evaluation of skin penetration *in vitro:* the influence of viability and metabolic status of the skin, *Toxicol. Appl. Pharmacol.,* 72, 272, 1984.

70. **Kao, J., Hall, J., and Holland, J. M.,** Quantitation of cutaneous toxicity: an *in vitro* approach using skin organ culture, *Toxicol. Appl. Pharmacol.,* 68, 206, 1983.

71. **Kao, J., Hall, Shugart, L. R., and Holland, J. M.,** An *in vitro* approach to studying cutaneous metabolism and disposition of topically applied xenobiotics, *Toxicol. Appl. Pharmacol.,* 75, 289, 1984.

72. **Kao, J., Patterson, F. K., and Hall, J.,** Skin penetration and metabolism of topically applied chemicals in six mammalian species, including man: an *in vitro* study with benzo(a)pyrene and testosterone, *Toxicol. Appl. Pharmacol.,* 81, 502, 1985.

73. **Kao, J. and Hall, J.,** Skin absorption and cutaneous first pass metabolism of topical steroids: *in vitro* studies with mouse skin in organ culture, *J. Pharmacol. Exper. Ther.,* 241, 482, 1987.

74. **Collier, S. W., Sheikh, N. M., Sakr, A., Lichtin, J. L., Stewart, R. F., and Bronaugh, R. L.,** Maintenance of skin viability during *in vitro* percutaneous absorption /metabolism studies, *Toxicol. Appl. Pharmacol.,* 99, 522, 1989.

75. **Boone, C. W.,** A surveillance procedure applied to sera, in *Tissue Culture: Methods and Applications,* Kruze, P., and Patterson, M., Eds., Academic Press, New York, 1973, 677.

76. **Doughty, M. J., Davis, M. H., and Gruenstein, E.,** Reversible change in the fibroblast lysomal enzyme dipeptidyl aminopeptidase-1 (Cathepsin C) related to the commercial source of fetal bovine serum in the culture medium, *In Vitro,* 21, 340, 1985.

77. **Hayden, R. E., Yeung, C. S. T., Paniello, R. C., Bello, S. L., and Dawson, S. M.,** The effect of glutathione and vitamins A, C, and E on acute skin flap survival, *Laryngoscope,* 97, 1176, 1987.

78. **Poole, C. A., Reilly, H. C., and Flint, M. H.,** The adverse effects of HEPES, TES, BES zwitterion buffers on the ultrastructure of cultured chick embryo epiphyseal chondrocytes, *In Vitro,* 18, 755, 1982.

79. **Parker, R. C., Ed.,** Balanced salt solutions and pH control, in *Methods of Tissue Culture,* Harper Row, New York, 1962, 53.

80. **Hawkins, G. S. and Reifenrath, W. G.,** Influence of skin source, penetration cell fluid, and partition coefficient on *in vitro* skin penetration, *J. Pharm. Sci.,* 75, 378, 1986.

81. **Smith, L. H. and Holland, J. M.,** Interaction between benzo[a]pyrene and mouse skin in organ culture, *Toxicology,* 21, 47, 1981.

Chapter 6

ASSESSMENT OF VEHICLE FACTORS INFLUENCING PERCUTANEOUS ABSORPTION

Joel L. Zatz

TABLE OF CONTENTS

I. INTRODUCTION

Although the usual application of percutaneous absorption data is to live human beings, there are a number of reasons for utilizing *in vitro* experiments, particularly in the elucidation of vehicle effects. A large number of formulations can be screened *in vitro* in a short period of time, and at relatively low cost. It is possible to measure the influence of potentially damaging or toxic additives without putting subjects at risk. The data that can be obtained *in vitro* are usually more complete and more precise than are available from *in vivo* studies. In many cases, such data can be used to prove absorption mechanisms. Finally, the application of models often makes it possible to perceive general patterns and extrapolate the information obtained to new compounds and new situations.

A. GENERAL EXPERIMENTAL CONSIDERATIONS

The experimental design must be selected with regard to the intended application. Most *in vitro* experiments have been of the infinite-dose type. The skin is occluded and the drug concentration remains constant during the experiment. This design is relatively easy to interpret in terms of apparent permeability, partition, and diffusion coefficients. It is thus possible to compare different vehicles in terms of these parameters. However, infinite dosing often exposes the stratum corneum to an excessive amount of the preparation. It certainly does not replicate conditions of use of dermatological products, which are typically applied to the skin in the form of an unoccluded thin film. In such cases, changes in composition of the vehicle may occur due to evaporation; for example, and in many cases, the drug concentration drops as permeation proceeds. If the *in vitro* experiment is supposed to mimic conditions of use of a dermatological product, a finite-dose design should be used. The data patterns are usually more complex and therefore more difficult to explain through a simple model.

Figure 1 shows the type of data that may be obtained from finite dose experiments.[1] In the examples shown, solutions of lidocaine in mixtures of water and propylene glycol were applied to the skin at a dosage of 11 µl/cm². Flux values at early times were of the same ranking as from an infinite dose application; the vehicle with the highest solvency (80% propylene glycol) exhibited the lowest flux while the vehicle in which lidocaine was least soluble (40% propylene glycol) had the highest flux. However, the curve representing the 40% propylene glycol solution behaved differently over time from the others. After an initial peak, the flux dropped, then rose again, and finally fell for a second time. The decrease in flux observed at 1 h is ascribed to a reduction in skin surface temperature due to evaporation of water from the solution. The second decrease is caused by depletion of lidocaine from the solution, which reduces the driving force for permeation during the course of the experiment.

B. MEMBRANE MODELS

An important consideration in designing the experimental procedure is the choice of membrane model. Excised human skin is preferred because there is frequently a good correlation between *in vitro* and *in vivo* data. Animal skin is usually more permeable than human skin and the extent varies from one compound to another, so that it is difficult to predict absolute absorption rates through human skin from animal experiments. However, in formulation development the rank order from various vehicles is often more significant than the rates themselves. In this instance excised animal skin has a theoretical advantage in that the experiments can be completed in a shorter period of time. There are additional advantages in terms of the reproducibility of experimental data. Human skin is usually obtained from either surgical procedures or autopsy, and the supply is limited. The donors may be of any age, some individuals may have been "sun worshippers", while others were

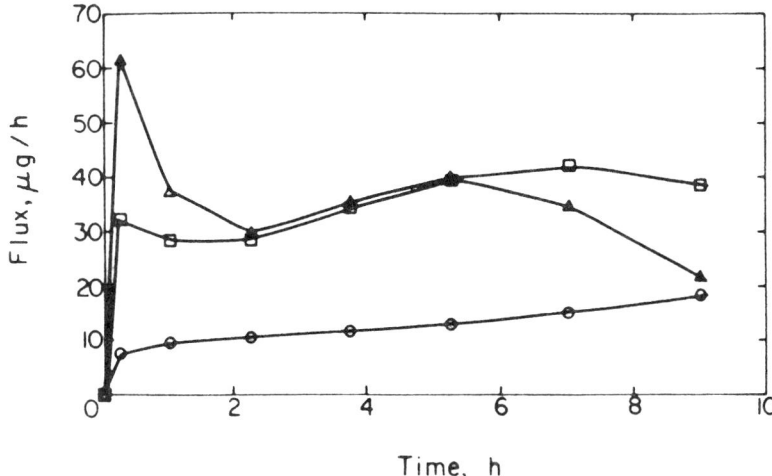

FIGURE 1. Mean rates of lidocaine penetration through hairless mouse skin from propylene glycol-water vehicles. Applied dose was 11 μl/cm². ○, 80% propylene glycol; □, 60% propylene glycol; △, 40% propylene glycol. (From Sarpotdar, P. P. and Zatz, J. L., *J. Pharm. Sci.*, 75, 176, 1986. With permission of the copyright owner, the American Pharmaceutical Association).

shut-ins. The variety of previous histories and diets is extensive, and so data obtained from studies of human skin tend to be quite variable. Laboratory animals, on the other hand, are generally available on demand. They are raised and fed under controlled conditions, so that their skin tends to exhibit less inter-subject variability than human skin.

An apparent contradiction to this principle with respect to nitroglycerin permeation was recently reported.[2] Flux from a transdermal system through skin of three species was measured. Hairless mouse penetration was more variable than that through excised human epidermis. The flux through dermatomed Yucatan pig skin was closer to human skin values, and the variability was less than with hairless mouse.

Extrapolation of permeation data comparing vehicles obtained with animal skin to human skin assumes that the rate-limiting step resides within the same location in both species. If a particular vehicle causes a shift in the rate-limiting step in one species but not the other, or if the test species is more sensitive to the effects of an interactive solvent than human skin, there will be problems in extrapolation.

In addition to permeation experiments utilizing excised skin, other types of *in vitro* studies can provide useful information regarding vehicle effects. These include release experiments, in which diffusion of active substances through the vehicle is measured, and permeation through polymer membranes. Release experiments do not utilize a skin membrane. The preparation under study is either in contact with a receptor fluid or else donor and receptor are separated by a simple, highly permeable membrane that prevents physical mixing of the two phases. Water or an aqueous fluid is frequently used as the receptor. In such cases, release from water-soluble and water-washable vehicles is usually favored. Some investigators have tried to mimic skin by using octanol or isopropyl myristate as a receptor. In general, results of release experiments do not parallel skin permeation behavior.

While some investigators have attempted to use polymer membranes as a substitute for skin, their major value appears to be in investigations of mechanism and the validation of theoretical concepts. Some examples will be presented below.

C. EXPERIMENTAL CONTROLS IN VEHICLE STUDIES

It is obviously necessary to maintain uniform experimental conditions in comparative studies, such as those used to evaluate simple vehicles or more complex formulations. A variable that is difficult to control is the inherent penetration resistance of the membrane, which is a function of many unknown and often uncontrollable factors. This is especially the case with human skin obtained from autopsy or surgery. There may be groups of individuals whose skin permeability tends to be relatively high, while the skin of others is more resistant. Furthermore, the amounts of skin available from a single donor may be sufficient for only a few experiments, so that an extensive comparison of vehicles or formulations often utilizes skin from more than one individual. The variability inherent in skin membranes from different sources often obscures small to moderate, but nevertheless real differences. Another problem is that certain formulations may falsely appear to be particularly effective in delivering a drug through the skin, having been tested on membranes with high permeability.

These problems can be minimized or eliminated by including suitable controls in the experimental design. Permeation of a reference substance under standard conditions can be used to normalize data from different treatments. One approach in which a control formulation was applied to each piece of skin is described later in this chapter.

Vehicles can be categorized according to several schemes based on physical properties.[3-5] Some studies on vehicles have focused on a comparison of different types of bases. For example, an oleaginous ointment may be compared to an absorption base, a cream, polyethylene glycol ointment, etc. While it is possible to obtain useful data by this approach, it is usually difficult to identify the reasons for any differences that are found without additional information. Among the variables are stratum corneum hydration, alteration in thermodynamic activity of the permeant in the vehicle, species present within the vehicle as a function of pH, and interaction of the vehicle with the stratum corneum. These factors will be addressed in the following sections.

II. SPECIFIC VEHICLE FACTORS

A. STRATUM CORNEUM HYDRATION

It is well known that percutaneous absorption of most substances is increased by raising the water content of the stratum corneum. Vehicles can increase stratum corneum hydration by providing water directly or occluding the skin surface, thus reducing the rate of water transpiration to the outside. Water that would normally be lost to the atmosphere is thus picked up by the stratum corneum. Among the vehicles in common use, petrolatum is the most occlusive.[6] As a general rule, ointments tend to be more occlusive than creams, while lotions and powders offer little resistance to moisture transport.

Wang et al.[7] evaluated hydrocortisone penetration from a group of oil-in-water emulsions using *in vitro* measurements through excised human skin as well as the vasoconstriction bioassay. In a series of emulsions whose total oil phase volume was maintained constant, replacement of mineral oil by petrolatum increased both the degree of occlusivity and the permeability coefficient of hydrocortisone.

The effect of humectants, such as glycerin, on skin hydration has been a matter of controversy. In a recent paper, Batt et al.[8] measured a series of instrumental properties of skin following application of glycerin solutions in water. At moderate values of relative humidity, transepidermal water loss (TEWL) values decreased as a result of glycerin treatment, and this effect persisted for several hours. These data, which suggested an increase in the degree of stratum corneum hydration, were in contradiction to earlier *in vitro* studies conducted at very low relative humidity.[6]

An error that is sometimes made is to assume that water evaporates completely following the application of an aqueous solution of a hydroscopic substance to the skin. As long as some material remains on the skin surface, its composition will be in equilibrium with the ambient relative humidity. Depending on relative humidity and the original composition, either evaporation will occur or else water may be extracted from the atmosphere until the mixture reaches equilibrium.[1]

B. PERMEANT ACTIVITY IN THE VEHICLE

In the usual situation in which transport across the stratum corneum is the rate limiting step in permeation, one major effect of vehicles is to determine permeant activity within the vehicle. This influences the activity gradient across the membrane irrespective of its nature. In other words, regardless of whether the membrane is skin, stratum corneum itself, or a polymer film, an increase in activity within the vehicle will result in an increase in penetration rate. This mechanism does not involve any interaction with the membrane and may therefore be characterized as noninteractive.

A second type of vehicle effect involves a change in membrane properties. This may be due to extraction of a component significant to maintaining barrier properties, a change in solvation of the membrane or some other perturbation that effectively changes the nature of the membrane. These interactions depend on the nature of the membrane; skin and polymer membranes would be expected to respond differently.

Interpretation of a simple comparison of flux values from different vehicles may be confounded by the simultaneous contribution of both effects. In order to assess stratum corneum-vehicle interactions, it is necessary to first factor out the nonspecific interactions. This may be done through experimental design or by performing a calculation to estimate the activity effect, and then determining the residual interactive influence.

The notion that a saturated solution should logically serve as the reference state in assessing activity within the vehicle was first put forth by T. Higuchi.[9] Since the activity of excess solid is assigned a value of unity, the activity of a solution may be estimated as C_v/S, where C_v is concentration within the vehicle and S is the solubility in the same vehicle. This principle leads to the proposition that saturated solutions containing the same permeant should have the same noninteractive contribution to flux, despite differences in the solute concentration. In cases where the interactive contribution is minor, the fluxes will be essentially the same.

The idea is illustrated in Figure 2 which shows the flux of methylparaben through polydimethyl-siloxane membranes from saturated solution in water, propylene glycol, polyethylene glycol 400, glycerin and polyol-water mixtures.[10] Despite large differences in solubility, the flux values are identical. The same result was obtained for other parabens. The solvents used are essentially inert toward the membrane; they are not imbibed to a significant extent and the membrane does not swell in their presence.

Substitution of alcohols or alcohol-water mixtures for these inert solvents leads to a large increase in permeability.[10,11] With pure alcohols, there is up to a 40-fold increase in flux. The flux enhancement is due principally to an increase in the membrane/vehicle partition coefficient. The data are consistent with a model in which alcohol molecules enter the membrane and form clusters capable of dissolving additional solute.[12] According to this model, flux is dependent both upon permeant activity and alcohol activity. An interesting observation is the existence of a peak in the flux-concentration profile with an alcohol solvent (Figure 3). At low concentrations, permeant activity is small, but alcohol activity is high. Increasing permeant concentration raises flux, but not proportionally. At concentrations approaching saturation, the permeant concentration is high; however, the alcohol activity is reduced, so that the interactive contribution is less than maximal. An optimum point occurs at an intermediate concentration, accounting for the peak in flux.

The noninteractive solvent effect described above gives rise to the rule of thumb that with a fixed concentration of a drug, permeation will be better from a vehicle that is a relatively poor solvent medium than one with high solvency. Furthermore, increasing the concentration of a solute in a system that is already in equilibrium with excess pure material

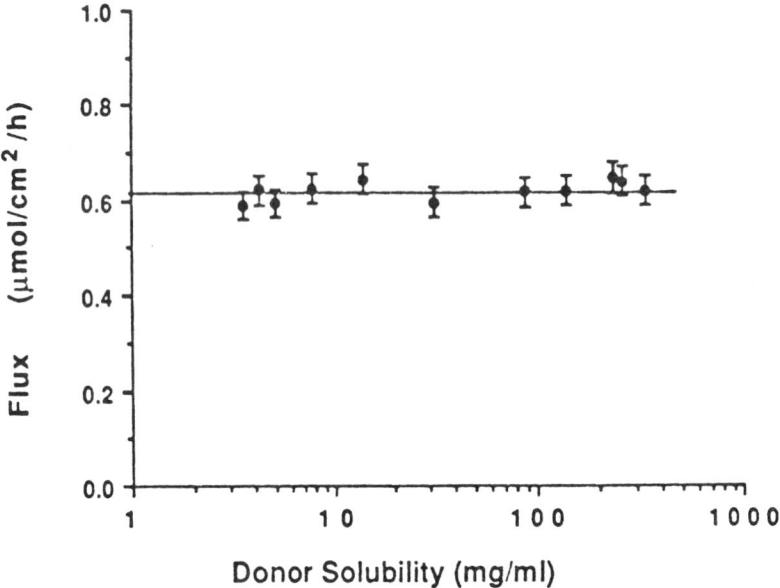

FIGURE 2.   Steady-state flux of methylparaben from saturated solution in water and polyols (polyethylene glycol 400, propylene glycol, glycerin) and water-polyol mixtures through polydimethylsiloxane membranes. (From Twist, J. N. and Zatz, J. L., *J. Soc. Cosmet. Chem.* 11, 85, 1960. With permission.)

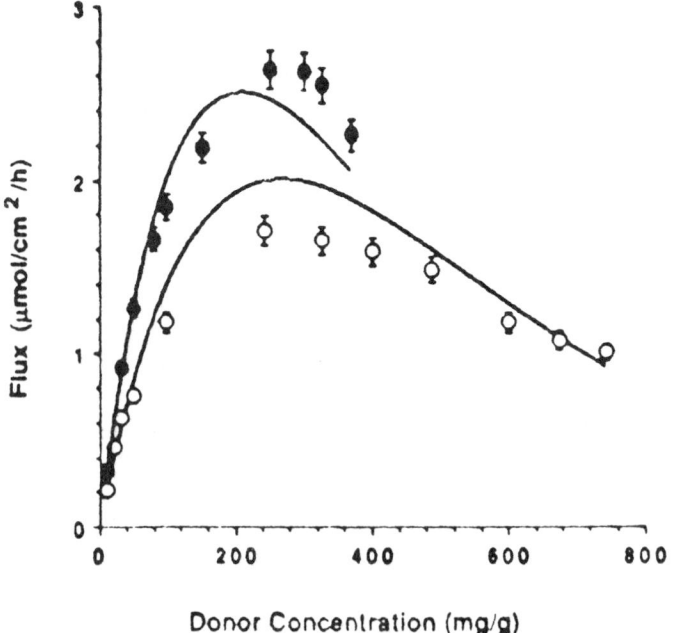

FIGURE 3.   Steady-state flux profiles for methylparaben (●) and butylparaben (○) through polydimethylsiloxane membranes from solution in 1-propanol. The points are measured values; the lines are calculated values. (From Twist, J. N. and Zatz, J. L., *J. Pharm. Sci.*, 79, 28, 1990. With permission of the copyright owner, the American Pharmaceutical Association.)

does not increase its activity; therefore, no increase in flux through skin is anticipated. On the other hand, if conditions within the vehicle are such that supersaturated solutions may form, flux values will be increased purely by noninteractive means. Coldman et al.[13] demonstrated this effect in studies of labeled fluocinolone acetonide through excised human skin. The vehicles were mixtures of isopropanol with either isopropyl myristate or propylene glycol. After application to the skin, evaporation of isopropanol concentrated the steroid in the nonvolatile component and its flux was increased. The largest increases occurred in those systems in which supersaturated solutions formed *in situ*. Interestingly, occlusion of the skin actually resulted in a decrease in drug flux with many of the vehicles employed in this study. Occlusion prevented evaporation of the isopropanol so that the increase in activity of fluocinolone acetonide within the vehicle did not take place.

C. pH OF THE VEHICLE

Extremes of pH are avoided in designing skin products so as to prevent overt skin damage. Strong acid and alkali act as primary irritants. The skin is tolerant to vehicles whose pH ranges from about 3 or 4 to about 10. Going beyond these suggested limits may result in irritation. Of course, *in vitro* experiments may be conducted at any pH. Within the range quoted above, pH has little, if any, effect on skin penetration of nonionic compounds.

With ionizable drugs, a major effect of pH is on the fraction of drug that is uncharged. While the water solubility of such drugs is greatest at pH values favoring the ionized form, charged species penetrate the skin poorly in comparison with their unionized counterpart. Although binding to skin proteins may be a significant factor, theoretical descriptions of the permeability of ionigenic drugs have assumed that both species permeate independently and the total flux is the sum of the flux values for the component species.[14] It is often further assumed that the diffusion coefficient for both species is the same, so that differences in permeability can be attributed to changes in partitioning.

The effect of ionization on skin penetration has generally been studied under infinite dose conditions using simple aqueous solutions as vehicles. One method involves working with vehicles of varying pH, so that the drug in question is essentially completely ionized in one test vehicle and almost completely uncharged in another. Steady-state flux values are obtained experimentally and the permeability coefficients, calculated from flux and vehicle concentration, may be compared. This approach was used in an investigation of scopolamine permeation through excised human skin.[14] As might be expected, the permeability coefficient, P, of the uncharged form was considerably higher than that of the charged specie. The permeability ratio, P(ionized)/P(unionized), was about 0.06

A second approach was employed by Swarbrick et al. in a study of permeation of chromone-2-carboxylic acids through human epidermis.[15] Making the usual assumption about the independence of permeation of the species present, these authors derived the following equation:

$$\frac{J}{C^{A^-}} = \left\{ \frac{P^{HA}}{K_a} \right\} C^{H_3O^+} + P^{A^-} \qquad (1)$$

in which J is the measured steady-state flux, K_a is the ionization constant, C^{A^-} and $C^{H_3O^+}$ are the vehicle concentration of ionized compound and hydrogen ion, respectively, and P^{HA} and P^{A^-} are permeability coefficients of the uncharged and charged species, respectively. A plot of J/C^{A^-} against $C^{H_3O^+}$ should be linear. Both permeability coefficients can be determined from the slope and intercept of the line. In the study cited,[15] the permeability ratio for the acidic compounds under investigation was approximately 10^{-4}.

This method permits evaluation of compounds whose water solubility may be negligible over a given range of pH, which would preclude experimental determination of flux values

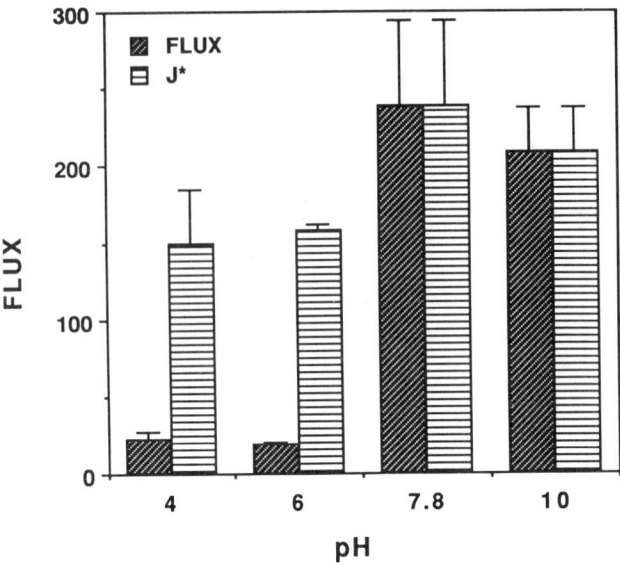

FIGURE 4.   Flux of lidocaine species through dermatomed human skin
from 40% propylene glycol at various donor pH values.

within that range. A disadvantage is that many data points are compressed in one side of
the figure if the pH range investigated is too large. All values do not then receive equal
weight in calculating the permeability coefficients.

In a variation on the above themes, Fleeker et al. measured flux through snakeskin at
two pH values and then solved simultaneous equations based on the usual assumptions to
calculate permeability coefficients of charged and uncharged forms of two model drugs,
clonidine and indomethacin.[16] The permeability ratio for clonidine was approximately 0.09,
while that for indomethacin was approximately 0.01. Pretreatment of the skins with dodecyl
N,N-dimethylaminoacetate resulted in a significant increase in flux of both species of clonidine; however, with indomethacin, only the ionized form was affected while permeability
of the unionized form remained the same.

The effect of pH on lidocaine permeation through excised hairless mouse and dermatomed
human skin from an aqueous medium containing 40% propylene glycol was measured[17]
(Figure 4). Data was obtained at four pH values ranging from 4.0 to 10.0 and two methods
were used to calculate indivudal permeability coefficients. One was a solution of simultaneous
equations, as described by Fleeker et al.[16] of the data obtained at the two extreme pH values.
The second was a variation of the method described by Swarbrick et al.[15] adapted to a basic
drug. The equation used was

$$\frac{J_{TOT}}{C_L} = P_L + \frac{P_{LH^+}C_{H^+}}{K_a} \qquad (2)$$

in which J_{TOT} is the measured flux, C_L and P_L refer to the concentration and permeability
coefficient, respectively, of the uncharged species, and P_{LH^+} is the permeability coefficient
of the ionized form. J_{TOT}/C_L is plotted against C_{H^+}; the slope and intercept are used to
calculate the respective permeability coefficients.

Both methods yielded permeability coefficient values that were comparable. Interestingly, there was a significant species effect on the permeability ratio. With hairless mouse
skin, the permeability ratio was about 0.06, but the value for dermatomed human skin was

0.02, one-third as much. Part of the explanation for this difference lies in the relatively large contribution of the viable tissues to hairless mouse membrane resistance toward lidocaine *in vitro*.[18] The aqueous tissues are expected to represent more of a barrier to lidocaine base than the ionized form, so that the passage of the un-ionized species would be retarded relative to the other. This finding points to a problem in the use of certain animal skin models when evaluating the penetration potential of ionized drugs. The data for lidocaine suggest that the relative permeability of ionized and un-ionized species of a given compound may vary with the membrane chosen.

In the preceding paragraphs, the effect of ionization on percutaneous absorption has been discussed in terms of the relevant permeability coefficients. However, the rate of permeation of dissolved substances is a function of both the permeability coefficient and concentration within the vehicle. The maximum flux is defined as $P \times S$, the product of permeability coefficient and solubility. The un-ionized form of a given drug would have a higher P value but a smaller value of S than the charged form. Since the rate of absorption is of practical interest, it is useful to calculate a flux ratio for different forms of the same drug, as well as a permeability ratio. The data required for this calculation can be obtained by working with saturated solutions (yielding J^{max} directly) or else adjusting the flux values from a knowledge of the actual drug concentration and its solubility. The theoretical maximum flux obtained from this computation is labeled "J^*", defined in Equation 3.

$$J^* = J\left(\frac{S}{C_v}\right) \tag{3}$$

In this equation, J is the experimental flux of a solution whose concentration is C_v. The flux ratio is therefore defined as J^{max} (ionized)/J^{max} (un-ionized). or J^*(ionized)/J^*(un-ionized). As mentioned above, the permeability ratio for scopolamine was about 0.06. The flux ratio for the same compound was considerably higher, about 0.4 (14).

Figure 4 shows the relationship between J^* through human skin and pH for lidocaine.[17] The fluxes for the component species, the total calculated flux, and the measured values are included. Maximal flux is not highly dependent on pH. At low pH values, at which the compound is nearly completely ionized, the high solubility accounts for the relatively high flux obtained. As the pH is raised and more of the un-ionized form is generated, the solubility decreases, but the increase in permeability coefficient compensates for the reduction in concentration of dissolved drug and the total flux remains approximately constant.

While it is clear that, as expected, the un-ionized form of a compound partitions better into the lipoidal matrix of the stratum corneum, the data quoted above affirm that at least some charged species also penetrate the skin. It has been argued that the finite penetration of certain ions and highly polar compounds (such as sugars) indicates the existence of a polar pathway through the stratum corneum.[19]

III. QUANTITATING THE INTERACTIVE PENETRATION COMPONENT

Differences in skin delivery from various vehicles depend on both the noninteractive and interactive contributions. By accounting for the affinity of the vehicle for a given permeant, the importance of vehicle-skin interactions can be ascertained. Compounds intended to modify stratum corneum properties for the purpose of increasing skin permeation of drugs are called penetration enhancers. Other chapters in this book discuss enhancers and the mechanisms by which they operate.

A straightforward approach to measurement of vehicle-skin interactions involves pretreatment of the skin for a specified period of time by solvents or more complex vehicles.

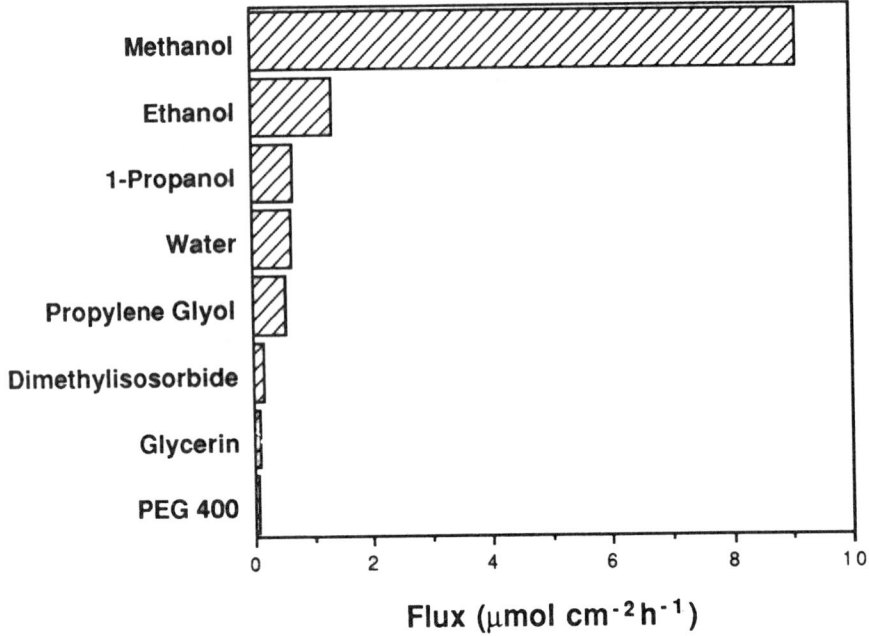

FIGURE 5.   Mean methylparaben flux through fuzzy rat skin from saturated donors. (Data from Twist, J. N. and Zatz, J. L., *J. Soc. Cosmet. Chem.*, 40, 231, 1989.

This is followed by application of a model permeant in a standard preparation, usually a simple solution. It may be assumed that pretreatment exerts no effect on the subsequently applied solution; the activity of the permeant is therefore constant. Differences in permeation behavior reflect changes in stratum corneum structure due to the pretreatment.

The interaction of stratum corneum with various solvents can also be compared by studying the penetration of compounds at the same activity. The usual reference point is the saturated solution. Thus, saturated solutions or solutions containing solute at the same degree of saturation may be compared.

Earlier in this chapter, the penetration of parabens through polydimethylsiloxane membranes was described. Despite large differences in solubility, flux was the same from saturated solution in several noninteractive solvents and solvent combinations. Different results were obtained with excised fuzzy rat skin.[20] Figure 5 contains flux data for saturated solutions of methylparaben. Propylene glycol, water, and 1-propanol yielded similar flux values. Polyethylene glycol 400 produced the lowst flux, about 7% that for water. Of the alcohols studied, methanol had the largest effect on the skin, perhaps due to extraction of intercellular lipid.

These data may be compared with results utilizing other solutes. With theophylline, flux values from the same group of solvents through fuzzy rat skin were substantially lower than for methylparaben,[20] but the same solvent rank order was obtained. In a study of permeation of benzocaine from saturated solutions through hairless mouse skin,[21] flux values for propylene glycol and water were about equal, but that for polyethylene glycol 400 was about one-tenth the other values. The flux of oxymorphone base through excised human skin was about 13 times greater from suspension in ethanol than propylene glycol.[22] Acyclovir flux from dispersions through excised rat skin was greater from isopropanol and ethanol than propylene glycol.[23] Data such as those depicted in Figure 5 allow a general categorization of solvents in terms of the interactive effect they are likely to have on skin permeation.

An important conclusion is that there is probably no vehicle that is totally inert toward

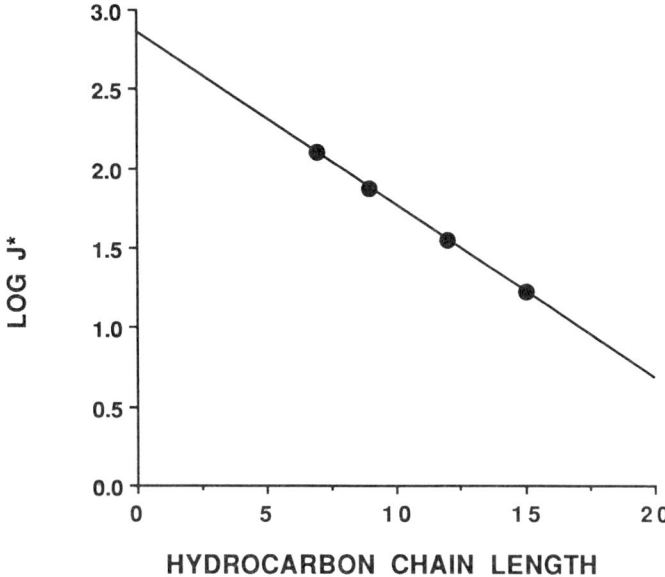

FIGURE 6. Effect of hydrocarbon chain length on normalized flux, J*, of caffeine through hairless mouse skin. (Data from Rahman, M. and Zatz, J. L., to be published.)

the stratum corneum. Water itself is known to be highly reactive, and the hydration level within the stratum corneum is a major factor affecting skin transport. We can anticipate some change in barrier properties whenever any vehicle system is applied to the skin.

In many instances, it is inconvenient or illogical to work with saturated solutions. This is true if the investigator wishes to use radiolabeled solutes or if the permeant is damaging to the skin at high concentrations. In such cases, it is possible to conduct the experiment by using a lower concentration and estimating the maximum flux as J* through Equation 3. As an example, the effect of several normal alkanes on caffeine J* for hairless mouse skin is shown in Figure 6.[24] Interaction of the solvents with the membrane decreases in magnitude as the alkane chain is lengthened.

An alternate method of separating interactive and noninteractive solvent effects was utilized by Roberts and Anderson in a study of phenol permeation.[25] The same solutions, at fixed phenol concentration, were applied to two membranes in parallel experiments. In one case, the membrane was a polyethylene film which was assumed to be inert toward the solvents. The second membrane was rat skin. Since phenol activity within a given vehicle was the same in both cases, the ratio of flux values provided an index of the degree of interaction with the skin membrane. Solvents which exhibited minimal skin interaction had a skin to polyethylene ratio of about two. The ratio for water and ethanol was approximately four. Dimethylformamide and dimethylsulfoxide had much higher ratios, 11 and 18, respectively, in line with their known tendency to disrupt the stratum corneum.

It should be noted that a simple comparison of rat skin flux values or permeability coefficients would not have been conclusive. For example, phenol flux from water was about 10 times that from dimethylsulfoxide, despite the fact that the latter disturbs the membrane barrier to a much greater extent. Dimethylsulfoxide is a much better solvent for phenol, so that solute activity was substantially lower in dimethylsulfoxide than water. This resulted in a higher escaping tendency from the water system which overshadowed the solvents' alteration of membrane properties. By cancelling the effect of differences in vehicle activity using the two membrane approach, the true interactive tendency towards skin was revealed.

Tiemessen et al.[26] described a novel method which may be useful in studying the effect of vehicles on skin penetration. They utilized a composite membrane which consisted of human stratum corneum sandwiched between silicone polymer membranes for *in vitro* studies under infinite dose conditions. Silicone adhesive was used to bond the three layers. Besides supplying mechanical strength, the silicone membranes prevent nonphysiological maceration of the skin in excess solvent. The flux obtained reflects the resistance of all the layers. The resistance due to the silicone membranes is determined by a separate experiment, and this is subtracted from the value for the composite membrane to yield the resistance for the stratum corneum itself. The technique requires that the major portion of the resistance reside within the stratum corneum, rather than the silicone layers. Therefore, drugs must permeate the polymer membrane well to be candidates for this technique.

When testing a series of formulations in which the vehicles utilized do not cause swelling of the silicone polymer, differences in penetration rate through this material alone reflect only the noninteractive contributions to flux. The flux through the composite membrane is a function of all of the factors discussed previously. Thus, the experiments utilized in the sandwich technique permit separate evaluation of the noninteractive and interactive contributions to flux.

IV. SURFACE ACTIVE AGENTS

Compounds with surface activity have traditionally been used in dermatological and cosmetic skin treatments for a variety of purposes. They serve as wetting agents, dispersants, solubilizers, and emulsifiers. Alteration of skin permeation by surface active compounds has been extensively studied and the literature has been reviewed.[27-31]

In general, nonionic surfactants are much less likely to affect skin permeability when applied in dilute solution in purely aqueous vehicles than are charged surfactants. However, alteration in the solvent can have a dramatic influence (see below).

An extensive *in vitro* study of the effect of cationic surfactants on lidocaine permeation utilized dermatomed human skin from six donors.[32] The vehicles, consisting essentially of 5% lidocaine suspensions in a 1:4 propylene glycol to water vehicle, were applied under infinite dose conditions; both water and lidocaine penetration were monitored simultaneously. The relationship between water and lidocaine flux from a vehicle containing no surfactants is shown in Figure 7. Similar correlation between *in vitro* skin penetration of water and other substances had been noted earlier.[33] Apparently, the skin membranes fell into two major groups with respect to relative permeation of both compounds.

To circumvent the variation in properties of skin from different donors, each piece of skin served as its own control in the experiments.[32] A reference formulation, containing lidocaine but no surfactant, was applied to the skin and penetration was monitored for 24 h. The skin was then washed, and a test formulation (containing surfactant), was applied. After an additional 24 h, the skin surface was washed once more and a reference formulation reapplied. The approach was validated by repeatedly applying the reference formulation; there was no significant change in flux of either water or lidocaine over the entire treatment period (72 h).

Figure 8 shows the effect of one of the surfactants, tetradecyl trimethylammonium bromide, at a concentration of 81 m*M*.[32] There is a significant increase in flux of both penetrants after 24 h. The flux remains at a high level following removal of the test donor (at 48 h) and substitution of a second control formulation.

The enhancement ratio, ER, is defined as the ratio of the flux from a test formulation to that of the initial reference. The damage ratio, DR, is defined as the flux from the second control divided by that of the initial control. The enhancement ratio is a measure of the effect of the surfactant on the stratum corneum. The damage ratio is a measure of the

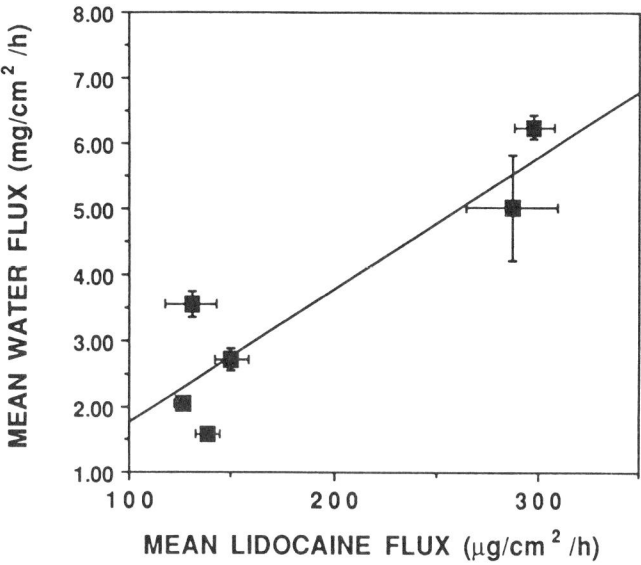

FIGURE 7. Correlation of water and lidocaine flux through dermatomed human skin. Bars indicate SEM. (From Kushla, G. P. and Zatz, J. L., *J. Pharm. Sci.*, in press. With permission.)

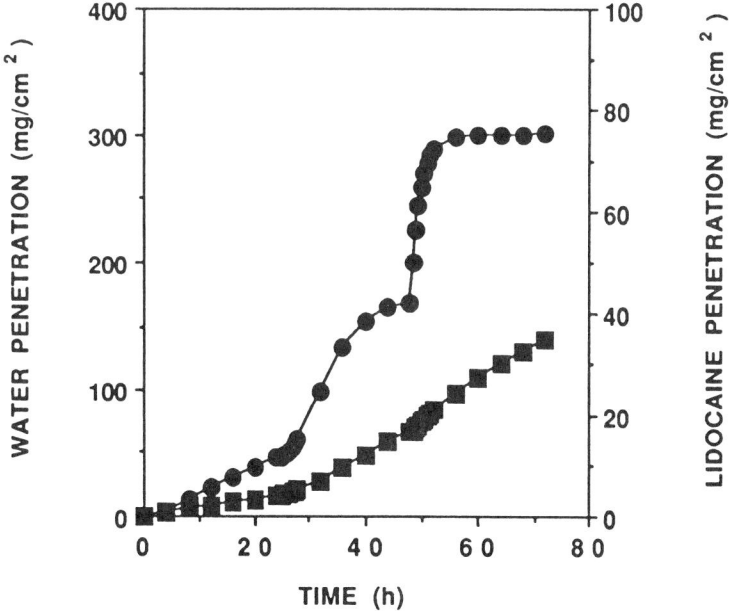

FIGURE 8. Cumulative amount penetrated through dermatomed human skin vs. time following sequential application of control; test formulation (containing 81 m*M* tetradecyl trimethylammonium bromide); control, 24-h applications; ●, lidocaine; ■, water. (From Kushla, G. P. and Zatz, J. L., *J. Pharm. Sci.*, in press. With permission.)

reversibility of the interaction under the experimental conditions. With the cationics employed in this study, the damage ratios were of approximately the same magnitude as the enhancement ratio values.

Some data from the study, including the standard errors, are shown in Figure 9.[32] The

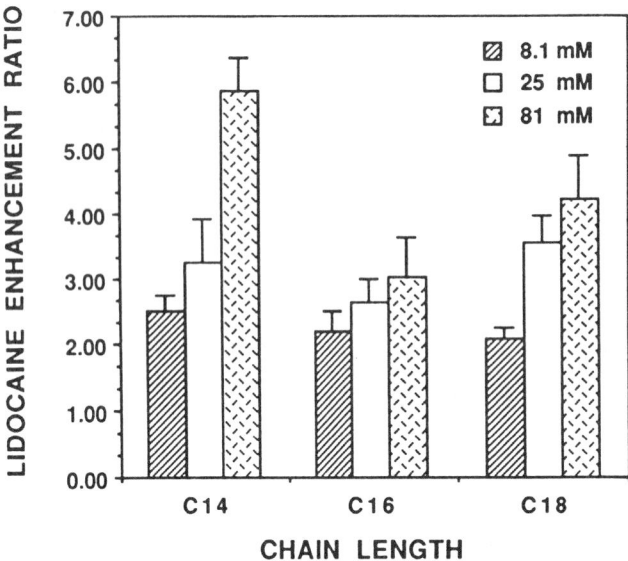

FIGURE 9.   Enhancement ratios for lidocaine as a function of alkyl chain length and surfactant concentration in 20% propylene glycol. The surfactants were members of the dimethylbenzyl ammonium chloride series. Bars indicate SEM. (From Kushla, G. P. and Zatz, J. L., *J. Pharm. Sci.*, in press. With permission.)

effect of surfactant concentration on lidocaine enhancement ratio, although relatively small, was statistically significant.

V. SOLVENTS AND SURFACE-ACTIVE PENETRATION ENHANCERS

It has become evident that the potency of surface-active compounds that increase the skin penetration of other substances (such as drugs) depends on the nature of the medium. In 1984, Cooper showed that the addition of polar lipids (fatty alcohols and acids) to a propylene glycol vehicle increased the penetration of salicylic acid across human epidermis *in vitro*.[34] Permeability from this combination was much greater than from an aqueous vehicle containing the same surface-active agents. The addition of high concentrations of propylene glycol to aqueous donors containing polysorbates dramatically increased the penetration of lidocaine[1] and hydrocortisone[35] through hairless mouse skin.

Okamoto et al. studied the penetration of acyclovir through excised hairless mouse and rat skin from vehicles containing several solvents and lipoidal additives, including 1-do-decylazacycloheptan-2-one, in combination.[23] All of the formulations contained suspended drug, to avoid differences in thermodynamic activity. None of the additives had any effect on skin penetration in an isopropyl myristate vehicle. Their activity was evident only in vehicles consisting of propylene glycol, ethanol, or isopropanol. Donor concentrations of the additives were also monitored in the same study. Although it was hypothesized that the low level of percutaneous absorption from isopropyl myristate vehicles was related to the tendency of the lipoidal additives to remain in this vehicle rather than enter the stratum corneum, data presented in the paper showed that the amounts of each additive leaving the donor from isopropyl myristate and propylene glycol, a highly effective vehicle, were much the same.

A number of combinations of myristic acid and various solvents were evaluated as potential vehicles for transdermal delivery of oxymorphone.[22] Several vehicles contained an excess of oxymorphone base, making it possible to judge the extent of their interaction with human stratum corneum. The addition of 10% myristic acid to a propylene glycol dispersion raised drug flux from 1.1 to 94 $\mu g\ cm^{-2}\ h^{-1}$. The flux from a simple ethanol dispersion was 13 $\mu g\ cm^{-2}\ h^{-1}$; there was a slight increase in flux due to addition of 10% myristic acid, to 19 $\mu g\ cm^{-2}\ h^{-1}$. Among the solvents used in combination with myristic acid, propylene glycol gave the highest oxymorphone penetration followed by ethanol. The combination of myristic acid and polyethylene glycol 400 exhibited the lowest flux of the group, 0.7 $\mu g\ cm^{-2}\ h^{-1}$.

The data suggest that the solvent itself must be a good penetrant if combinations with other agents are to increase skin penetration. However, the mechanism by which solvents and surface-active substances interact to enhance drug permeation has not been clearly defined.

VI. CONCLUSIONS

In vitro skin penetration experiments are useful in exposing the role of vehicles. Solvents and other vehicle components interact with the stratum corneum in a variety of ways. With proper experimental design, noninteractive and membrane-specific vehicle effects can be separated, increasing our understanding of the mechanisms by which vehicles and vehicle components operate to influence skin permeation. The effectiveness of surface active compounds which influence stratum corneum structure and thereby modify the permeability of other compounds is also a function of vehicle composition. The ionized form of certain drugs is capable of permeating stratum corneum, although the permeability coefficient of the unionized form is always greater.

REFERENCES

1. **Sarpotdar, P. P. and Zatz, J. L.,** Evaluation of penetration enhancement of lidocaine by nonionic surfactants through hairless mouse skin *in vitro, J. Pharm. Sci.,* 75, 176, 9186.
2. **Roberts, M. E. and Mueller, K. R.,** Comparisons of *in vitro* nitroglycerin (NTG) flux across Yucatan pig, hairless mouse, and human skins, *Pharm. Res.,* 7, 673, 1990.
3. **Block, L. H.,** Medicated applications, in *Remington's Pharmaceutical Sciences,* 17th ed., Gennaro, Ar. R., Ed., Mack Printing, Easton, PA, 1985, 1567.
4. **Furher, C.,** The classification of agents used in dermatology, in *Dermal and Transdermal Absorption,* Brandau, R. B. and Lippold, H., Eds., Eissenschaftliche Verlagsgeselischaft mbH, Stuttgart, Germany, 1982, 14.
5. **Flynn, G. L.,** Topical drug absorption and topical pharmaceutical systems, in *Modern Pharmaceutics,* 2nd ed., Banker, G. S. and Rhodes, C. T., Eds., Marcel Dekker, New York, 1990, 263.
6. **Powers, D. H. and Fox, C.,** The effect of cosmetic emulsions on the stratum corneum, *J. Soc. Cosmet. Chem.,* 10, 109, 1959.
7. **Wang, J. C. T., Winarna, S. R., Lichtin, J. L., and Patel, B. G.,** Effect of phase-volume ratio of o/w emulsion vehicles on the activity of a topically applied vasoconstrictor, *J. Soc. Cosmet. Chem.,* 39, 139, 1988.
8. **Batt, M. D., Davis, W. B., Fairhurst, E., Gerrard, W. A., and Ridge, B. D.,** Changes in the physical properties of the stratum corneum following treatment with glycerol, *J. Soc. Cosmet. Chem.,* 39, 367, 1988.
9. **Higuchi, T.,** Physical chemical analysis of percutaneous absorption process from creams and ointments, *J. Soc. Cosmet. Chem.,* 11, 85, 1960.
10. **Twist, J. N. and Zatz, J. L.,** Influence of solvents on paraben permeation through idealized skin model membranes, *J. Soc. Cosmet. Chem.,* 37, 429, 1986.
11. **Twist, J. N. and Zatz, J. L.,** Membrane-solvent-solute interaction in a model permeation system, *J. Pharm. Sci.,* 77, 536, 1988.

12. **Twist, J. N. and Zatz, J. L.,** A model for alcohol-enhanced permeation through polydimethylsiloxane membranes, *J. Pharm. Sci.,* 79, 28, 1990.

13. **Coldman, M. F., Poulsen, B. J., and Higuchi, T.,** Enhancement of percutaneous absorption by the use of volatile:nonvolatile systems as vehicles, *J. Pharm. Sci.,* 58, 1098, 1969.

14. **Michaels, A. S., Chandresakharan, S. K. and Shaw, J. S.,** Drug permeation through human skin: theory and *in vitro* experimental measurement, *A. I. Ch. E. J.,* 21, 985, 1975.

15. **Swarbrick, J., Lee, G., Brom, J., and Gensmantel, N. P.,** Drug permeation through human skin. II. Permeability of ionizable compounds, *J. Pharm. Sci.,* 73, 1352, 1984.

16. **Fleeker, C., Wong, O., and Rytting, J. H.,** Facilitated transport of basic and acidic drugs in solutions through snakeskin by a new enhancer-dodecyl N,N-dimethylamino acetate, *Pharm. Res.,* 6, 443, 1989.

17. **Kushla, G. P. and Zatz, J. L.,** Influence of pH on lidocaine penetration through human and hairless mouse skin *in vitro, Int. J. Pharm.,* in press.

18. **Kushla, G. P. and Zatz, J. L.,** Lidocaine penetration through human and hairless mouse skin *in vitro, J. Soc. Cosmet. Chem.,* 40, 41, 1989.

19. **Flynn, G. L.,** Mechanism of percutaneous absorption from physicochemical evidence, in *Percutaneous Absorption,* 2nd ed., Bronaugh, R. L. and Maibach, H. I., Eds., Marcel Dekker, New York, 1989, 27.

20. **Twist, J. N. and Zatz, J. L.,** The effect of solvents on solute penetration through fuzzy rat skin, *J. Soc. Cosmet. Chem.,* 40, 231, 1989.

21. **Zatz, J. L. and Dalvi, U. G.,** Evaluation of solvent-skin interaction in percutaneous absorption, *J. Soc. Cosmet. Chem.,* 34, 327, 1983.

22. **Aungst, B. J., Blake, J. A., Rogers, N. J., and Hussain, M. A.,** Transdermal oxymorphone formulation development and methods for evaluating flux and lag times for two skin permeation-enhancing vehicles, *J. Pharm. Sci.,* in press.

23. **Okamato, H., Muta, K., Hashida, M., and Sezaki, H.,** Percutaneous penetration of acyclovir through excised hairless mouse and rat skin: effect of vehicle and percutaneous penetration enhancer, *Pharm. Res.,* 7, 64, 1990.

24. **Rahman, M. and Zatz, J. L.,** to be published.

25. **Roberts, M. S. and Anderson, R. A.,** The percutaneous absorption of phenolic compounds: the effect of vehicles on the penetration of phenol, *J. Pharm. Pharmacol.,* 27, 599, 1975.

26. **Tiemessen, H. L. G. M., Bodde, H. E., Mollee, H. and Junginger, H. E.,** A human stratum corneum-silicone membrane sandwich to simulate drug transport under occlusion, *Int. J. Pharm.,* 53, 119, 1989.

27. **Zatz, J. L.,** Percutaneous Absorption, in *Controlled Drug Bioavailability,* Vol. 3, Smolen, V. F. and Ball, L. A., Eds., John Wiley & Sons, New York, 1985, 185.

28. **Zatz, J. L. and Sarpotdar, P. P.,** Influence of vehicles on skin penetration in *Transdermal Delivery of Drugs,* Vol. 2, Kydonieus, A. F. and Berner, B., Eds., CRC Press, Boca Raton, 1987, 85.

29. **Idson, B. and Behl, C. J.,** Drug structure vs. penetration, in *Transdermal Delivery of Drugs,* Vol. 3, Kydonieus, A. F. and Berner, B., Eds., CRC Press, Boca Raton, 1987, 85.

30. **Walters, K. A.,** Penetration enhancers and their use in transdermal therapeutic systems, in *Transdermal Drug Delivery,* Hadgraft, J. and Guy, R. H., Eds., Marcel Dekker, New York, 1989, 197.

31. **Walters, K. A.,** Surfactants and percutaneous absorption, in *Prediction of Percutaneous Absorption,* Scott, R. C., Guy, R. H., and Hadgraft, J., Eds., IBC Technical Services, London, 1990, 148.

32. **Kushla, G. P. and Zatz, J. L.,** Correlation of water and lidocaine flux enhancement by cationic surfactants *in vitro, J. Pharm. Sci.,* in press.

33. **Bronaugh, R. L.,** Percutaneous absorption: *in vitro* techniques, in *Percutaneous Absorption,* 2nd ed., Bronaugh, R. L. and Maibach, H. I., Eds., Marcel Dekker, New York, 1989, 239.

34. **Cooper, E. R.,** Increased skin permeability for lipophilic molecules, *J. Pharm. Sci.,* 73, 1153, 1984.

35. **Sarpotdar, P. P. and Zatz, J. L.,** Percutaneous absorption enhancement by nonionic surfactants, *Drug Dev. Ind. Pharm.,* 12, 1625, 1986.

Chapter 7

CUTANEOUS METABOLISM

Steven W. Collier, Jan E. Storm, and Robert L. Bronaugh

TABLE OF CONTENTS

I. INTRODUCTION

Metabolism in skin has been studied by investigators wishing to understand the processes which regulate its growth, differentiation, and function; and by investigators interested in characterizing its xenobiotic metabolizing capacity. The outer layer, the epidermis, is a differentiated tissue weighing approximately 225 g and completely regenerates itself on an average of every 45 d.[1,2] The metabolism of endogenous substrates by skin to replicate and differentiate is inextricably linked to its production of a stratum corneum with barrier functionality. Thus, the study of metabolism of carbohydrates, lipids, and proteins has begun to elucidate the role metabolism plays in maintaining the development and health of skin.

The skin's metabolic activity is also manifest in its ability to biotransform a variety of xenobiotic substances. Xenobiotic metabolism by skin has been extensively reviewed.[3-5] Knowledge of the requirements to maintain endogenous metabolism is essential to the study of xenobiotic metabolism. This chapter describes some of the techniques we use in our investigations, some of the data obtained from these studies, and how these data are related to the processes of cutaneous xenobiotic metabolism during percutaneous absorption.

II. WHY CUTANEOUS METABOLISM SHOULD BE CONSIDERED

Xenobiotic metabolic activity of the skin should be considered in the development and dispensation of transdermally delivered drugs. Although most xenobiotic metabolizing enzyme activities in the skin are substantially lower than in the liver, it is possible that the activity of some enzymes may equal or exceed that in the liver. For example, quinone reductase has been reported to be approximately 1.5 times more active in the skin than in the liver.[7] Thus the metabolizing capacity of the skin for each new drug should be examined. If biotransformation of a transdermally delivered drug is great enough, altered potency or therapeutic index may result. Induction of cutaneous enzymes over the course of the transdermal therapy could also result and manifest itself in toxic effects not present during initial stages of therapy. In addition, certain disease states and lifestyles might also alter the enzymatic activity of the skin resulting in altered patterns of biotransformation. Finally, metabolism of systemically absorbed compounds, and, thus, their pharmacokinetic behavior, might be influenced by the extent of their metabolism by skin.[8,9]

Xenobiotic metabolism in the skin should also be considered when attempting to characterize the health risks resulting from dermal exposure to chemicals present in air, water or soil; present in the workplace; or, present in consumer products such as cosmetics. Many chemicals, such as mutagens and carcinogens, may be biotransformed in the skin to reactive metabolites which form covalent bonds with tissue proteins initiating the mutagenic or carcinogenic process. Even minute levels of metabolic activity could initiate mutagenic events if a carcinogen is percutaneously absorbed.

III. DETERMINATION OF METABOLISM IN SKIN

In vivo, the extent of cutaneous metabolism is difficult to differentiate from systemic metabolism. Frequently the rates of cutaneous metabolism are low when compared to the liver and the detection of metabolites is sometimes precluded by the quantity of blood which must be collected. When venous effluents are collected for analysis, the contribution of skin metabolism is confounded with that of metabolism in the blood. A better qualitative and quantitative determination of cutaneous metabolism can often be performed *in vitro*. *In vitro* experimental techniques employed for the study of skin metabolism include the use of cytosolic and microsomal fractions or homogenates, culture of isolated skin cells, and organ culture of explanted primary skin tissue. Some important considerations must be made in the experimental design and data interpretation using any of these techniques. For example, enzymatic assays utilizing skin subcellular fractions or homogenates are usually optimized for the enzyme process(es) which is(are) anticipated for the specific chemical under study. Unexpected biotransformations which would result in unusual metabolites may therefore be overlooked. Cells in culture undergo a selection process with each passage which may include changes in their metabolizing capacities. Moreover, the phenotype which they express is that of the cell(s) which thrive under the culture conditions provided and not necessarily the phenotype of the primary tissue from which they were originally isolated. There are also special considerations for organ explant cultures, such as maintaining viability and preventing microbial overgrowth, which are discussed in this section and in Chapter 5.

The predominant experimental technique for cutaneous enzyme assays has been through the use of epidermal homogenates[10-13] and epidermal cells in culture.[14] Microsomal and cytosolic fractions have been used in our laboratory to study the activities and kinetic parameters of some enzymatic processes and compare activities between skin and liver.[6,15,16] A species and organ comparison of aryl hydrocarbon hydroxlyase (AHH), 7-ethoxycoumarin deethylase (ED) and glutathione-S-transferase (GST) activity was conducted in rodents. AHH and ED activities were determined in microsomal fractions prepared from liver and skin.[15] Microsomal fractions were prepared according to a method derived from Mukhtar and Bickers,[17] and Rettie et al.[18]

Interspecies differences are apparent in AHH activity in both skin and liver (Figure 1). AHH activity in mouse skin is two to six times greater than in the skin of rats. Cutaneous AHH activity in SENCAR mice is somewhat greater than in BALB/c mice in these experiments while hepatic AHH activity was equivalent in both strains (Figure 1). AHH activity in liver of both strains of mice is equivalent and about seven times greater than AHH activity in liver of either rat strain.

Interspecies differences are also apparent in ED activity of both skin and liver (Figure 2). Both mouse strains exhibit similar cutaneous activity, which is about 20 times greater than either rat strain. A similar pattern of ED activity exists in liver (Figure 2). Both mouse strains exhibit similar hepatic activity which is about 7 times greater than either rat strain.

Interspecies differences are also observed in glutathione-S-transferase activity of both skin and liver (Table 1). Because GST activity of cytosolic fractions is so much greater than either AHH or ED activity of microsomal fractions, it was possible to derive kinetic parameters for this enzyme. The liver V_{max} in either strain of mice is about twice that found in rats; while the skin V_{max} in either strain of mice is more than five times greater than it is in rats. No interspecies differences in the affinity of this enzyme for the substrate used (CDNB) were observed.

This study demonstrates a markedly greater metabolic capacity of mouse skin as compared to rat skin. Activity of two Phase I enzymes, AHH and ED is several times higher in BALB/c and SENCAR mice than in fuzzy and Osborne Mendel rats. Also the V_{max} of the Phase II enzyme, GST, was several times higher in both mouse strains than in either rat strain.

The overall metabolizing capacity of skin was less than that of liver by nearly 2 orders of magnitude. The AHH activity in skin is only 0.3% (BALB/c), 0.6% (SENCAR), 0.8% (Osborne Mendel), or 1.0% (fuzzy) of its activity in liver. ED activity in skin is only 0.2% (BALB/c), 0.3% (SENCAR), or 0.1% (Osborne Mendel, Fuzzy) of its activity in liver. Similarly, the maximal velocity of GST in skin is only 1% (rats) or 3% (mice) of its maximal velocity in liver.

The microsomal metabolism of caffeine, butylated hydroxytoluene (BHT), acetyl ethyl tetramethyl tetralin (AETT), and salicylic acid were also studied in the skin and liver of fuzzy rats.[6] The results of these determinations and other literature values for skin and liver

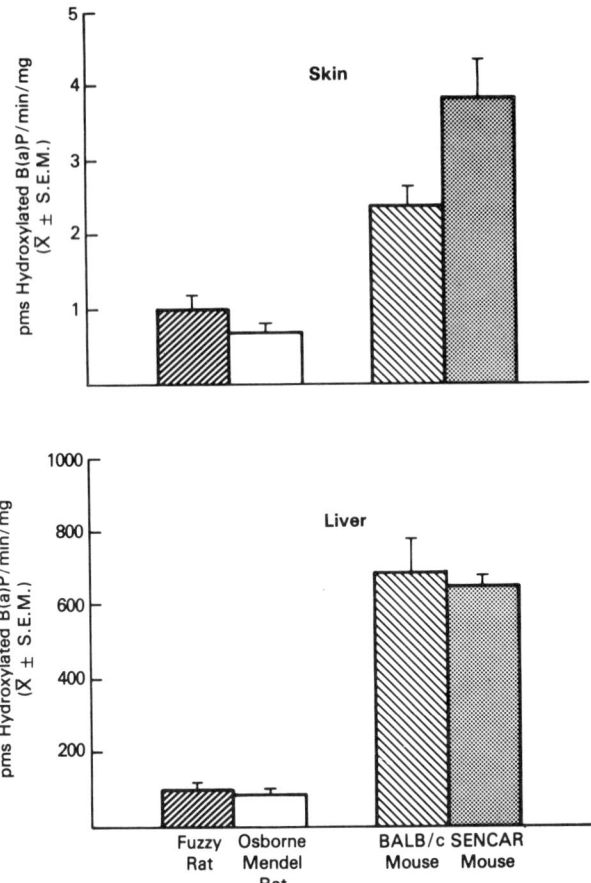

FIGURE 1. Aryl hydrocarbon hydroxylase activity in skin and liver microsomes. Tissues from two rats or mice were pooled prior to initial homogenization, and samples were assayed in duplicate. Three or four samples were assayed for each tissue and enzyme. Liver assays contained approximately 0.5 mg protein and were incubated at 37°C for 10 min. Skin assays contained approximately 1.0 mg protein and were incubated at 37°C for 30 min. AHH activity was determined by measuring the hydroxylated B[a]P formed using B[a]P as substrate according to a method slightly modified from Nebert and Gelboin.[19] All values are Means ± SEM of three or four determinations.

activities are given in Table 2. A similar pattern is seen in the relative ratios of skin and liver activities, i.e., skin activity may be as much as one or two orders of magnitude less than liver activity.

It is possible that the lower enzyme activity in skin compared to liver results from the fact that there is a tenfold lower microsomal protein concentration in skin as compared to the liver of rats.[22] It is clear that even if a compound is metabolized extensively following systemic administration, (such as some of the compounds listed in Table 2) it may still undergo little or no biotransformation during passage through the skin.

IV. EXPERIMENTAL DETERMINATION OF CUTANEOUS METABOLISM DURING PERCUTANEOUS ABSORPTION

Skin is not a homogeneous tissue. From the basal layer to the stratum corneum, lies a continuum of epidermal differentiation with water and ionic gradients.[23,24] The low pH of

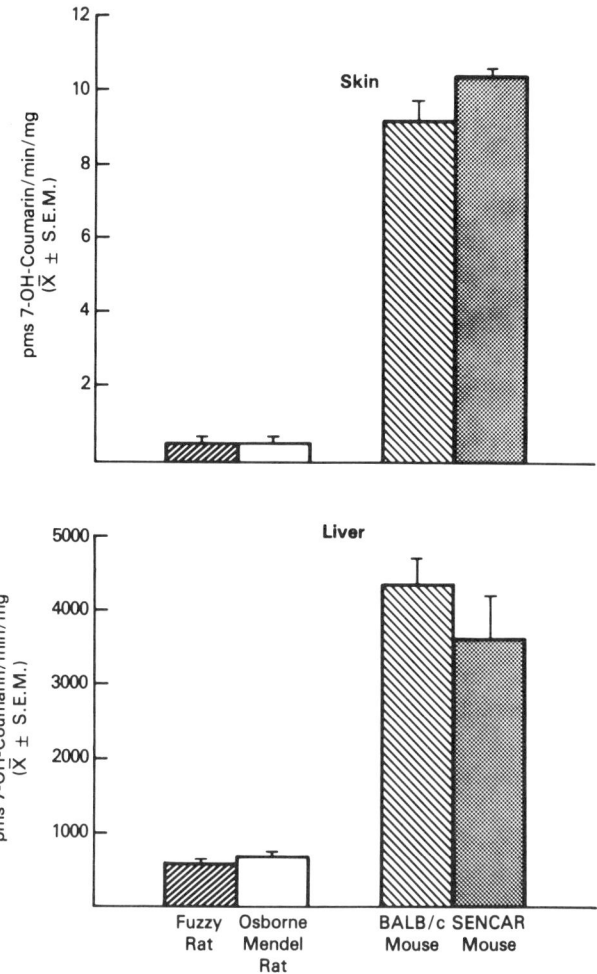

FIGURE 2. 7-Ethoxycoumarin deethylase activity in skin and liver microsomes. Tissues from two rats or mice were pooled prior to initial homogenization and samples were assayed in duplicate. Three or four samples were assayed for each tissue and enzyme. Liver assays contained approximately 0.5 mg protein and were incubated at 37°C for 10 min. Skin assays contained approximately 1.0 mg protein and were incubated at 37°C for 40 min. ED activity was determined by measuring the hydroxycoumarin formed using 7-ethoxycoumarin as substrate according to a method slightly modified from Greenlee and Poland.[20] All values are Means ± SEM of three or four determinations.

the epidermis may be due to its highly anaerobic nature[25] and this may govern certain cutaneous metabolic processes. The dermal portion of skin is structurally, functionally, and embryologically distinct from the epidermis. The use of fresh viable skin for the measurement of metabolism during percutaneous absorption preserves the barrier and architectural properties of the organ and makes it a useful model for estimating metabolic rates and processes encountered *in vivo*.

The determination of metabolism in skin is predicated on the assumption that viable tissue is used in the study. A discussion of what conditions are suitable for maintaining viable skin in flow-through diffusion cells has been presented earlier in this book (Chapter 5). Investigators have used skin in organ culture for the measurement of metabolism during

TABLE 1
Kinetic Parameters of Cytosolic Glutathione-S-Transferase Activity in Two Strains of Mice and Rats

Strain/species	Parameter	Liver	Skin
Fuzzy rat	V_{max} (nm/min/mg)	1432	14
	K_m (mM)	0.17	0.13
Osborne Mendel rat	V_{max}	963	10
	K_m	0.18	0.11
Balb/c mouse	V_{max}	2181	65
	K_m	0.36	0.51
SENCAR mouse	V_{max}	1984	67
	K_m	0.24	0.33

Note: GST activity was determined in cytosolic fractions prepared from liver and skin according to a method slightly modified from Habig et al.[21] using chlorodinitrobenzene (CDNB) as substrate. Liver assays contained 0.05 mg protein and skin assays contained approximately 2.0 mg protein; both were incubated for 5 min at 37°C. Conjugation of CDNB with glutathione was determined at 4 concentrations ranging from 0.08 to 0.64 mM. Values from 3 to 4 samples at each substrate concentration were averaged and kinetic parameters (V_{max} and K_m) were derived using Lineweaver Burk plots.

TABLE 2
Cutaneous and Systemic Metabolism in the Rat

Compound	Activity of microsomal enzymes	
	Liver	Skin
	(pmol/min/mg protein)	
Caffeine	4.1	N.D.[a]
BHT	17000	113
AETT	667	2.5
Salicylic acid	15.3	N.D.[a]

Other Compounds Metabolized During Percutaneous Absorption

Benzo(a)pyrene	1670[36]	4.7[37]
Testosterone	4200[38]	10[39]
Estradiol	16840[40]	25[39]

[a]   N.D. = not detected.

percutaneous absorption.[6,16,26-30] The condition of the skin during the experiment remains a preoccupation with investigators correlating *in vitro* skin metabolism results to the *in vivo* situation. Receptor fluids such as distilled water, normal saline, phosphate buffered saline, and organic solvent solutions are not compatible with the continued metabolic functioning of freshly prepared skin. In static diffusion cells with physiological receptor solutions, skin loses viability quickly if the receptor fluid is not frequently replaced.[31] An incubation study of skin sections demonstrates how skin rapidly loses its glycolytic ability.

Dermatome sections of hairless guinea pig skin, 200 μm thick, 14 mm diameter, were floated in 3 ml of HHBSS or HHBSS with 1.5% of poloxamer 188 or poloxamer 338. The

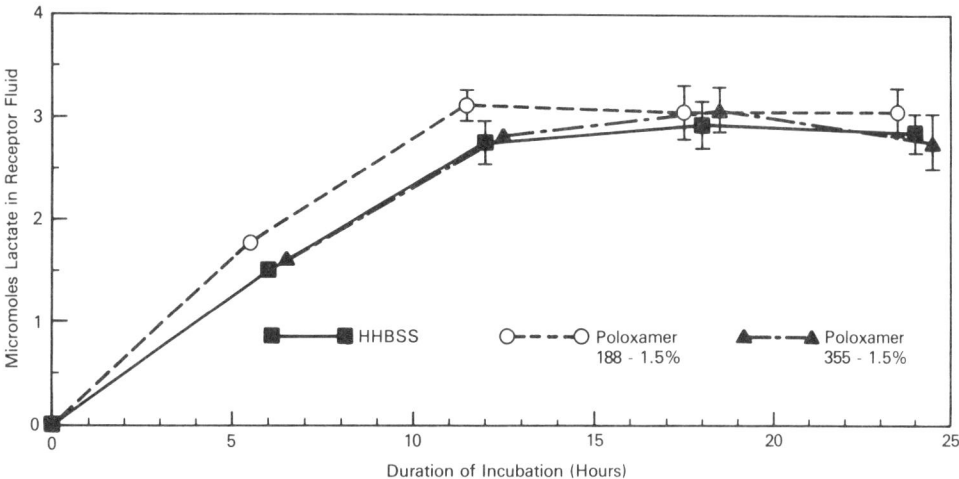

FIGURE 3. Lactate accumulation in incubation medium. Fuzzy rat skin, 200 μm thick, 14 mm diameter, was floated in 3 ml HHBSS or HHBSS containing 1.5% poloxamer 188 or 1.5% poloxamer 338. All values are Means ± SEM of four determinations.

HHBSS solutions contained 1 g dextrose/liter (5.6 mmol). Inclusion of 1.5% surfactant showed no differences in initial rates of glycolysis. Lactate production ceased after 2.5 to 3 μmoles lactate accumulated in the medium (approximately 12 h of incubation) for all receptor fluids (Figure 3). This represents a 9% utilization of the total available glucose in the incubation medium. For long duration studies utilizing viable skin in static diffusion cells, periodic replacement for the diffusion cell contents is necessary. The use of flow-through diffusion cells eliminates this problem by continually replacing receptor fluid. The skin flap system described by Riviere (Chapter 16) can be successfully configured to use a recirculating receptor fluid with a reservoir of suitable size and with proper monitoring and adjustment of receptor fluid glucose levels.

A. VEHICLE EFFECTS

The viability of skin *in vitro* is dependent on the environment into which it is placed. While receptor fluid effects are considered to be most important, the effects of topical vehicles or test compounds should not be ignored. The use of acetone as a volatile vehicle for surface disposition of test materials has been scrutinized for potential membrane alteration.[32] Recently, a high and low dose acetone application was tested for its effects on glycolysis by dermatomed hairless guinea pig skin sections maintained in flow-through diffusion cells. Using a HEPES-buffered Hanks' balanced salt solution with glucose[33] (pH 7.4) as a receptor fluid, fractions were collected at the initial and terminal portions of a 48-h perfusion study. The rate of lactate formation for either treatment was not significantly different from the untreated skin at the initial or terminal portions of the study (Figure 4). Furthermore, glycolytic activity remained at initial rates for the duration of the 48-h perfusion. The high dose (50 μl/0.64 cm²) of acetone was considered to be in great excess of any amount which would ever be used as a vehicle and that viability is not greatly affected by the use of this or lower surface densities of application. Acetone is a very volatile solvent and is perhaps the most common volatile vehicle for surface disposition of test compounds in percutaneous absorption studies. If other vehicles are used in a metabolism study, particularly less volatile ones, their effects on tissue should be determined.

B. MAMMALIAN CELL VS. MICROBIAL METABOLISM

Roth and James have reviewed skin microflora and the conditions affecting their growth.[34]

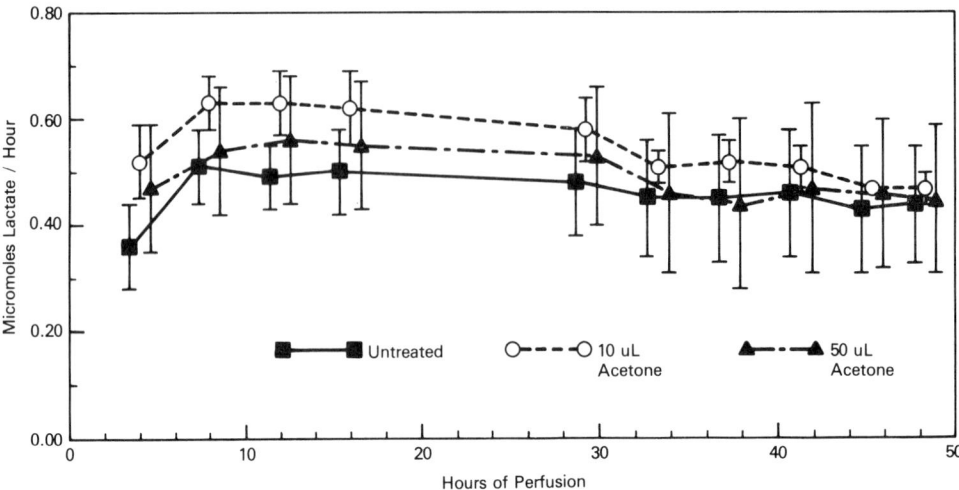

FIGURE 4.   Rates of lactate production by Fuzzy rat skin in flow-through diffusion cells. Skin sections were allowed to equilibrate for 1 h before topical dosing by pipette with either 10 or 50 µl of acetone per 0.64 cm² or left untreated. Anaerobic glucose utilization was determined by the lactate dehydrogenase method. Values are Means ± SEM of three or four determinations.

The skin is colonized by a wide variety of microorganisms. Micrococcaceae, coryneform organisms, propionibacteria, mycoflora, and occasionally Gram-negative rods populate normal skin. Their relative populations and proportions vary with species, climate, body location, age, and sex. The normal host defenses against bacterial overpopulation include an intact stratum corneum, exfoliation, an acid mantle, topically secreted IgA antibodies, and antibacterial compounds produced by resident organisms. Bacterial adherence to skin is affected by an intact stratum corneum and rapid corneocyte turnover. An unbroken barrier prevents binding of the bacteria to unexposed attachment factors such as fibronectin and fibrinogen. The low pH of the skin (approximately 5.5) is caused in part by fatty acids liberated from sebaceous triglycerides by microflora. This acid mantle is inhibitory to some organisms and stimulatory to propionibacteria. Perturbation of the normal skin lipids or pH results in increased bacterial colonization.[35]

The normal *in vivo* condition of skin in percutaneous absorption studies is one of an intact stratum corneum with a constant, nonpathogenic state of microflora colonization. It is possible under these conditions to have a contribution of surface microbial metabolism to topically absorbed compounds. This scenario suggests the possibility of microbial alteration of percutaneous absorption by biotransformation of the parent compound. *In vitro* studies have been aimed at elucidation of the contributions of mammalian cell metabolism. The placement of nonsterile primary tissue into a diffusion cell apparatus with a physiological receptor fluid is basically an inoculation of the receptor fluid with skin microflora. Appropriate techniques are required to prevent microbial overgrowth of the skin or receptor fluid.

An investigation of estradiol biotransformation to estrone demonstrates two potential problems with measurement of *in vitro* cutaneous metabolism.[33] Table 3 shows the percentage of metabolized estradiol penetrating the skin and partitioning into the receptor phase as a function of receptor fluid. The percentage of estrone appearing in the receptor fluid remained constant at approximately 20% throughout the 24-h duration of the experiment for skin sections perfused with MEM or HHBSS with average ethyl acetate extraction efficiencies of 93.1% and 94.4%, respectively. PBSG perfused skin showed declining rates of metabolism consistent with its inability to sustain glucose metabolism (Chapter 5). Extraction efficiency was 99.1% for the PBS fractions. No estrone was recovered from distilled water control

TABLE 3
Effect of the Receptor Fluid on the Measurement of Estradiol
Metabolism in Diffusion Cells

Fraction receptor fluid	Estrone formation (% of sample radioactivity)			
	0—6 h	6—12 h	12—18 h	18—24 h
MEM + FBS	1.1 ± 0.04	1.4 ± 0.04	2.1 ± 2.1	3.0 ± 2.4
MEM	19.4 ± 2.6	20.3 ± 1.9	22.4 ± 1.8	23.8 ± 2.5
HHBSS	20.8 ± 1.3	20.1 ± 1.4	21.3 ± 1.3	20.6 ± 1.6
PBSG	17.2 ± 2.0	16.2 ± 3.3	7.7 ± 1.1	5.3 ± 1.3
Water	2.3 ± 1.0	3.1 ± 0.8	3.2 ± 0.7	6.7 ± 0.8
Water + gentamicin	0	0	0	0

Note: Values are the Mean ± SEM of 4 to 6 determinations. MEM = minimum essential medium; FBS = fetal bovine serum; HHBSS = HEPES-buffered Hanks' balanced salt solution; PBSG = phosphate-buffered saline with 0.1% glucose. Estradiol was applied to fuzzy rat skin (5 $\mu g/cm^2$).

perfusions when gentamicin sulfate was added and high recoveries (99.4%) were obtained. Exclusion of gentamicin sulfate from the distilled water perfusate showed baseline recovery of approximately 3% estrone from the receptor fluid. The final distilled water fraction (18 to 24 h) contained twice the amount of estrone as the preceding fractions. Untreated skin sections perfused with gentamicin-deficient receptor fluids showed the presence of Gram-positive and Gram-negative rods under histopathological examination. It seems likely that the estrone formation in the gentamicin-deficient distilled water was due to bacterial metabolism rather than mammalian cell metabolism. The supplementation of MEM with 10% FBS resulted in an apparent reduction of metabolism. A corresponding reduction in the efficiency of the ethyl acetate extraction (89.7%) was noted and believed to be the result of binding of estrone to the serum protein, thereby masking metabolism.

These results demonstrate that adequate measures must be taken to insure the microbiological integrity of the diffusion cell system so as to be able to differentiate between mammalian cell and bacterial metabolism. The apparatus must be sterilized prior to each study. This is routinely done by dipping the diffusion cells in 70% ethanol (EtOH) solution and assembling them with a section of Parafilm in place of the skin section. The EtOH solution is placed in the receptor fluid reservoir and the entire system flushed after assembly. The EtOH solution is drained and replaced with a receptor fluid which has been sterilized by passage through a sterile 0.2 μm filter. Antibiotics such as gentamicin or a penicillin and streptomycin mixture are added to the receptor fluid to suppress bacterial growth. Receptor fluid fractions are either extracted immediately after collection, or refrigerated, or frozen until they can be extracted.

The handling of the skin after harvest and before mounting on the diffusion cell may affect its microbial integrity. Sterilization of a primary tissue is not practical without producing cytotoxicity. Vigorous scrubbing may disrupt the barrier layer of the skin. We have found that a mild rinse with a 1% detergent solution in distilled water to remove loose surface debris before preparing a dermatome section to be sufficient preparation. The skin is pinned to a Styrofoam block and lightly patted dry with a paper towel. A moist skin surface lubricates the passage of the dermatome over the skin. The dermatome surfaces should be wiped clean with a 70% EtOH solution before and after use. All instruments and surfaces which contact the skin are also sterilized with an EtOH solution rinse prior to their use. The skin may then be punched into circles and mounted in the sterilized diffusion cell apparatus.

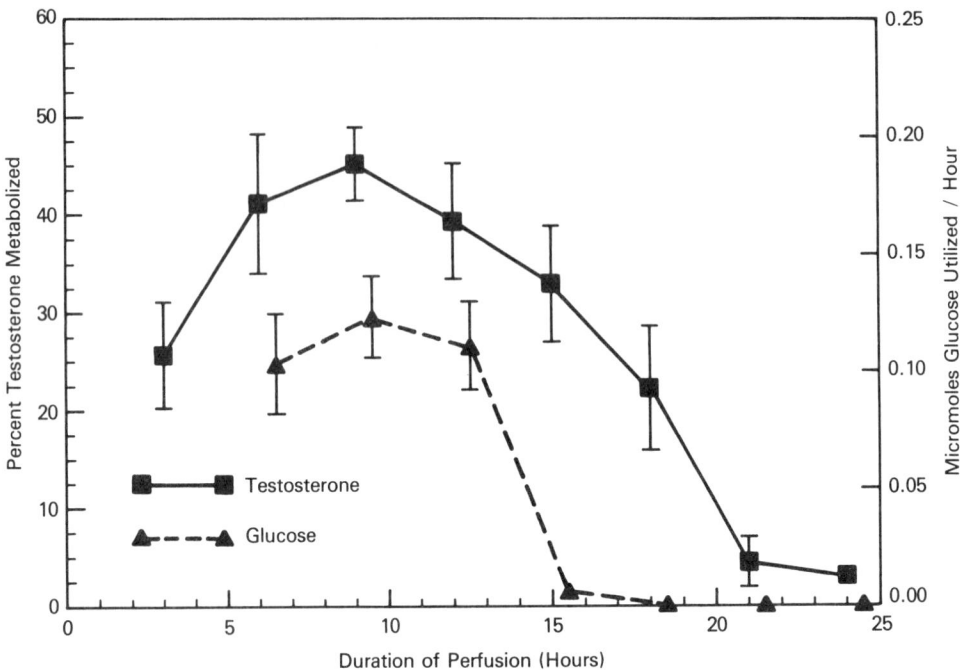

FIGURE 5.   Inhibition of lactate formation and testosterone metabolism by sodium fluoride. Fuzzy rat skin sections (200 μm thick) were placed in flow-through diffusion cells. [^{14}C]-testosterone was applied to the stratum corneum in 10 μl acetone vehicle at a surface density of 5 μg/cm^2. Initial receptor fluid was HHBSS (1.5 ml/h). HHBSS with sodium fluoride (10 mM) was initiated after 12 h. Parent compound and metabolites were extracted from receptor fluid fractions two times with 10 ml ethyl acetate, dried with 10 g anhydrous sodium sulfate and concentrated under a nitrogen stream for thin-layer chromatography (TLC) analysis. The receptor fluid extracts were spotted on silica gel TLC plates and chromatographed with a cyclohexane/ethyl acetate (4:3) solvent system for testosterone and metabolites. Following chromatography, the location and concentration of the parent compound and metabolite(s) were determined using a TLC imaging plate scanner. Lactate production was used to determine the rate of anaerobic glucose utilization and was measured by the lactate dehydrogenase method. Testosterone metabolites co-chromatographed with standards for 5α-androstane-3,17-dione, and 4-androstane-3,17-dione or 5α-dihydrotestosterone. All values are Means ± SEM of three or four determinations.

C. INHIBITION OF CUTANEOUS METABOLISM

As with any metabolic study, proper controls are necessary to establish to what extent the appearance of a compound is attributable to metabolism. Studies with tissue homogenates frequently use parallel samples in which the protein has been denatured by heating as controls. This is not practical with intact skin sections as the barrier properties may be affected. Freezing skin to destroy its viability has been employed[27] as well as perfusion of skin with receptor fluids which do not maintain viability.[6,16] Any penetrating material which is not the parent compound is ascribed to be the product of nonenzymatic processes and subtracted from the quantity appearing in the receptor fluid and skin section at the end of an experiment with viable skin. The inhibition of metabolism by the addition of an inhibitor to the receptor fluid may be suitable for metabolic controls. The inhibition of testosterone metabolism and glycolysis in fuzzy rat skin by sodium azide and sodium fluoride was studied. The amounts of metabolite(s) recovered in the receptor fluid fractions are expressed as a percentage of the total radioactivity penetrating into the fraction (Figures 5 and 6).

The results show that 10 mM sodium fluoride, effectively eliminates glycolysis shortly after infusion. Testosterone metabolism was essentially eliminated by the end of the 24-h experiment. The infusion of 10 mM sodium azide had a small effect on glycolysis by the end of the 24-h experiment and a moderate inhibitory effect on testosterone metabolism.

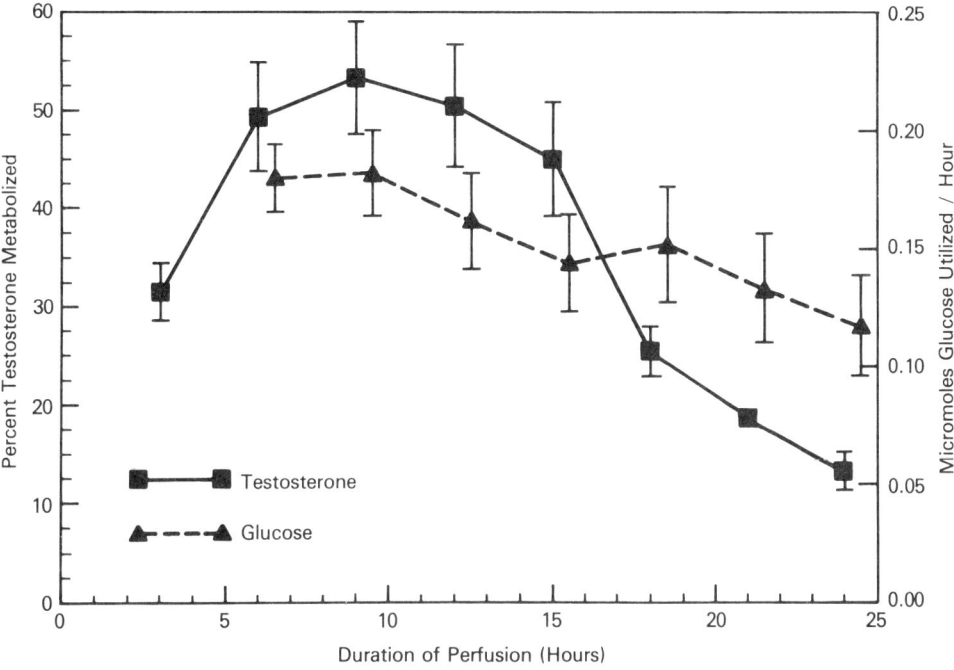

FIGURE 6. Inhibition of lactate formation and testosterone metabolism by sodium azide. Fuzzy rat skin sections (200 μm thick) were placed in flow-through diffusion cells. [^{14}C]-testosterone was applied to the stratum corneum in 10 μl acetone vehicle at a surface density of 5 μg/cm². Initial receptor fluid was HHBSS (1.5 ml/h). HHBSS with sodium azide (10 mM) was initiated after 12 h. Parent compound and metabolites were extracted from receptor fluid fractions two times with 10 ml ethyl acetate, dried with 10 g anhydrous sodium sulfate and concentrated under a nitrogen stream for TLC analysis. The receptor fluid extracts were spotted on silica gel TLC plates and chromatographed with a cyclohexane/ethyl acetate (4:3) solvent system for testosterone and metabolites. Following chromatography, the location and concentration of the parent compound and metabolite(s) were determined using a TLC imaging plate scanner. Lactate production was used to determine the rate of anaerobic glucose utilization and was measured by the lactate dehydrogenase method. Testosterone metabolites co-chromatographed with standards for 5α-androstane-3,17-dione, and 4-androstane-3,17-dione or 5α-dihydrotestosterone. All values are Means ± SEM of three or four determinations.

The absolute kinetics of testosterone metabolism inhibition might be confounded by the continued outward diffusion of testosterone metabolites formed before introduction of the inhibitors. The difference in extent of metabolic inhibition resulting from the addition of sodium azide or fluoride to the receptor fluid can be explained by their mechanism of action. Azide is a potent cytochrome oxidase inhibitor. Fluoride in the presence of phosphate forms the phosphofluoridate ion which strongly inhibits the glycolytic enzyme, enolase. Inhibition of the glycolytic cycle prevents the formation of reducible substrates resulting in an abolition of ATP formation from glucose. Cytochrome oxidase inhibition blocks the respiratory chain but does not immediately affect glycolysis. Since anaerobic glucose utilization predominates in skin, a sizable portion of the total cutaneous ATP is generated in this manner. Inhibition of the respiratory chain, while greatly decreasing ATP production, does not completely block it. Cellular processes requiring ATP may retain their function for a longer period of time with respiratory chain blockade than with inhibition of glycolysis.

The leaching of an esterase and a deaminase from skin into the receptor fluid in static diffusion cells has been reported.[41,42] In such a situation, retention of enzymatic activity in the receptor fluid results in an overestimation of cutaneous metabolism due to continued biotransformation of the parent compound in the receptor fluid. Hydrolytic reactions are the only Phase I reactions that do not utilize energy. Such enzymes lack the requirements of

TABLE 4
Activity of Microsomal Enzymes in Skin[a]

Species	Aryl hydrocarbon hydroxylase (pmol/min/mg protein)	Ethoxycoumarin deethylase (pmol/min/mg protein)
Hairless guinea pig	2.51 ± 0.35[b]	3.8 ± 2.7[c]
Human	0.24 ± 0.08[b]	Not detectable
SENCAR mouse	3.35 ± 0.07[b]	10.4 ± 1.4

[a]   Values are Means $\pm$ SEM of three or four determinations. Hairless guinea pig and human skin were from 200 μm sections: mouse skin was full thickness.
[b]   Significantly different from both species (LSD test, $p < 0.01$).
[c]   Less than SENCAR mouse (*t* test, $p = 0.06$).

From Storm, J.E., Collier, S.W., Stewart, R.F., and Bronaugh, R.L., *Fundam. Appl. Toxicol.*, 15, 132, 1990.

cofactors for their activity and so might be expected to function in a dilute solution. Other metabolic events requiring a source of biochemical energy are dependent on the intact cell, its mitochondria, and a catabolic substrate, or the artificial introduction of an electron carrier for continued metabolism. Hydrolytic enzymes such as esterases occur in the cytosol and the endoplasmic reticulum. They are also present extracellularly in the stratum corneum. While it may seem unnecessary to maintain skin section viability for the study of certain specific hydrolytic biotransformations, cell lysis might increase leakage of hydrolytic enzymes into receptor fluid fractions thereby exaggerating cutaneous metabolic effects. Serum supplements and purified albumin fractions added to receptor fluids to increase partitioning of penetrating compounds may also possess hydrolytic enzymatic activity. Incubation of the test compound in aliquots of the receptor fluid and extraction and analysis for metabolites can be used to control extracutaneous nonenzymatic or enzymatic reactions.

D. METABOLITE KINETICS DURING PERCUTANEOUS ABSORPTION

For percutaneously absorbed compounds with biologically active metabolites (toxic or therapeutic), metabolic rates determined during percutaneous absorption provide the intermediate kinetic data for linking pharmacokinetic parameters such as rates of absorption with pharmacodynamic effects. Saturability of cutaneous metabolic pathways may be common for moderately and quickly absorbed compounds. To investigate some of these relationships, activities for AHH and ED were determined in mouse, guinea pig, and human skin[16] (Table 4). Benzo[a]pyrene (B[a]P) is very lipophilic and resists partitioning into aqueous receptor fluids of *in vitro* systems (Chapter 5). The more polar metabolites of B[a]P are able to partition into receptor fluids where the majority of metabolites were recovered in *in vitro* diffusion cell studies. Figure 7 shows the cumulative metabolic disposition of B[a]P in Sencar mouse and hairless guinea pig skin from these studies and the rates of metabolism based upon the appearance of metabolites in the receptor fluid fractions. Metabolism in human skin was not reliably detectable in these experiments. Since B[a]P did not readily partition into receptor fluid its concentration at the enzyme sites in skin was presumably saturating. Thus the rates of metabolism illustrated in Figure 7 (inset) probably represent the maximum possible metabolic capability of skin for this substrate. Sencar mouse skin metabolized more B[a]P than hairless guinea pigs which in turn metabolized more B[a]P than human skin. This pattern coincides with the pattern of AHH activity illustrated in Table 4, where Sencar mouse cutaneous AHH activity is greater than hairless guinea pig AHH activity which was in turn greater than human cutaneous AHH activity, and suggests that cutaneous AHH activity was a limiting factor in the amount of B[a]P metabolites formed.

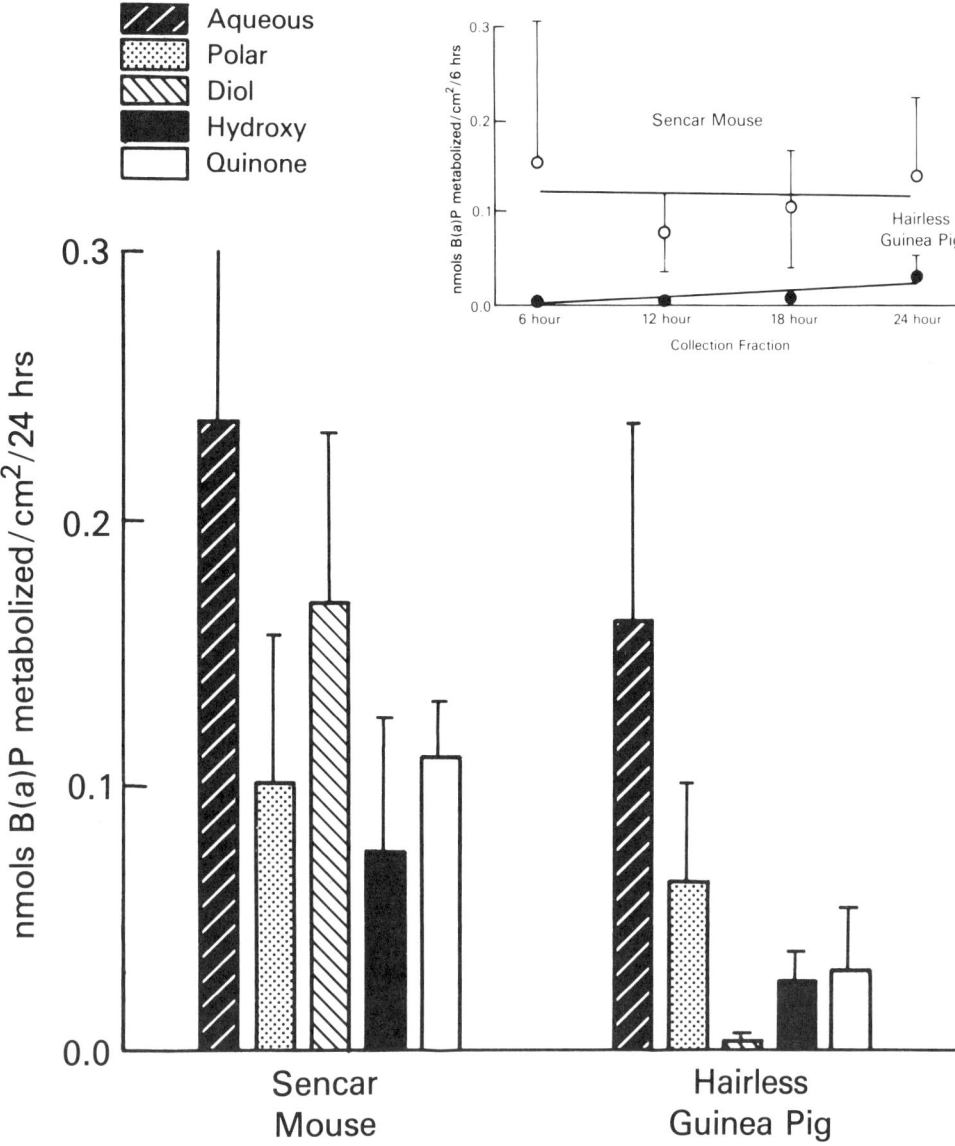

FIGURE 7. B[a]P metabolized (nmol/cm²/24 h and nmol/cm²/6 h [inset]) by Sencar mouse and hairless guinea pig skin in flow-through diffusion cells. Values are Means ± SEM of four or five determinations. (From Storm, J.E., Collier, S.W., Stewart, R.F., and Bronaugh, R.L., *Fundam. Appl. Toxicol.*, 15, 132, 1990.)

The hydrophilic compounds, 7-ethoxycoumarin and 7-hydroxycoumarin, readily partition into the receptor fluid. This makes it possible to quantitate parent and metabolite in the receptor fluid at different times during the course of 7-ethoxycoumarin penetration, and the relationship between rate of penetration and extent of metabolism could be more fully elucidated. Ethoxycoumarin deethylase is most active in Sencar mouse skin, less active in hairless guinea pig skin, and was not reliably detectable in human skin (Table 4). The higher activity in mouse and guinea pig skin is reflected in the inability to saturate the deethylase by percutaneously penetrating 7-ethoxycoumarin (Figure 8). With an ever-increasing penetration of 7-ethoxycoumarin, there was an increased formation and penetration of its metabolite 7-OH-coumarin in these species (Figure 8). Whereas, in human skin, the enzyme

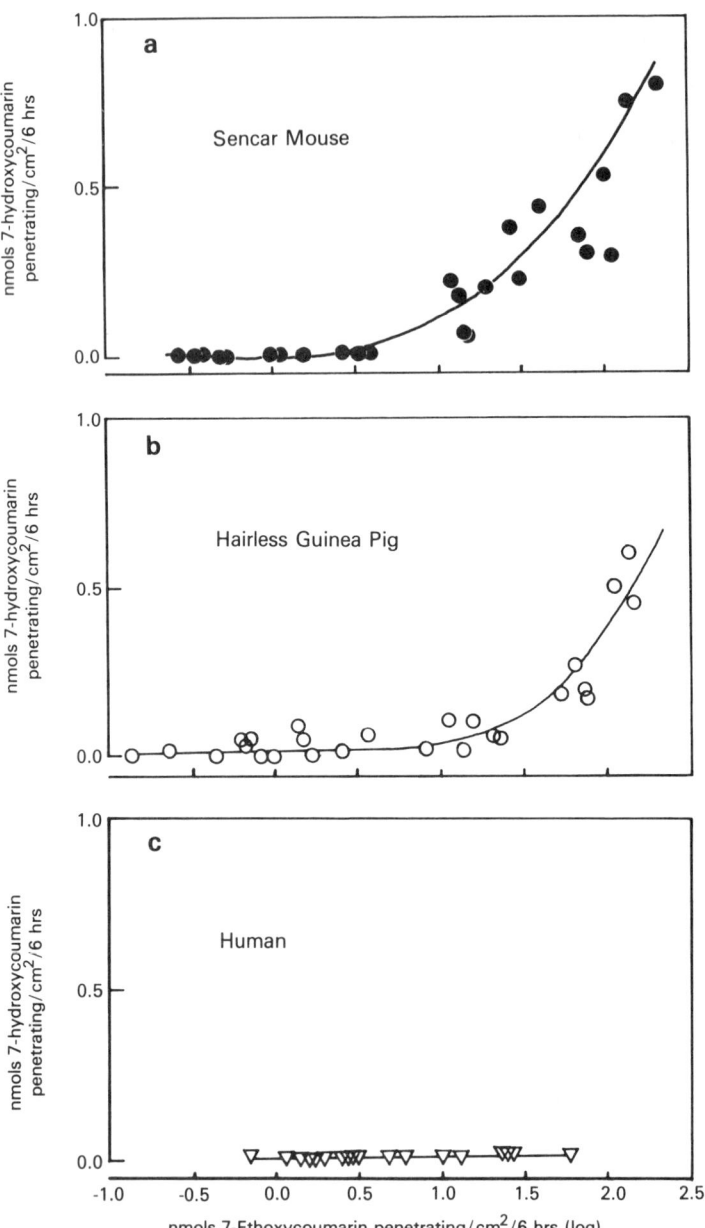

FIGURE 8.   Relationship between rate of 7-EC percutaneous penetration and 7-EC cutaneous metabolism in Sencar mouse (a), hairless guinea pig (b), and human (c) skin. Values represent individual determinations of rate of 7-EC penetration (log) and simultaneous rate of 7-hydroxycoumarin penetration for each 6-h collection period after application of 7-EC in 10 µl acetone at doses ranging from 5 to 250 µg/cm². (From Storm, J.E., Collier, S.W., Stewart, R.F., and Bronaugh, R.L., *Fundam. Appl. Toxicol.*, 15, 132, 1990.)

appears to be saturated at a relatively low rate of penetration regardless how great the penetration of 7-ethoxycoumarin, the same rate of 7-OH-coumarin formation and penetration is observed.

V. CONCLUSIONS

Rates of xenobiotic metabolism by skin are usually a small fraction of that observed in the liver. Measurement of cutaneous metabolism requires rigorous attention to the methodologies employed. A well characterized and validated system with an adequately prepared viable skin membrane is crucial to obtaining credible and reproducible results. Sterile technique and inhibition of microbial growth must be used in order to ascribe measured metabolism to the skin. Sensitive analytical assays with adequate controls must be employed to measure low levels of metabolites.

In vitro skin metabolism studies can be used to identify metabolites and measure metabolite kinetics of toxicological or therapeutic importance. With highly toxic or carcinogenic compounds or potent therapeutic agents, even minute levels of cutaneous metabolism may be of consequence. The consideration of cutaneous metabolism in topically delivered drug product development and risk-assessment of percutaneously absorbed compounds reduces the uncertainty in linking equivalent dose to pharmacodynamic response.

REFERENCES

1. **Leider, M. and Bunce, C. M.,** Physcial dimensions of the skin, Determination of the specific gravity of skin, hair and nail. *Arch. Dermatol.,* 69, 563, 1954.
2. **Bergstresser, P. R. and Taylor, J. R.,** Epidermal 'turnover time'— a new examination, *Br. J. Dermatol.,* 96, 503, 1977.
3. **Kao, J. and Carver, M.,** Cutaneous metabolism of xenobiotics, *Drug Metab. Rev.,* 22, 336, 1990.
4. **Kappus, H.,** Drug metabolism in the skin, in *Pharmacology of the Skin II.* Greaves, M. W. and Shuster, S., Eds., Springer-Verlag, New York, 1989, 123.
5. **Pannatier, A., Jenner, P., Testa, B., and Etter, J. C.,** The skin as a drug-metabolizing organ, *Drug Metab. Rev.,* 8, 319, 1978.
6. **Bronaugh, R. L., Stewart, R. F., and Storm, J. E.,** Extent of cutaneous metabolism during percutaneous absorption of xenobiotics, *Toxicol. Appl. Pharmacol.,* 99, 534, 1989.
7. **Kahn, W. A., Das, M., Stick, S., Javed, S., Bickers, D. R., and Mukhtar, H.,** Induction of epidermal NAD(P)H:quinone reductase by chemical carcinogens: a possible mechanism for the detoxification. *Biochem. Biophys. Res. Comm.,* 146, 126, 1987.
8. **West, D. P., Fischer, J. H., Barbour, M. M., Cwik, M. J., Micali, G., and Fiedler, V. C.,** Altered theophylline metabolism in patients with psoriasis, *DICP, The Ann. of Pharmacotherapy,* 24, 464, 1990.
9. **Riviere, J. E.,** *In vitro* absorption—Skin flap model, Chapter 16, This book.
10. **Bickers, D. R., Dutta-Choudhury, T., and Mukhtar, H.,** Epidermis: a site of drug metabolism in neonatal rat skin. Studies on cytochrome P-450 content and mixed-function oxidase and epoxide hydrolase activity, *Mol. Pharmacol.,* 21, 239, 1982.
11. **Andersson, P., Eddsbäcker, S., Ryrfeldt, A., and Von Bahr, C.,** *In vitro* biotransformation of glucocorticoids in liver and skin homogenate fraction from man, rat, and hairless mouse, *J. Steroid Biochem.,* 16, 787, 1982.
12. **Cheung, Y. W., Li Wan Po, A., and Irwin, W. J.,** Cutaneous biotransformation as a parameter in the modulation of the activity of topical corticosteroids, *Int. J. Pharmaceutics,* 26, 175, 1985.
13. **Kulkarni, A. P., Nelson, J. L., and Radulovic, L. L.,** Partial purification and some biochemical properties of neonatal rat cutaneous glutathione S-transferases, *Comp. Biochem. Physiol.,* 87B, 1005, 1987.
14. **Coomes, M. W., Norling, A. H., Pohl, R. J., Müller, D. and fouts, J. R.,** Foreign compound metabolism by isolated skin cells from the hairless mouse, *J. Exp. Pharmacol. Therapeutics,* 225, 770, 1983.
15. **Storm, J. E., Stewart, R. F., and Bronaugh, R. L.,** Interstrain and species differences in xenobiotic metabolizing capacity in skin: evidence of enhanced activity in SENCAR mice, *Toxicologist,* 8, 126, 1988.

16. **Storm, J. E., Collier, S. W., Stewart, R. F., and Bronaugh, R. L.,** Metabolism of xenobiotics during percutaneous penetration: role of absorption rate and cutaneous enzyme activity, *Fundam. Appl. Toxicol.,* 15, 132, 1990.

17. **Mukhtar, H. and Bickers, D. R.,** Age related changes in benzo(a)pyrene metabolism and epoxide-metabolizing enzyme activities in rat skin, *Drug Metab. Dispos.,* 11, 562, 1983.

18. **Rettie, A. E., Williams, F. M., Rawlins, M. D., Mayer, R. T., and Burke, M. D.,** Major differences between lung, skin, and liver in the microsomal metabolism of homologous series of resorufin and coumarin ethers, *Biochem. Pharmacol.,* 35, 3495, 1986.

19. **Nebert, D. W. and Gelboin, H. V.,** Substrate inducible microsomal aryl hydroxylase in mammalian cell culture. I. Assay and properties of induced enzyme, *J. Biol. chem.,* 243, 6242, 1968.

20. **Greenlee, W. F. and Poland, A.,** An improved assay of 7-ethoxycoumarin O-deethylase activity: induction of hepatic enzyme activity in C57BL/6J and DBA/2J mice by phenobarbital, 3-methylcholanthrene and 2,3,7,8-tetrachlorodibenzo-*p*-dioxin, *J. Pharmacol. Exp. Ther.,* 205, 596, 1978.

21. **Habig, W. H., Pabst, M. J., and Jakoby, W. B.,** Glutathione-S-transferase. The first enzymatic step in mercapturic acid formation, *J. Biol. Chem.,* 249, 7130, 1974.

22. **Bickers, D. R.,** The skin as a site of drug and chemical metabolism, in *Current Concepts in Cutaneous Toxicity*. Drill, V. A. and Lazar, P., Eds., Academic Press, New York, 1980, 95.

23. **Warner, R. R., Myers, M. C., and Taylor, D. A.,** Electron probe analysis of human skin: element concentration profiles, *J. Invest. Dermatol.,* 90, 78, 1988.

24. **Warner, R. R., Myers, M. C., and Taylor, D. A.,** Electron probe analysis of human skin: determination of the water concentration profile, *J. Invest. Dermatol.,* 90, 218, 1988.

25. **Cohen, R. D. and Iles, R. A.,** Intracellular pH: measurement, control, and metabolic interrelationships, *CRC Crit. Rev. Clin. Lab. Sci.,* 6, 101, 1975.

26. **Holland, J. M., Kao, J. Y., and Whitaker, M. J.,** A multisample apparatus for kinetic evaluation of skin penetration *in vitro*: the influence of viability and metabolic status of the skin, *Toxicol. Appl. Pharmacol.,* 72, 272, 1984.

27. **Kao, J., Hall, J., and Holland, J. M.,** Quantitation of cutaneous toxicity: an *in vitro* approach using skin organ culture, *Toxicol. Appl. Pharmacol.,* 68, 206, 1983.

28. **Kao, J., Hall, J., Shugart, L. R., and Holland, J. M.,** An *in vitro* approach to studying cutaneous metabolism and disposition of topically applied xenobiotics, *Toxicol. Appl. Pharmacol.,* 75, 289, 1984.

29. **Kao, J., Patterson, F. K., and Hall, J.,** Skin penetration and metabolism of topically applied chemicals in six mammalian species, including man: an *in vitro* study with benzo(a)pyrene and testosterone, *Toxicol. Appl. Pharmacol.,* 81, 502, 1985.

30. **Kao, J. and Hall, J.,** Skin absorption and cutaneous first pass metabolism of topical steroids: *in vitro* studies with mouse skin in organ culture, *J. Pharmacol. Exper. Ther.,* 241, 482, 1987.

31. **Macpherson, S. E., Scott, R. C., and Williams, F. M.,** Metabolism of pesticides during percutaneous absorption *in vitro*, in *Prediction of Percutaneous Penetration; Methods, Measurements, Modelling,* Scott, R. C., Guy, R. H., and Hadgraft, J., Eds., IBC, London, 1990, 135.

32. **Hinz, R. S., Hodson, C. D., Lorence, C. R., and Guy, R. H.,** *In vitro* percutaneous penetration: evaluation of utility of hairless mouse skin, *J. Invest. Dermatol.,* 93, 87, 1989.

33. **Collier, S. W., Sheikh, N. M., Sakr, A., Lichtin, J. L., Stewart, R. F., and Bronaugh, R. L.,** Maintenance of skin viability during *in vitro* percutaneous absorption/metabolism studies, *Toxicol. Appl. Pharmacol.,* 99, 522, 1989.

34. **Roth, R. R. and James, W. D.,** Microbial ecology of the skin, *Annu. Rev. Microbiol.,* 42, 441, 1988.

35. **Korting, H. C., Kober, M., Mueller, M., and Braun-Falco, O.,** Influence of repeated washings with soap and synthetic detergents on pH and resident flora of the skin of forehead and forearm, *Acta. Derm. Venereol.,* 67, 41, 1987.

36. **Van Cantfort, J., DeGraeve, J., and Gielen, J. E.,** Radioactive assay for aryl hydrocarbon hydroxylase. Improved method and biological significance, *Biochem. Biophys. Res. Comm.,* 79, 505, 1977.

37. **Bickers, D. R., Mukhtar, H., and Yang, S. K.,** Cutaneous metabolism of benzo(a)pyrene: comparative studies in C57BL/6N and DBA/2N mice and neonatal Sprague-Dawley rats, *Chem. Biol. Interact.,* 43, 263, 1983.

38. **Cheng, K. C. and Schenkman, J. B.,** Testosterone metabolism by cytochrome P-450 isozymes RLM_3 and RLM_5 and by microsomes, *J. Biol. Chem.,* 258, 11738, 1983.

39. **Davis, B. P., Rampini, E., and Hsia, S. L.,** 17β-Hydroxysteroid dehydrogenase of rat skin, *J. Biol. Chem.,* 247, 1407, 1972.

40. **Cheng, K. C. and Schenkman, J. B.,** Metabolism of progesterone and estradiol by microsomes and purified cytochrome P-450 RLM_3 and RLM_5, *Drug Metab. Dispos.,* 12, 222, 1984.

41. **Yu, C. D., Fox, J. L., Ho, H. F. H., and Higuchi, W. I.,** Physical mode evaluation of topical prodrug delivery-Sinultaneous transport and bioconversion of vidarabine-5' valerate. II. Parameter determinations, *J. Pharm. Sci.,* 68, 1347, 1979.

42. **Bundgaard, H., Hoelgaard, A., and Mollgaard, B.,** Leaching of hydrolytic enzymes from human skin in cutaneous permeation studies as determined with metronidazole and 5-fluorouracil prodrugs, *Int. J. Pharm.,* 15, 285, 1983.

Chapter 8

EFFECTS OF OCCLUSION*

D. Bucks, R. Guy, and H. Maibach

TABLE OF CONTENTS

* Sections of this chapter have been adapted from the 2nd edition in this series on Percutaneous Penetration[11] and from the doctoral thesis entitled "Prediction of Percutaneous Absorption".[12]

I. INTRODUCTION

Mammalian skin provides a relatively efficient barrier to the ingress of exogenous materials and the egress of endogenous compounds, particularly water. Loss of this vital function results in death from dehydration; compromised function is associated with complications seen in several dermatological disorders. Stratum corneum intercellular lipid domains form a major transport pathway for penetration.[14-16,22] Perturbation of these lamellar lipids causes skin permeation resistance to fall and has implicated their crucial role in barrier function. Indeed, epidermal sterologenesis appears to be modulated by the skin's barrier requirements.[31] Despite the fact that the skin is perhaps the most impermeable mammalian membrane, it is semipermeable; as such, the topical application of pharmaceutical agents has been shown to be a viable route of entry into the systemic circulation as well as an obvious choice in the treatment of dermatological ailments. Of the various approaches employed to enhance the percutaneous absorption of drugs, occlusion (defined as the complete impairment of passive transepidermal water loss at the application site) is the simplest and most common method in use.

The increased clinical efficacy of topical drugs caused by covering the site of application was first documented by Garb.[21] Subsequently, Scholtz[36] using fluocinolone acetonide, and Sulzberger and Witten[37] using hydrocortisone, reported enhanced corticoid activity with occlusion in the treatment of psoriasis. The enhanced pharmacological effect of topical corticosteroids under occlusion was further demonstrated by the vasoconstriction studies of McKenzie[29] and McKenzie and Stoughton.[30] Occlusion has also been reported to increase the percutaneous absorption of various other topically applied compounds.[9,18,26-27] However, as will be shown below, short term occlusion does not necessarily increase the percutaneous absorption of all chemicals.

II. PERCUTANEOUS ABSORPTION OF *p*-PHENYLENEDIAMINE (PPDA) IN GUINEA PIGS

The *in vivo* percutaneous absorption of PPDA from six occlusive patch test systems was investigated by Kim et al.[27] The extent of absorption was determined using ^{14}C radiotracer methodology. The ^{14}C-PPDA was formulated as 1% PPDA in petrolatum (USP) and applied from each test system at a skin surface dose of 2 mg/cm^2. Thus, the amount of PPDA was normalized with respect to the surface area of each patch test system (and, hence, to the surface area of treated skin). A sixfold difference in the level of skin absorption ($p < 0.02$) was found (Table 1).

The rate of ^{14}C excretion following topical application of the radiolabelled PPDA in the various patch test systems is shown in Figure 1. Clearly, the rate and extent of PPDA absorption was dependent upon the occlusive patch test system employed. It should be noted that a nonocclusive control study was not conducted.

III. PERCUTANEOUS ABSORPTION OF VOLATILE COMPOUNDS IN RHESUS MONKEYS

The *in vivo* percutaneous absorption of two fragrances (safrole and cinnamyl anthranilate) and two chemical analogs (cinnamic alcohol and cinnamic acid) were measured under nonoccluded and plastic wrap (Saran Wrap®—a chlorinated hydrocarbon polymer) occluded conditions by Bronaugh et al.[3] The extent of absorption following single dose administration was determined using ^{14}C radiotracer methodology. Each compound was applied at a topical dose of 4 μg/cm^2 from a small volume of acetone. The fragrance materials were well absorbed through monkey skin. Plastic wrap occlusion of the application site resulted in large increases

TABLE 1
Percutaneous Absorption of PPDA from Patch
Test Systems[a]

Patch test system	mg PPDA in chamber	Mean % dose absorbed (SD)
Hill Top chamber	40	53 (21)
Teflon (control)	16	49 (9)
Small Finn chamber	16	30 (9)
Large Finn chamber	24	23 (7)
AL-Test chamber	20	8 (1)
Small Finn chamber with paper disc insert	16	34 (20)

Note: The rate of ^{14}C excretion following topical application of the radiolabeled
PPDA in the various patch test systems is shown in Figure 1. Clearly,
the rate and extent of PPDA absorption was dependent upon the occlusive
patch test system employed. It should be noted that a nonocclusive control
study was not conducted.

[a] 2 mg/mm^2 PPDA for 48 h on the dorsal mid-lumbar region of the guinea
pig.

Data from Kim, H. O., Wester, R. C., McMaster, J. R., Bucks, D. A. W.,
and Maibach, H. I., *Contact Dermatitis*, 17, 178, 1987.

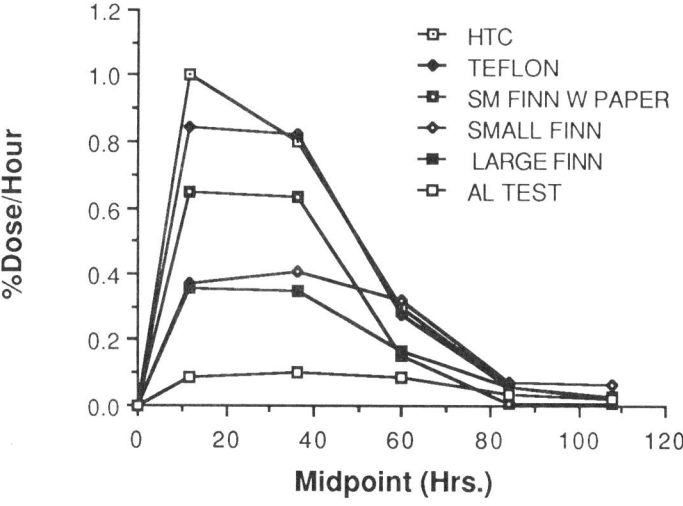

FIGURE 1. *In vivo* percutaneous absorption of PPDA (2 mg/mm^2) following a
48 h exposure on the dorsal lumbar region of guinea pigs (Redrawn from Kim, H.
O., Wester, R. C., McMaster, J. R., Bucks, D. A. W., and Maibach, H. I., *Contact
Dermatitis,* 17, 178, 1987.)

in absorption (see Table 2). The authors also presented *in vitro* data documenting the
significant increase in percutaneous absorption of these chemicals under occluded compared
to nonoccluded conditions.

Investigation of the effect of occlusion on the percutaneous absorption of six additional
volatile compounds (benzyl acetate, benzamide, benzoin, benzophenone, benzyl benzoate,
and benzyl alcohol) was conducted using the same *in vivo* methodology. These studies
included occlusion of the site of application with a glass cylinder (secured to the skin by

TABLE 2
***In Vivo* Percutaneous Absorption of
Fragrances in Monkeys**

	% Dose absorbed[a]	
	Nonprotected	Plastic wrap occlusion
Cinnamyl anthranilate	26.1 (4.6)	39.0 (5.6)
Safrole	4.1 (1.6)	13.3 (4.6)
Cinnamic alcohol	25.4 (4.4)	74.6 (14.4)
Cinnamic acid	38.6 (16.6)	83.9 (5.4)

Note: 24-h exposure at 4 $\mu g/cm^2$ prior to soap and water washing.

[a] Single dose application; values corrected for incomplete renal elimination. Mean ± SD (N = 4).

Data from Bronaugh, R. L., Stewart, R. F., Wester, R. C., Bucks, D., and Maibach, H. I., *Fd. Chem. Toxicol.*, 23, 111, 1985.

TABLE 3
***In Vivo* Percutaneous Absorption of Benzyl Derivatives in Monkeys**

	% Dose absorbed[a]			
	Nonprotected	Plastic wrap occlusion	Glass chamber occlusion	Log Ko/w
Benzamide	47 (14)	85 (8)	73 (20)	0.64
Benzyl alcohol	32 (9)	56 (29)	80 (15)	0.87
Benzoin	49 (6)	43 (12)	77 (4)	1.35
Benzyl acetate	35 (19)	17 (5)	79 (15)	1.96
Benzophenone	44 (15)	69 (12)	69 (10)	3.18
Benzyl benzoate	57 (21)	71 (9)	65 (20)	3.97

Note: 24-h exposure at 4 $\mu g/cm^2$ prior to soap and water washing.

[a] Single dose application; values corrected for incomplete renal elimination. Mean ± SD (N = 4).

Data from Bronaugh, R. L., Wester, R. C., Bucks, D. A. W., and Maibach, H. I., *Fd. Chem. Toxicol.*, 28, 369, 1990.

silicone glue) capped with Parafilm, occlusion with plastic wrap, and nonprotected conditions.[4] As shown in Table 3, occlusion, in general, enhances the percutaneous absorption of these compounds. However, differences in percutaneous absorption were observed between plastic wrap and "glass chamber" occlusive conditions. The absorption of benzyl acetate was lower under plastic wrap compared to the nonprotected condition, whereas glass chamber occlusion resulted in the greatest bioavailability. This discrepancy might be due to compound sequestration by the plastic wrap.

IV. PERCUTANEOUS ABSORPTION OF STEROIDS IN MAN

The earliest attempt to correlate the increased pharmacological effect of hydrocortisone under occlusive conditions with the pharmacokinetics of absorption was reported by Feldmann and Maibach.[18] In this study, the rate and extent of ^{14}C-label excretion into the urine following topical application of ^{14}C-hydrocortisone to the ventral forearm of normal human volunteers were measured. Radiolabeled hydrocortisone (75 μg) was applied in acetone solution (1000 μl) as a surface deposit over 13 cm^2 of skin. The authors estimated that this was equivalent to a sparing application of a 0.5% hydrocortisone topical preparation (5.8 $\mu g/cm^2$). The site of application was either nonprotected or occluded with plastic wrap (Saran Wrap®). When the skin was unprotected, the dosing site was washed 24 h post application. On the other hand, when the skin was occluded, the plastic wrap remained in place for 96 h (4 d) post application before the application site was washed. The % of the applied dose excreted into the urine, corrected for incomplete renal elimination, was (mean ± SD) 0.46 ± 0.20 and 5.9 ± 3.5 under nonprotected and occluded conditions, respectively (see Tables 4, 5, and Figure 2). These numbers differ from those in the original paper which were calculated incorrectly. A paired t test of the results indicates a significant difference ($p = 0.01$) in cumulative absorption of hydrocortisone between two exposure conditions. Quantitatively, the occlusive condition employed increased the cumulative absorption of hydrocortisone by about an order of magnitude. However, note that the occlusive system retained the drug in contact with the skin for 96 h compared to the 24 h exposure period under nonprotected conditions.

Guy et al.[26] investigated the effect of occlusion on the percutaneous absorption of steroids *in vivo* following single and multiple application. The extent of absorption of four steroids (progesterone, testosterone, estradiol, and hydrocortisone), using radiotracer elimination into the urine following topical application to the ventral forearm of male volunteers, was reported. The chemical dose was 4 $\mu g/cm^2$ and application area 2.5 cm^2. The ^{14}C-labeled chemicals were applied in 20 μl acetone. In the occlusive studies, after evaporation of the vehicle, the site of application was covered with a plastic (polyethylene-vinyl acetate copolymer) chamber.[34] In all cases, after 24 h, the site of application was washed with soap and water using a standardized procedure.[5] In the occlusive studies, the administration site was then recovered with a new chamber. An essentially identical protocol was also performed following a multiple dosing regimen.[6] Daily topical doses of three of the steroids (testosterone, estradiol, and hydrocortisone) were administered over a 14 d period. The first and eighth doses were ^{14}C-labeled and urinary excretion of radiolabel was followed. As above, the 24 h washing procedure was performed daily and a new chamber applied. Occlusive chambers and washes were collected and assayed for residual surface chemical. The results of this study are in Table 6. Steroid percutaneous absorption as a function of penetrant octanol-water partition coefficient (Ko/w) is shown in Figure 3. The studies indicate that:

1. The single-dose measurements of the percutaneous absorption of hydrocortisone, estradiol, and testosterone are predictive of percutaneous absorption following a comparable multiple dose regimen (see chapter on the effect of repetitive application), under both occluded and nonoccluded conditions.
2. Occlusion significantly ($p < 0.05$) increased the percutaneous absorption of estradiol, testosterone, and progesterone, but not that of hydrocortisone.
3. Percutaneous absorption increases with increasing Ko/w up to testosterone but declines for progesterone, under occluded and nonoccluded conditions.
4. The occlusive procedure generally permits excellent dose accountability (Table 7).

The percutaneous absorption of these same four steroids under "protected" (i.e., covered, but nonocclusive) conditions has also been measured *in vivo*[9,10] using the same methodology. The data obtained from these later experiments permitted the effect of occlusion to be rigorously assessed (since complete mass balance of the applied dose was possible). With the exception of hydrocortisone (Table 8), occlusion significantly increased the percutaneous absorption ($p < 0.01$) of the steroids. These results were in excellent agreement with the comparable nonprotected studies described above. As stated before, excellent dose accountability was reported (Table 9).

To investigate the apparent discrepancy between the effect of plastic wrap occlusion[18]

TABLE 4

Percutaneous Absorption[a] of ^{14}C after Topical Administration of ^{14}C Hydrocortisone under Nonprotected[b] Conditions (% of Dose)

Sbj	0—12 H	12—24 H	Day 2	Day 3	Day 4	Day 5	Day 6	Day 7	Day 8	Day 9	Day 10	Total %D
1	0.11	0.11	0.23	0.07	0.17	0.08	0.09	0.07	0.04	0.07	0.03	1.08
2	0.11	0.06	0.11	0.09	0.09	0.07	0.05	0.05	0.04	0.00	0.00	0.67
3	0.03	0.16	0.19	0.15	0.12	0.12	0.18	0.04	0.05	0.05	0.03	1.14
4	0.05	0.05	0.05	0.01	0.04	0.09	0.05	0.03	0.06	0.03	0.00	0.46
5	0.00	0.01	0.11	0.16	0.07	0.11	0.12	0.10	0.01	0.00	0.04	0.73
6	0.11	0.03	0.07	0.04	0.03	0.03	0.02	0.00	0.00	0.00	0.00	0.33
Mean	0.07	0.07	0.13	0.09	0.09	0.08	0.09	0.05	0.04	0.02	0.02	0.74
SD	0.05	0.05	0.07	0.06	0.06	0.03	0.06	0.03	0.02	0.03	0.02	0.32

[a] Topical urinary excretion values calculated using an IV correction factor of 62.9%.
[b] Nonprotected, site of application washed 24-h postapplication with soap and water.

Data from Feldmann, R. and Maibach, H. I., *Arch. Dermatol.*, 91, 661, 1965.

TABLE 5
Percutaneous Absorption[a] of ^{14}C after Topical Administration of ^{14}C Hydrocortisone under Occluded[b] Conditions (% of Dose)

Sbj	0—12 H	12—24 H	Day 2	Day 3	Day 4	Day 5	Day 6	Day 7	Day 8	Day 9	Day 10	Total %D
1	1.69	0.98	1.37	1.51	1.62	1.66	0.37	0.41	0.28	0.30	0.21	10.39
2	0.14	0.60	2.00	2.26	1.19	1.19	0.82	0.59	0.39	0.27	0.21	9.66
3	0.72	0.16	1.78	1.93	1.38	1.08	0.64	0.57	0.45	0.21	0.18	9.09
4	0.13	0.95	0.45	0.74	1.35	0.58	0.64	0.38	0.32	0.10	0.01	5.65
5	0.78	0.19	0.51	0.24	0.21	0.24	0.22	0.35	0.10	0.03	0.00	2.86
6	3.86	1.86	3.12	3.09	2.53	1.99	1.44	0.42	0.39	0.23	0.17	19.10
Mean	1.22	0.79	1.54	1.63	1.38	1.12	0.69	0.45	0.32	0.19	0.13	9.46
SD	1.41	0.63	1.01	1.04	0.75	0.65	0.43	0.10	0.13	0.10	0.10	5.52

[a] Topical urinary excretion values calculated using an IV correction factor of 62.9%.

[b] Occluded with plastic wrap for 4 d, site of application washed 96-h postapplication with soap and water.

Data from Feldmann, R. and Maibach, H. I., *Arch. Dermatol.*, 91, 661, 1965.

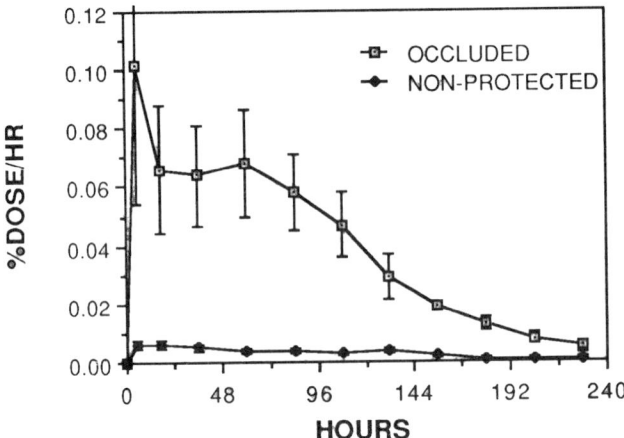

FIGURE 2. Percutaneous absorption of hydrocortisone in man. Human
96 h occluded versus 24-h nonprotected exposure of hydrocortisone at 4 µg/
cm² prior to soap and water washing. Occlusion was with plastic wrap. (Data
from Feldmann, R. J. and Maibach, H. I., *Arch Dermatol.*, 91, 661, 1965.)

TABLE 6
Percutaneous Absorption of Steroids in Man

	Mean % applied dose absorbed (± SD)	
	Nonprotected	**Occlusion**
Hydrocortisone		
Single application	2 ± 2[a]	4 ± 2
Multiple application:		
1st Dose	3 ± 1	4 ± 1
8th Dose	3 ± 1	3 ± 1
Estradiol		
Single application	11 ± 5[a]	27 ± 6
Multiple application:		
1st Dose	10 ± 2	38 ± 8
8th Dose	11 ± 5	22 ± 7
Testosterone		
Single application	13 ± 3[a]	46 ± 15
Multiple application:		
1st Dose	21 ± 6	51 ± 10
8th Dose	20 ± 7	50 ± 9
Progesterone		
Single application	11 ± 6[a]	33 ± 9

[a]   Data from Feldmann, R. and Maibach, H. I., *J. Invest.
       Dermatol.*, 52, 89, 1969.

and that of the plastic chamber on hydrocortisone absorption,[26] we repeated the measurements
of penetration using plastic wrap (Saran Wrap®) with the experimental protocol of Guy et
al.[26] Under these circumstances, we found no difference between plastic wrap and plastic
chamber occlusion on the percutaneous absorption of hydrocortisone (Table 10).

V. PERCUTANEOUS ABSORPTION OF PHENOLS IN MAN

We subsequently investigated the effect of occlusion on the *in vivo* percutaneous absorption of phenols following single dose application. The occlusive and protective chamber

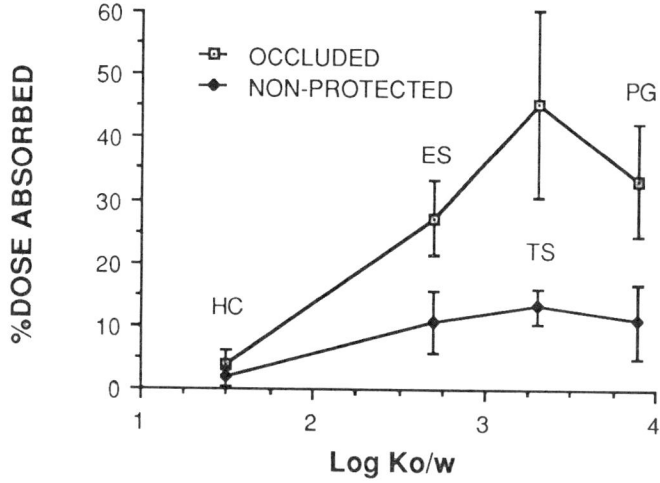

FIGURE 3. Percutaneous absorption of 4 steroids (HC = hydrocortisone, ES = estradiol, TS = testosterone, PG = progesterone) in man as a function of penetrant octanol/water partition coefficient. Exposure period 24-h at 4 μg/cm² prior to soap and water washing. (Redrawn from Guy et al., in *Skin Pharmacokinetics*, Shroot, B. and Schaefer, H., Eds., Karger, Basel, 1987, 70.)

TABLE 7
Accountability of Applied Dose in Occluded Studies[a]

	Observed PA	% Removed from skin	Total % dose
Hydrocortisone			
Single dose[b]	4 ± 2	64 ± 5	68 ± 4
1st MD[c]	4 ± 1	82 ± 5	85 ± 4
8th MD[d]	3 ± 1	78 ± 2	81 ± 3
Estradiol			
Single dose[b]	27 ± 6	60 ± 12	87 ± 13
1st MD[c]	38 ± 8	62 ± 6	100 ± 4
8th MD[d]	22 ± 7	59 ± 8	81 ± 6
Testosterone			
Single dose[b]	46 ± 15	44 ± 7	90 ± 8
1st MD[c]	51 ± 10	48 ± 9	99 ± 4
8th MD[d]	50 ± 9	42 ± 9	92 ± 17
Progesterone			
Single dose[b]	33 ± 9	47 ± 10	80 ± 6

Note: Mean (in % applied dose) ± SD

[a] 24-h exposure at 4 μg/cm² prior to soap and water washing. Occlusion was with a plastic (Hilltop) chamber.
[b] Single dose study.
[c] First dose of a 14-d multiple-dose study.
[d] Eighth dose of a 14-d multiple-dose study.

Data from Bucks, D. A. W., et al., Unpublished observations.

TABLE 8

Percutaneous Absorption of Steroids in Man
Single Dose Application for 24 h @ 4 μg/cm²

| | Mean % dose absorbed (± SD; N ≥ 5) | |
	Protected[a]	Occluded[b]
Hydrocortisone	4 ± 2	4 ± 2
Estradiol	3 ± 1	27 ± 6
Testosterone	18 ± 9	46 ± 15
Progesterone	13 ± 6	33 ± 9

[a]   Ventilated plastic chamber.
[b]   Occlusive plastic chamber.

Data from Guy, R. H., Bucks, D. A. W., McMaster, J. R., Villaflor, D. A., Roskos, K. V., Hinz, R. S., and Maibach, H. I., in *Skin Pharmacokinetics,* Shroot, B. and Schaefer, H., Eds., Karger, Basel, 70, 1987.; Bucks, D. A. W., Maibach, H. I., and Guy, R. H., in *Percutaneous Absorption,* Vol. 2, Bronaugh, R. and Maibach, H., Eds., Marcel Dekker, New York, 1989, 77.; and, Bucks, D. A. W., McMaster, J. R., Maibach, H. I., and Guy, R. H., *J. Invest. Dermatol.,* 90, 29, 1988.

TABLE 9

Accountability of Applied Dose in Protected Studies using
Ventilated Plastic Chambers[a]

	Observed PA	% Removed from skin	Total % dose
Hydrocortisone	4 ± 2	85 ± 6	89 ± 6
Estradiol	3 ± 1	96 ± 1	100 ± 1
Testosterone	18 ± 9	77 ± 8	96 ± 2
Progesterone	13 ± 6	82 ± 7	96 ± 3

Note: Mean % dose (± SD, N ≥ 5).

[a]   Single dose application for 24 h at 4 μg/cm²

Data from Bucks et al.[9,10]

TABLE 10

Percutaneous Absorption of
Hydrocortisone in Man

	% Dose absorbed[a]
Plastic wrap occlusion	4.7 (2.1)[b]
Plastic chamber occlusion[c]	4.0 (2.4)
"Protected" condition[d]	4.4 (1.7)

[a]   Single dose application for 24 h at 4 μg/cm²; values corrected for incomplete renal elimination.
[b]   Mean ± SD (N = 6)
[c]   Guy et al.[26]
[d]   Bucks et al.[9,10]

methodology described by Bucks et al.[7,10] was utilized. Nine ^{14}C ring labelled para-substituted phenols (4-aminophenol, 4-acetamidophenol, 4-propionylamidophenol, phenol, 4-cyano-phenol, 4-nitrophenol, 4-iodophenol, 4-heptyloxyphenol and 4-pentyloxyphenol) were used. As in the earlier steroid studies, the site of application was the ventral forearm of male volunteers and the area of application 2.5 cm[2]. Penetrants were applied in 20 μl ethanol (95%). The chemical dose was 2 to 4 μg/cm[2]. After vehicle evaporation, the application site was covered with either an occlusive or protective device. After 24 h, the patch was removed and the site washed with a standardized procedure.[5] The application site was then recovered with a new chamber of the same type. Urine was collected for seven days. On the seventh day: (1) the second chamber was removed, (2) the dosing site was washed with the same procedure, and (3) the upper layers of stratum corneum from the application site were removed by cellophane tape stripping. Urine, chambers, washes, and skin tape strips were collected and assayed for radiolabel. Percutaneous absorption of each compound under protected and occluded conditions is presented in Tables 11 through 19 and Figures 4 through 12. Phenol percutaneous absorption as a function of the penetrant octanol-water partition coefficient (Ko/w) is shown in Figure 13. Phenol percutaneous absorption is summarized in Table 20. The methodology permitted excellent dose accountability (Tables 21 and 22). The studies indicate that:

1. Occlusion significantly increased (unpaired t test, $p < 0.05$) the penetration of phenol, heptyloxyphenol and pentyloxyphenol.
2. Occlusion did not enhance the absorption of aminophenol, 4-acetamidophenol, pro-pionylamidophenol, cyanophenol, nitrophenol, and iodophenol.
3. The methodology employed again permitted excellent dose accountability.

VI. DISCUSSION

A predominant effect of occlusion is to increase hydration of the stratum corneum, thereby swelling the corneocytes, and promoting the uptake of water into intercellular lipid domains. The normal water content of stratum corneum is 5 to 15%, a value which can be increased up to 50% by occlusion.[1,33] Upon removal of a plastic occlusive dressing after 24 h of contact, transepidermal water loss values are increased by an order of magnitude;[10] the elevated rate then returns rapidly (~ 15 minutes) to normal with extraneous water dissipation. With occlusion, skin temperature generally increases from 32°C to as much as 37°C.[28] Faergemann et al.[17] showed that occlusion: (1) increases the transepidermal flux of chloride and carbon dioxide, (2) increases microbial counts on skin, and (3) increases the surface pH of skin from a preoccluded value of 5.6 to 6.7. Anhidrosis results from occlu-sion.[32,23] Plastic chamber occlusion can also cause skin irritation (personal observation). Occlusion-induced increases in mitotic rate of skin and epidermal thickening have been documented by Fisher and Maibach.[20]

With respect to percutaneous absorption, occlusion (or a protective cover) prevents loss of the surface deposited chemical by friction and/or exfoliation; bioavailability may, thereby, be increased. However, comparison of the data in Tables 6 and 8, for the percutaneous absorption of steroids under nonprotected and protected conditions, shows clearly that the potential increase in bioavailability from protection of the site of application does not explain the increase in steroid absorption under occluded conditions.

Occlusion does not necessarily increase percutaneous absorption. Hydrocortisone ab-sorption under occluded conditions was not enhanced in single dose or multiple dose ap-plication studies (Table 23).

This lack of penetration enhancement under occluded conditions has been observed with certain para-substituted phenols. However, a trend of occlusion-induced absorption enhance-

TABLE 11A
Percutaneous Absorption of Aminophenol Under Occluded Conditions[a] in Man (% of Dose)

Subjects	0—4 H	4—8 H	8—12 H	12—24 H	Day 2	Day 3	Day 4	Day 5	Day 6	Day 7	Total %D
A	0.04	0.08	0.19	0.51	2.36	1.09	0.67	0.54	0.50	0.45	6.43
B	0.03	0.03	0.08	1.05	2.10	1.54	0.59	0.38	0.23	0.12	6.15
C	0.00	0.06	0.17	2.55	6.16	1.78	0.31	0.48	0.18	0.10	11.78
D	0.02	0.32	0.32	2.33	3.31	1.95	0.88	0.38	0.27	0.15	9.95
E	0.00	0.16	0.18	1.08	2.40	0.96	0.52	0.35	0.22	0.20	6.07
Mean	0.02	0.13	0.19	1.50	3.27	1.47	0.60	0.42	0.28	0.20	8.08
SD	0.02	0.12	0.09	0.89	1.68	0.43	0.21	0.08	0.13	0.14	2.63

TABLE 11B
Percutaneous Absorption of Aminophenol Under Protected Conditions[a] in Man (% of Dose)

Subjects	0—4 H	4—8 H	8—12 H	12—24 H	Day 2	Day 3	Day 4	Day 5	Day 6	Day 7	Total %D
A	0.03	0.04	0.29	1.60	2.97	1.49	0.68	0.32	0.29	0.29	7.99
B	0.17	0.26	0.03	0.05	0.33	0.69	0.46	0.41	0.44	0.28	3.11
C	0.01	0.04	0.07	0.21	0.58	0.67	0.52	0.41	0.22	0.16	2.89
D	0.02	0.07	0.19	0.63	1.66	0.85	0.59	0.22	0.25	0.13	4.60
E	0.05	0.13	0.30	1.25	2.70	1.68	0.73	0.45	0.12	0.19	7.60
F	0.02	0.13	0.46	2.14	3.63	1.72	0.43	0.59	0.26	0.12	9.50
Mean	0.05	0.11	0.22	0.98	1.98	1.18	0.57	0.40	0.26	0.20	5.95
SD	0.06	0.08	0.16	0.82	1.34	0.50	0.12	0.12	0.11	0.07	2.78

[a]   Site of application washed 24-h postapplication with soap and water.

From Bucks, D. A. W., Lee, M., Maibach, H. I., and Guy, R. H., in preparation.

TABLE 12A
Percutaneous Absorption of Acetaminophen Under Occluded Conditions[a] in Man (% of Dose)

Subjects	0—4 H	4—8 H	8—12 H	12—24 H	Day 2	Day 3	Day 4	Day 5	Day 6	Day 7	Total %D
A	0.00	0.01	0.05	0.26	0.42	0.56	0.37	0.47	0.40	0.48	3.01
B	0.02	0.09	0.01	0.01	0.02	0.09	0.09	0.09	0.11	0.09	0.61
C	0.01	0.09	0.19	0.12	0.77	1.52	1.06	0.77	0.68	0.79	6.00
D	0.00	0.06	0.08	0.07	0.17	0.17	0.16	0.16	0.11	0.15	1.11
E	0.00	0.00	0.02	0.09	0.41	0.56	0.48	0.30	0.39	0.35	2.60
F	0.02	0.27	0.43	0.27	1.16	0.77	0.66	0.84	0.73	0.65	5.80
Mean	0.01	0.09	0.13	0.14	0.49	0.61	0.47	0.44	0.40	0.42	3.19
SD	0.01	0.09	0.16	0.11	0.42	0.52	0.36	0.31	0.27	0.28	2.28

TABLE 12B
Percutaneous Absorption of Acetaminophen Under Protected Conditions[a] in Man (% of Dose)

Subjects	0—4 H	4—8 H	8—12 H	12—24 H	Day 2	Day 3	Day 4	Day 5	Day 6	Day 7	Total %D
A	0.00	0.06	0.02	0.20	0.77	1.12	1.55	0.96	0.97	1.14	6.79
B	0.00	0.01	0.02	0.14	0.24	0.13	0.32	0.29	0.28	0.29	1.71
C	0.06	0.08	0.01	0.14	0.76	0.56	1.13	0.60	0.73	0.70	4.77
D	0.00	0.01	0.04	0.08	0.13	0.16	0.23	0.23	0.24	0.32	1.45
E	0.01	0.01	0.01	0.02	0.11	0.18	0.20	0.18	0.21	0.19	1.13
F	0.02	0.03	0.07	0.36	0.57	0.91	1.37	0.84	1.10	1.34	6.62
Mean	0.01	0.04	0.03	0.15	0.43	0.51	0.80	0.52	0.59	0.66	3.74
SD	0.02	0.03	0.02	0.12	0.31	0.43	0.62	0.33	0.40	0.48	2.64

[a] Site of application washed 24-h postapplication with soap and water.

From Bucks, D. A. W., Lee, M., Maibach, H. I., and Guy, R. H., in preparation.

TABLE 13A

Percutaneous Absorption of Propionylamidophenol Under Occluded Conditions[a] in Man (% of Dose)

Subjects	0—4 H	4—8 H	8—12 H	12—24 H	Day 2	Day 3	Day 4	Day 5	Day 6	Day 7	Total %D
A	0.00	0.02	0.27	2.14	7.17	4.50	3.64	1.75	0.67	0.98	21.15
B	0.00	0.02	0.19	1.25	4.42	3.72	2.81	1.85	1.56	0.91	16.73
C	0.40	0.56	1.29	1.12	3.56	3.91	3.53	6.30	3.20	1.99	25.85
D	0.00	0.11	0.08	0.43	2.22	2.60	1.45	1.43	0.53	0.58	9.43
E	0.00	0.06	0.13	0.71	2.74	2.31	1.02	0.54	0.64	0.39	8.55
F	0.39	2.55	4.52	3.77	11.9	4.04	2.06	1.19	0.48	0.65	31.54
Mean	0.13	0.56	1.08	1.57	5.33	3.51	2.42	2.18	1.18	0.92	18.88
SD	0.20	1.00	1.74	1.23	3.66	0.86	1.09	2.07	1.07	0.57	9.11

TABLE 13B

Percutaneous Absorption of Propionylamidophenol Under Protected Conditions[a] in Man (% of Dose)

Subjects	0—4 H	4—8 H	8—12 H	12—24 H	Day 2	Day 3	Day 4	Day 5	Day 6	Day 7	Total %D
A	0.04	0.05	0.06	0.39	2.14	1.68	1.83	0.77	1.28	0.78	9.03
B	0.03	0.02	0.06	0.17	0.77	1.36	0.96	0.52	0.73	1.23	5.85
C	0.07	0.12	0.10	0.47	1.02	1.58	1.35	1.06	0.66	0.34	6.77
D	0.00	0.00	0.21	2.36	6.52	6.34	2.96	1.75	1.79	1.14	23.08
E	0.01	0.12	1.11	2.17	3.36	1.93	1.03	0.51	0.31	0.23	10.77
Mean	0.03	0.06	0.31	1.11	2.76	2.58	1.63	0.92	0.95	0.74	11.10
SD	0.03	0.05	0.45	1.06	2.34	2.12	0.82	0.51	0.58	0.45	6.97

[a] Site of application washed 24-h postapplication with soap and water.

From Bucks, D. A. W., Lee, M., Maibach, H. I., and Guy, R. H., in preparation.

TABLE 14A
Percutaneous Absorption of Phenol Under Occluded Conditions[a] in Man (% of Dose)

Subjects	0—4 H	4—8 H	8—12 H	12—24 H	Day 2	Day 3	Day 4	Day 5	Day 6	Day 7	Total %D
A	1.25	4.36	8.19	12.8	11.1	0.75	0.22	0.26	0.05	0.11	39.08
B	1.76	9.25	8.74	13.5	2.39	0.12	0.06	0.03	0.07	0.05	35.99
C	3.00	7.34	7.31	12.4	4.98	0.16	0.15	0.05	0.08	0.00	35.47
D	2.27	5.36	5.50	8.35	4.99	0.39	0.16	0.05	0.04	0.02	27.14
E	3.30	4.24	4.97	10.8	7.83	1.32	0.16	0.07	0.03	0.04	32.78
F	7.26	6.23	4.05	6.38	8.17	0.48	0.19	0.09	0.07	0.00	32.93
Mean	3.14	6.13	6.46	10.71	6.58	0.53	0.16	0.09	0.06	0.04	33.90
SD	2.16	1.92	1.89	2.81	3.07	0.45	0.06	0.08	0.02	0.04	4.04

TABLE 14B
Percutaneous Absorption of Phenol Under Protected Conditions[a] in Man (% of Dose)

Subjects	0—4 H	4—8 H	8—12 H	12—24 H	Day 2	Day 3	Day 4	Day 5	Day 6	Day 7	Total %D
A	0.22	7.08	6.11	10.5	7.26	0.61	0.13	0.11	0.16	0.10	32.27
B	3.85	2.14	2.07	2.20	2.48	0.38	0.32	0.20	0.21	0.25	14.09
C	3.53	1.30	4.25	6.99	4.29	0.34	0.19	0.11	0.15	0.10	21.26
D	2.10	5.63	5.34	6.57	2.85	0.52	0.15	0.08	0.09	0.08	23.42
E	2.69	3.41	3.34	6.99	4.62	0.39	0.13	0.07	0.15	0.10	21.89
F	4.20	8.49	5.04	7.51	2.57	0.20	0.24	0.15	0.11	0.08	28.59
Mean	2.76	4.68	4.36	6.79	4.01	0.41	0.19	0.12	0.15	0.12	23.59
SD	1.47	2.85	1.47	2.66	1.83	0.14	0.07	0.05	0.04	0.07	6.31

Note: Data corrected for 27.2% of the applied dose evaporating from the skin surface upon application.

[a] Site of application washed 24-h postapplication with soap and water.

From Bucks, D. A. W., Lee, M., Maibach, H. I., and Guy, R. H., in preparation

TABLE 15A
Percutaneous Absorption of Cyanophenol Under Occluded Conditions[a] in Man (% of Dose)

Subjects	0—4 H	4—8 H	8—12 H	12—24 H	Day 2	Day 3	Day 4	Day 5	Day 6	Day 7	Total %D
A	2.17	8.92	11.1	12.7	3.08	0.33	0.14	0.19	0.34	0.17	39.17
B	5.56	15.3	11.9	8.49	1.18	0.29	0.38	0.40	0.21	0.21	43.85
C	31.8	3.28	0.27	3.62	4.59	0.44	0.29	0.24	0.17	0.30	44.97
D	17.6	17.5	8.47	9.00	1.22	0.20	0.22	0.12	0.10	0.11	54.56
E	11.3	13.1	7.01	6.53	1.82	0.43	0.38	0.13	0.15	0.16	40.99
F	3.32	26.9	12.0	5.35	2.21	0.25	0.21	0.23	0.31	0.15	50.88
Mean	12.0	14.2	8.45	7.62	2.35	0.32	0.27	0.22	0.21	0.18	45.74
SD	11.3	8.01	4.47	3.20	1.30	0.09	0.10	0.10	0.09	0.06	5.90

TABLE 15B
Percutaneous Absorption of Cyanophenol Under Protected Conditions[a] in Man (% of Dose)

Subjects	0—4 H	4—8 H	8—12 H	12—24 H	Day 2	Day 3	Day 4	Day 5	Day 6	Day 7	Total %D
A	0.34	11.8	11.5	12.4	2.90	0.28	0.08	0.07	0.07	0.08	39.60
B	0.67	0.68	0.02	0.38	4.86	0.90	0.24	0.20	0.23	0.15	8.35
C	3.03	6.40	6.78	6.19	4.56	0.29	0.20	0.17	0.14	0.00	27.76
D	1.83	8.43	10.4	9.77	2.19	0.16	0.10	0.06	0.03	0.06	33.07
E	1.25	3.90	4.03	7.05	4.19	0.12	0.11	0.09	0.11	0.01	20.86
F	0.07	25.8	15.9	12.4	1.77	0.00	0.06	0.04	0.01	0.08	56.18
Mean	1.20	9.51	8.11	8.04	3.41	0.29	0.13	0.11	0.10	0.06	30.97
SD	1.10	8.85	5.68	4.58	1.30	0.32	0.07	0.07	0.08	0.05	16.36

[a]   Site of application washed 24-h postapplication with soap and water.

From Bucks, D. A. W., Lee, M., Maibach, H. I., and Guy, R. H., in preparation.

TABLE 16A
Percutaneous Absorption of Nitrophenol Under Occluded Conditions[a] in Man (% of Dose)

Subjects	0—4 H	4—8 H	8—12 H	12—24 H	Day 2	Day 3	Day 4	Day 5	Day 6	Day 7	Total %D
A	1.21	10.4	5.41	5.58	2.29	0.14	0.09	0.09	0.07	0.04	25.30
B	0.01	4.00	6.75	6.04	1.77	0.29	0.05	0.05	0.04	0.09	19.08
C	2.66	15.3	9.40	7.96	2.75	0.13	0.13	0.09	0.05	0.09	38.57
D	0.10	16.3	7.33	4.59	2.25	0.19	0.14	0.12	0.09	0.17	31.22
E	9.99	13.8	5.97	6.22	2.08	0.17	0.17	0.05	0.08	0.08	38.62
F	11.8	36.5	14.9	5.64	1.38	0.12	0.07	0.06	0.03	0.08	70.64
Mean	4.30	16.0	8.30	6.01	2.09	0.18	0.11	0.08	0.06	0.09	37.24
SD	5.25	11.0	3.53	1.11	0.47	0.06	0.04	0.03	0.02	0.04	18.04

TABLE 16B
Percutaneous Absorption of Nitrophenol Under Protected Conditions[a] in Man (% of Dose)

Subjects	0—4 H	4—8 H	8—12 H	12—24 H	Day 2	Day 3	Day 4	Day 5	Day 6	Day 7	Total %D
A	0.07	6.67	11.2	21.1	4.45	0.59	0.65	0.40	0.25	0.54	45.88
B	1.46	6.19	12.9	17.4	8.71	0.99	0.37	0.16	0.29	0.14	48.54
C	4.45	5.30	4.58	8.95	4.98	0.51	0.41	0.19	0.21	0.15	29.72
D	0.04	3.80	5.29	5.43	3.56	0.33	0.29	0.23	0.17	0.17	19.31
E	2.86	0.00	19.3	9.92	3.52	1.37	0.25	0.16	0.12	0.18	37.69
F	1.37	10.2	9.15	20.7	2.46	0.15	0.30	0.11	0.20	0.29	44.92
Mean	1.71	5.35	10.4	13.9	4.61	0.66	0.38	0.21	0.21	0.24	37.68
SD	1.70	3.36	5.43	6.68	2.19	0.45	0.14	0.10	0.06	0.15	11.30

[a] Site of application washed 24-h postapplication with soap and water.

From Bucks, D. A. W., Lee, M., Maibach, H. I., and Guy, R. H., in preparation.

TABLE 17A

Percutaneous Absorption of Iodophenol Under Occluded Conditions[a] in Man (% of Dose)

Subjects	0—4 H	4—8 H	8—12 H	12—24 H	Day 2	Day 3	Day 4	Day 5	Day 6	Day 7	Total %D
A	0.97	4.14	3.48	5.59	7.63	3.13	2.52	1.32	0.84	0.75	30.39
B	0.40	10.2	2.31	6.86	6.47	2.17	1.73	0.53	0.35	0.05	31.02
C	3.86	7.01	1.73	6.16	4.69	2.38	2.27	0.80	0.59	0.71	30.21
D	1.72	2.51	1.36	5.24	5.27	2.77	2.52	1.31	0.66	0.36	23.72
E	5.87	2.11	0.25	3.60	6.94	6.35	4.60	3.70	1.75	0.58	35.75
F	1.13	1.98	1.28	2.50	4.79	2.80	1.96	1.30	0.82	0.50	19.04
Mean	2.33	4.65	1.74	4.99	5.97	3.27	2.60	1.49	0.83	0.49	28.36
SD	2.11	3.29	1.09	1.64	1.22	1.55	1.03	1.13	0.48	0.26	5.96

TABLE 17B

Percutaneous Absorption of Iodophenol Under Protected Conditions[a] in Man (% of Dose)

Subjects	0—4 H	4—8 H	8—12 H	12—24 H	Day 2	Day 3	Day 4	Day 5	Day 6	Day 7	Total %D
A	4.03	5.48	2.13	6.06	3.96	3.84	2.42	1.57	1.48	0.91	31.87
B	1.21	0.94	1.20	6.92	2.21	2.88	1.66	0.39	0.30	0.22	17.93
C	2.15	2.36	1.21	1.75	3.78	2.17	2.02	1.07	5.71	0.41	22.63
D	1.30	5.57	3.05	3.61	4.33	2.79	1.37	0.79	0.46	0.27	23.54
E	2.54	1.99	0.98	2.48	4.03	1.88	1.71	0.93	0.68	0.35	17.57
F	4.10	5.78	5.01	4.71	4.90	1.44	2.07	0.59	0.38	0.16	29.13
Mean	2.56	3.69	2.26	4.25	3.87	2.50	1.87	0.89	1.50	0.38	23.78
SD	1.27	2.16	1.55	2.02	0.90	0.85	0.37	0.41	2.10	0.27	5.80

[a] Site of application washed 24-h postapplication with soap and water.

From Bucks, D. A. W., Lee, M., Maibach, H. I., and Guy, R. H., in preparation.

TABLE 18A
Percutaneous Absorption of Heptyloxyphenol Under Occluded Conditions[a] in Man (% of Dose)

Subjects	0—4 H	4—8 H	8—12 H	12—24 H	Day 2	Day 3	Day 4	Day 5	Day 6	Day 7	Total %D
A	0.13	12.85	8.27	9.44	3.49	1.74	0.17	0.18	0.11	0.08	36.45
B	1.55	7.10	3.99	4.24	3.79	1.10	0.85	0.21	0.28	0.15	23.25
C	4.66	0.00	22.5	5.99	5.20	1.80	0.70	0.40	0.30	0.16	41.76
D	2.91	5.75	1.71	6.03	6.08	4.46	1.14	0.65	0.33	0.29	29.36
E	20.7	5.90	7.15	6.82	4.52	1.14	0.37	0.25	2.26	0.10	49.17
F	6.12	6.06	2.41	10.2	6.46	2.05	1.11	0.84	0.28	0.20	35.69
Mean	6.00	6.28	7.68	7.12	4.93	2.05	0.72	0.42	0.59	0.16	35.95
SD	7.49	4.09	7.73	2.26	1.21	1.24	0.39	0.27	0.82	0.08	9.10

TABLE 18B
Percutaneous Absorption of Heptyloxyphenol Under Protected Conditions[a] in Man (% of Dose)

Subjects	0—4 H	4—8 H	8—12 H	12—24 H	Day 2	Day 3	Day 4	Day 5	Day 6	Day 7	Total %D
A	0.02	1.76	3.88	4.40	2.20	0.41	0.36	0.10	0.09	0.11	13.33
B	0.91	1.34	7.38	0.05	5.61	0.89	0.66	0.50	0.16	0.15	17.64
C	0.50	4.81	8.19	7.49	3.45	1.48	0.36	0.44	0.22	0.05	27.00
D	0.25	1.73	2.14	3.35	3.60	1.68	0.65	0.29	0.16	0.08	13.92
E	0.04	8.44	10.6	8.41	5.38	1.68	0.81	0.27	0.19	0.13	35.97
F	3.31	4.34	4.13	6.97	9.20	1.64	1.27	0.65	0.36	0.11	31.98
Mean	0.84	3.74	6.06	5.11	4.91	1.30	0.69	0.37	0.20	0.11	23.31
SD	1.26	2.73	3.19	3.14	2.46	0.53	0.34	0.20	0.09	0.03	9.68

a   Site of application washed 24-h postapplication with soap and water.

From Bucks, D. A. W., Lee, M., Maibach, H. I., and Guy, R. H., in preparation.

TABLE 19A

Percutaneous Absorption of Pentyloxyphenol Under Occluded Conditions[a] in Man (% of Dose)

Subjects	0—4 H	4—8 H	8—12 H	12—24 H	Day 2	Day 3	Day 4	Day 5	Day 6	Day 7	Total %D
A	9.59	9.79	5.90	6.45	6.61	0.63	0.10	0.05	0.11	0.10	39.35
B	6.85	4.45	2.68	2.75	2.64	1.16	0.44	0.23	0.14	0.12	21.46
C	0.64	3.03	7.83	4.41	2.92	1.92	0.61	0.22	0.10	0.08	21.77
D	12.1	4.92	4.37	3.15	1.26	0.34	0.12	0.04	0.02	0.08	26.43
E	4.89	9.65	5.74	4.01	2.53	0.60	0.20	0.16	0.08	0.06	27.92
F	10.7	8.35	4.45	6.06	4.87	1.35	0.78	0.26	0.15	0.07	37.01
Mean	7.46	6.70	5.16	4.47	3.47	1.00	0.38	0.16	0.10	0.09	28.99
SD	4.25	2.92	1.75	1.51	1.93	0.59	0.28	0.09	0.05	0.02	7.59

TABLE 19B

Percutaneous Absorption of Pentyloxyphenol Under Protected Conditions[a] in Man (% of Dose)

Subjects	0—4 H	4—8 H	8—12 H	12—24 H	Day 2	Day 3	Day 4	Day 5	Day 6	Day 7	Total %D
A	0.33	1.42	2.41	1.64	0.85	0.15	0.05	0.02	0.07	0.04	6.98
B	0.30	2.62	2.52	2.02	1.73	0.54	0.20	0.11	0.10	0.13	10.28
C	1.41	3.12	3.13	1.56	1.84	0.69	0.36	0.12	0.11	0.09	12.43
D	1.32	5.14	1.03	2.62	2.02	0.23	0.05	0.05	0.08	0.07	12.62
E	0.14	9.36	3.33	2.74	1.42	0.33	0.22	0.15	0.08	0.11	17.88
F	2.56	3.63	2.43	4.01	1.55	1.03	0.56	0.22	0.12	0.07	16.18
Mean	1.01	4.22	2.48	2.43	1.57	0.49	0.24	0.11	0.09	0.09	12.73
SD	0.93	2.80	0.81	0.91	0.41	0.33	0.20	0.07	0.02	0.03	3.94

[a]   Site of application washed 24-h postapplication with soap and water.

From Bucks, D. A. W., Lee, M., Maibach, H. I., and Guy, R. H., in preparation.

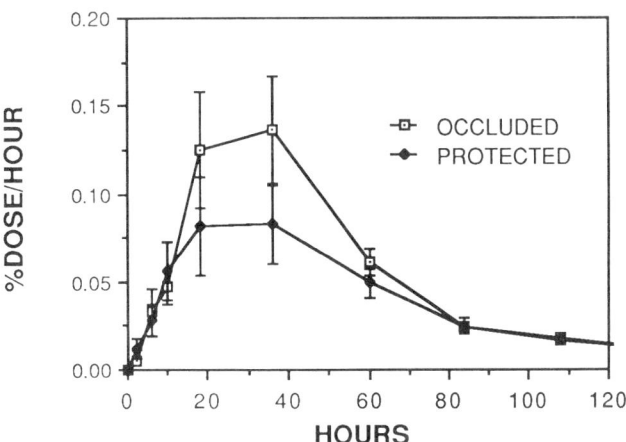

FIGURE 4. Percutaneous absorption of aminophenol in man under occluded and protected conditions (Mean ± SEM, N = 6). Exposure 24-h prior to soap and water washing.

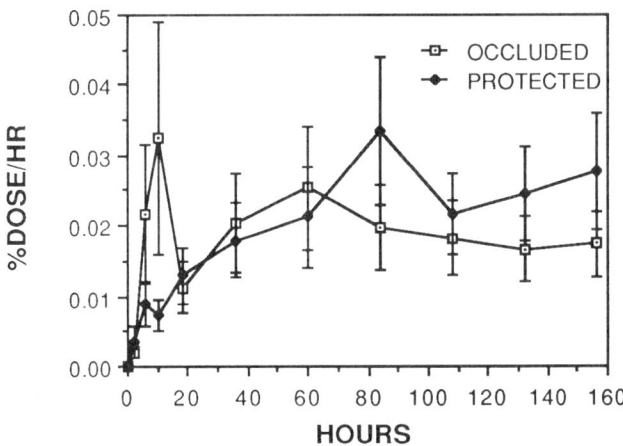

FIGURE 5. Percutaneous absorption of acetaminophen in man under occluded and protected conditions (Mean ± SEM, N = 6). Exposure 24-h prior to soap and water washing.

ment with increasing penetrant lipophilicity is apparent. But, insufficient data is yet available to quantify the degree of lipophilicity (as measured by octanol/water partition coefficient), which must be exhibited by a penetrant in order for occlusion-induced enhanced skin permeation to be manifest.

The increase in percutaneous absorption of hydrocortisone under occlusive conditions observed by Feldmann and Maibach[18] may be due to an acetone solvent effect. In this early work, the chemical was applied in 1.0 ml of acetone over an area of 13 cm². Might the pretreatment of the skin with a large volume of acetone compromise stratum corneum barrier function? It has been reported that acetone can damage the stratum corneum.[35,2] It is conceivable that the large volume of acetone used (76.9 μl/cm²) may be responsible for the observed increase in hydrocortisone penetration under plastic wrap occlusion. More likely,

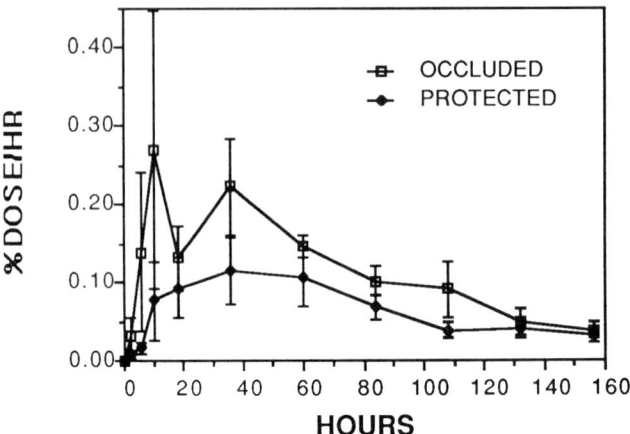

FIGURE 6. Percutaneous absorption of propionylamidophenol in man under occluded and protected conditions (Mean ± SEM, N = 6). Exposure 24-h prior to soap and water washing.

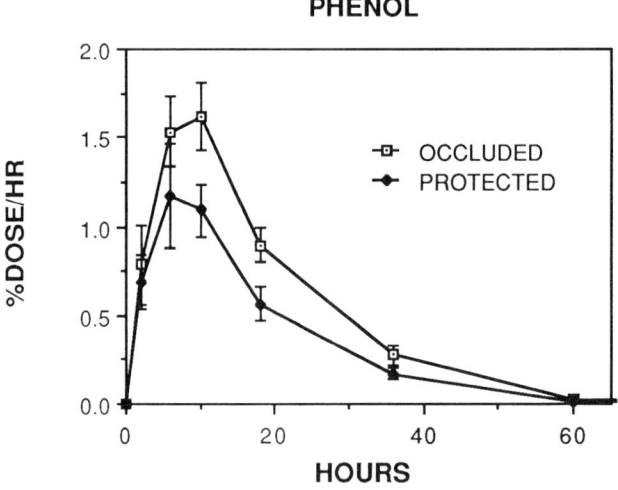

FIGURE 7. Percutaneous absorption of phenol in man under occluded and protected conditions (Mean ± SEM, N = 6). Exposure 24-h prior to soap and water washing.

it is reasonable to suggest that the increased duration of exposure (96 h compared to 24 h) may contribute to the increase in observed hydrocortisone percutaneous absorption. This enhancement in absorption was not observed in the experiments with hydrocortisone under occlusion[10] when the acetone surface concentration was only 8.0 μl/cm^2 (20 μl over 2.5 cm^2) and skin surface exposure was limited to 24 h.

The occlusion-induced enhancement of lipophilic compounds may be understood by considering the steps involved in the percutaneous absorption process. Minimally, after application in a volatile solvent, the penetrant must (1) dissolve/partition into the surface lipids of the stratum corneum, (2) diffuse through the lammelar lipid domains of the stratum corneum, (3) partition from the stratum corneum into the more hydrophilic viable epidermis, (4) diffuse through the epidermis and upper dermis, and (5) encounter a capillary of the cutaneous microvasculature and gain access to the systemic circulation (Figure 14).

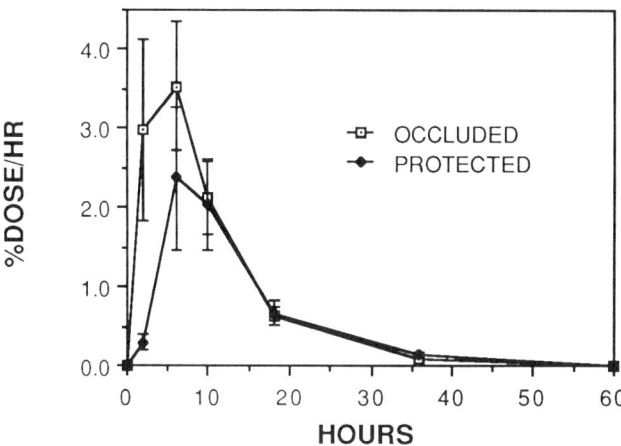

FIGURE 8. Percutaneous absorption of cyanophenol in man under occluded and protected conditions (Mean ± SEM, N = 6). Exposure 24-h prior to soap and water washing.

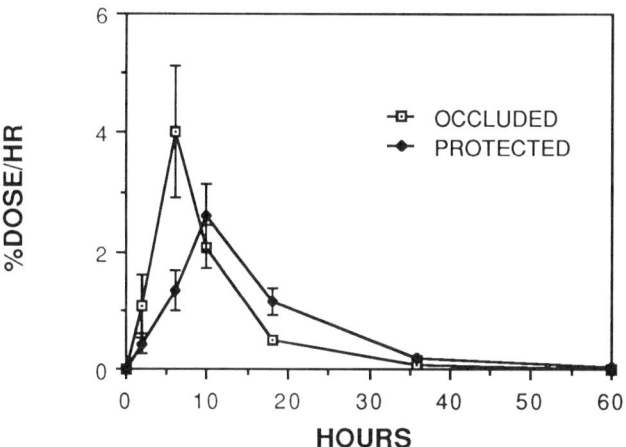

FIGURE 9. Percutaneous absorption of nitrophenol in man under occluded and protected conditions (Mean ± SEM, N = 6). Exposure 24-h prior to soap and water washing.

As stated above, occlusion hydrates the stratum corneum and, if the effect of hydration was simply to decrease the viscosity of the stratum corneum intercellular domain, then the penetration of all chemicals should be equally enhanced by occlusion. In other words, the relative increase in the effective diffusion coefficient of the penetrant across the stratum corneum would be independent of the nature of the penetrant. But this is not the situation observed; the degree of enhancement is compound specific. To account for this effect, we postulate that stratum corneum hydration alters the stratum corneum-viable epidermis partitioning step. Occlusion hydrates the keratin in corneocytes and increases the water content between adjacent intercellular lipid lamellae. A penetrant diffusing through the intercellular lipid domains will distribute between the hydrophobic bilayer interiors and the aqueous regions separating the head-groups of adjacent bilayers. Stratum corneum hydration magnifies

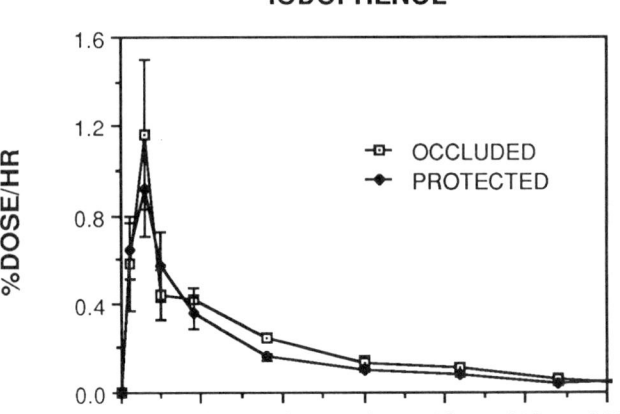

FIGURE 10. Percutaneous absorption of iodophenol in man under occluded and protected conditions (Mean ± SEM, N = 6). Exposure 24-h prior to soap and water washing.

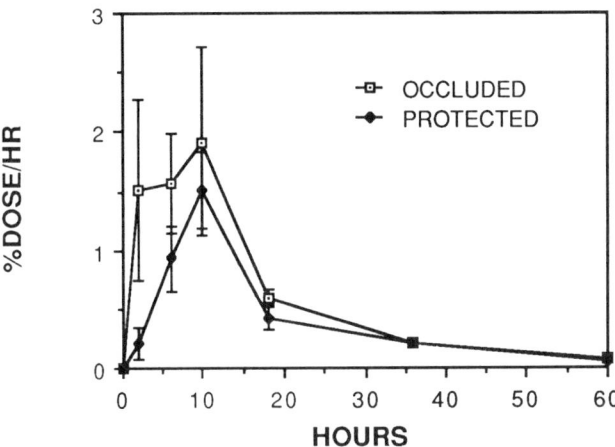

FIGURE 11. Percutaneous absorption of heptyloxyphenol in man under occluded and protected conditions (Mean ± SEM, N = 6). Exposure 24-h prior to soap and water washing.

the latter environment and increases the ''hydrophilic'' character of the stratum corneum somewhat. It follows that this leads, in turn, to a reduction in the stratum corneum-viable epidermis partition coefficient of the penetrant (because the two tissue phases now appear more similar). The decrease should facilitate the kinetics of transfer of penetrant through the stratum corneum and from the stratum corneum to the viable epidermis, and the relative effect on this rate should become greater as the lipophilicity of the absorbing molecule increases.[25] The limit of this mechanism of enhancement would occur when penetrant is either (1) completely insoluble in the aqueous phases of the stratum corneum or (2) sterically hindered from penetrating the skin at a measurable rate due, e.g., to large molecular size.

The importance of the partitioning step is implied by the dependence of percutaneous absorption on steroid lipophilicity (Figure 3). Penetration does not continue to increase with

PENTYLOXYPHENOL

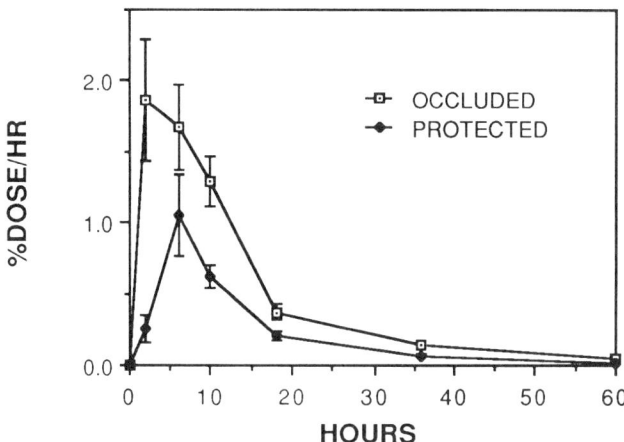

FIGURE 12. Percutaneous absorption of pentyloxyphenol in man under occluded and protected conditions (Mean ± SEM, N = 6). Exposure 24-h prior to soap and water washing.

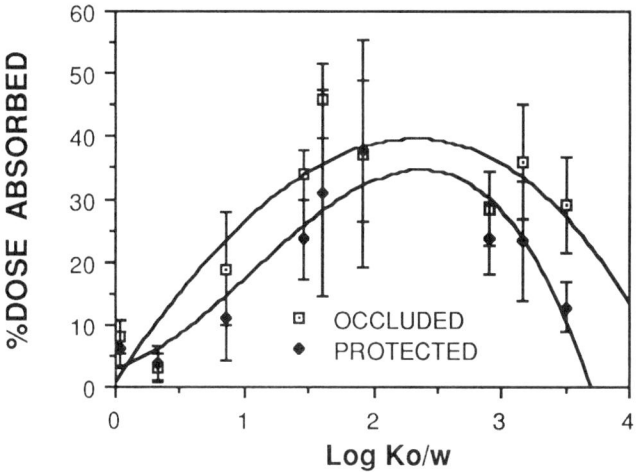

FIGURE 13. Percutaneous absorption of phenols in man under occluded and protected conditions as a function of penetrant octanol/water partition coefficient (Ko/w).

increasing lipophilicity (as would be predicted if the skin behaved as a simple lipid membrane). The attenuation in absorption may be explained by a shift in the rate-limiting step from diffusion through the stratum corneum to the transfer across the stratum corneum-viable epidermis interface. The effect should be most apparent when the penetrant's aqueous solubility is extremely low; thus, it follows that this transfer process should become slower as penetrant lipophilicity increases. This suggested mechanism is further supported by results obtained with the para-substituted phenols described above[7,8] (Figure 13).

Restricting evaporative loss of volatile compounds using plastic wrap occlusion enhances percutaneous absorption.[3] Clearly, this effect may increase the extent of absorption of these lipophilic volatiles in addition to the possible enhancement afforded by occlusion-induced hydration of the stratum corneum.

As noted above, occlusion does not always increase the percutaneous absorption of

TABLE 20
Percutaneous Absorption of Phenols in Man[a]

Compound	Log Ko/w	Mean % dose absorbed (SD)	
		Occluded	Protected
Aminophenol	0.04	8 (3)	6 (3)
Acetaminophen	0.32	3 (2)	4 (3)
Propionylamidophenol	0.86	19 (9)	11 (7)
Phenol[b]	1.46	34 (4)	24 (6)[c]
Cyanophenol	1.60	46 (6)	31 (16)
Nitrophenol	1.91	37 (18)	38 (11)
Iodophenol	2.91	28 (6)	24 (6)
Heptyloxyphenol	3.16	36 (9)	23 (10)[b]
Pentyloxyphenol	3.51	29 (8)	13 (4)[c]

[a]   Single dose application from 95% EtOH (N = 6) to the ventral forearm; 24-h exposure at 2 to 4 $\mu g/cm^2$ prior to soap and water washing.

[b]   Data analysis accounts for 27.2% of applied dose evaporating off skin surface during application.

[b]   Significant difference @ $p < 0.05$

[c]   Significant difference @ $p < 0.01$

From Bucks, D. A. W., Lee, M., Maibach, H. I., and Guy, R. H., in preparation.

TABLE 21
Accountability of Applied Dose in Occluded Studies[a]

Compound	Observed PA	% Removed from skin	Total % dose
Aminophenol	8 (3)	55 (18)	63 (17)
Acetaminophen	3 (2)	61 (24)	64 (24)
Propionylamidophenol	19 (9)	77 (9)	96 (2)
Phenol[b]	34 (4)	61 (13)	95 (10)
Cyanophenol	46 (6)	41 (9)	87 (7)
Nitrophenol	37 (18)	50 (11)	87 (13)
Iodophenol	28 (6)	63 (4)	91 (3)
Heptyloxyphenol	36 (9)	59 (7)	95 (3)
Pentyloxyphenol	29 (8)	71 (8)	100 (2)

Note: Mean % dose absorbed (SD)

[a]   Single dose application from 95% EtOH (N = 6) to the ventral forearm; 24-h exposure at 2 to 4 $\mu g/cm^2$ prior to soap and water washing.

[b]   Data analysis accounts for 27.2% of applied dose evaporating off skin surface during application. (Bucks et al., 1990)

From Bucks, D. A. W., Lee, M., Maibach, H. I., and Guy, R. H., in preparation.

topically applied agents. Furthermore, the extent of penetration may depend upon the method of occlusion. This finding has important implications in the design of a transdermal drug delivery system (TDS) for which the duration of application exceeds 24 h. We have found that about $^1/_3$ of normal, healthy, male volunteers experience plastic chamber occlusion-induced irritation following contact periods greater than 24 h; however, we have not observed any irritation of the skin using the nonocclusive patch system on the same volunteers following identical contact periods with the same penetrant. In those situations for which occlusion does not significantly increase the percutaneous absorption of a topically applied

TABLE 22
Accountability of Applied Dose in Protected Studies[a]

Compound	Observed PA	% Removed from skin	Total % dose
Aminophenol	6 (3)	85 (4)	91 (2)
Acetaminophen	4 (3)	93 (5)	97 (4)
Propionylamidophenol	11 (7)	84 (7)	95 (2)
Phenol[b]	24 (6)	68 (15)	92 (18)
Cyanophenol	31 (16)	70 (12)	101 (5)
Nitrophenol	38 (11)	65 (12)	103 (4)
Iodophenol	24 (6)	73 (7)	97 (2)
Heptyloxyphenol	23 (10)	71 (10)	95 (4)
Pentyloxyphenol	13 (4)	85 (3)	98 (2)

Note: Mean % dose absorbed (SD)

[a] Single dose application from 95% EtOH (N = 6) to the ventral forearm; 24-h exposure at 2 to 4 $\mu g/cm^2$ prior to soap and water washing.

[b] Data analysis accounts for 27.2% of applied dose evaporating off skin surface during application.

From Bucks, D. A. W., Lee, M., Maibach, H. I., and Guy, R. H., in preparation.

drug, or an occlusion-induced enhancement in percutaneous absorption is not required, a nonocclusive TDS is an approach worthy of consideration.

Conclusions of the above discussion are as follows:

1. Studies indicate that the extent of percutaneous absorption may depend upon the occlusive system used.[3,4,27]
2. Occlusion does not necessarily increase percutaneous absorption. Penetration of hydrophilic compounds, in particular, may not be enhanced by occlusion.
3. Mass balance (dose accountability) has been demonstrated using occlusive and nonocclusive patch systems *in vivo* in man. Dose accountability rigorously quantifies percutaneous absorption measured using radiotracer methodology and allows objective comparison between different treatment modalities.
4. Occlusion, per se, can cause local skin irritation and the implication of this observation in the design of TDS should be considered.

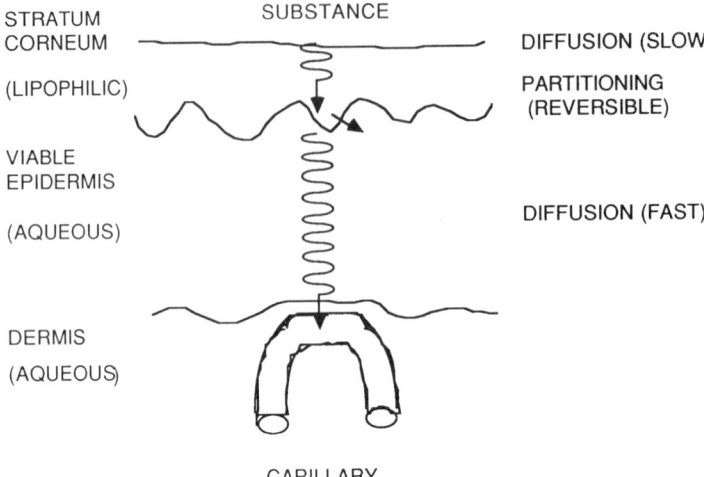

FIGURE 14.   Schematic depiction of percutaneous absorption.

TABLE 23
Percutaneous Absorption of Hydrocortisone in Man Following
Application from Acetone Solution

	% Dose absorbed[a]	Applied dose (μg/cm²)
Plastic wrap occlusion; single dose[b]	9 (6)	5.8
Plastic wrap occlusion; single dose[c]	5 (2)	4.0
Plastic chamber occlusion[d]	4 (2)	4.0
Plastic chamber occlusion[e]	4 (1)	4.0
Plastic chamber occlusion[f]	3 (1)	4.0
"Protected" Condition; single dose[g]	4 (2)	4.0
Nonprotected; single dose[h]	1 (0.3)	5.8
Nonprotected; single dose[i]	2 (2)	4.0
Nonprotected; multiple dose[j]	3 (1)	4.0
Nonprotected; multiple dose[k]	3 (1)	4.0

[a]   Values, mean (SD), corrected for incomplete renal elimination.

[b]   Occluded for 4 d, washed 96 h post-application with soap and water, see Table 5
      (Feldmann, R. J. and Maibach, H. I., *Arch. Dermatol.*, 91, 661, 1965.)

[c]   Occluded for 1 d, washed 24 h post-application with soap and water (Bucks et al.,
      unpublished results).

[d]   Single dose occluded for 7 d, washed 24 h post-application with soap and water (Guy
      et al., 1987).

[e]   % of first dose absorbed following daily doses. Occluded for 14 d, washed 24 h post-
      application with soap and water (Guy et al., 1987).

[f]   % of eighth dose absorbed following daily doses. Occluded for 14 d, washed 24 h
      post-application with soap and water (Guy et al., 1987).

[g]   Site covered for 7 d with a ventilated plastic chamber, washed 24 h post-application
      with soap and water (Bucks et al., 1988).

[h]   Site washed 24 h post-application with soap and water; see Table 4 (Feldmann, R.
      J. and Maibach, H. I., *Arch. Dermatol.*, 91, 661, 1965.)

[i]   Site washed 24 h post-application with soap and water (Feldmann, R. J. and Maibach,
      H. I., *J. Invest. Dermatol.*, 52, 89, 1969.)

[j]   % of first dose absorbed following daily doses for 14 d, site washed 24 h post-
      application with soap and water (Bucks, D. A. W., Maibach, H. I., and Guy, R.
      H., *J. Pharm. Sci.*, 74, 1337, 1985.)

[k]   % of eighth dose absorbed following daily doses for 14 d, site washed 24 h post-
      application with soap and water (Bucks, D. A. W., Maibach, H. I., and Guy, R.
      H., *J. Pharm. Sci.*, 74, 1337, 1985.)

REFERENCES

1. **Blank, I. H. and Scheuplein, R. J.,** The epidermal barrier, in *Progress in the Biological Sciences in Relation to Dermatology,* Vol. 2, Rook, A. J. and Champion, R. H., Eds., Cambridge University Press, Cambridge, 1964, 245.

2. **Bond, J. R. and Barry, B. W.,** Damaging effect of acetone on the permeability barrier of hairless mouse skin compared with that of human skin, *Int. J. Pharmaceut.,* 41, 91, 1988.

3. **Bronaugh, R. L., Stewart, R. F., Wester, R. C., Bucks, D. A. W., and Maibach, H. I.,** Comparison of percutaneous absorption of fragrances by humans and monkeys, *Fd. Chem. Toxicol.,* 23, 111, 1985.

4. **Bronaugh, R. L., Wester, R. C., Bucks, D. A. W., Maibach, H. I., and Sarason, R.,** *In vivo* percutaneous absorption of fragrance ingredients in rhesus monkeys and humans, *Fd. Chem. Toxicol.,* 28, 369, 1990.

5. **Bucks, D. A. W., Marty, J. P. L., and Maibach, H. I.,** Percutaneous absorption of malathion in the guinea pig: effect of repeated skin application, *Fd. Chem. Toxicol.,* 23, 919, 1985.

6. **Bucks, D. A. W., Maibach, H. I., and Guy, R. H.,** Percutaneous absorption of steroids: effect of repeated application, *J. Pharm. Sci.,* 74, 1337, 1985.

7. **Bucks, D. A. W., McMaster, J. R., Maibach, H. I., and Guy, R. H.,** Percutaneous absorption of phenols *in vivo, Clin. Res.,* 35, 672A, 1987.

8. **Bucks, D. A. W., McMaster, J. R., Maibach, H. I., and Guy, R. H.,** Prolonged residence of topically applied chemicals in the stratum corneum, *J. Pharm. Sci.,* 76, S125, 1987.

9. **Bucks, D. A. W., McMaster, J. R., Maibach, H. I., and Guy, R. H.,** Bioavailability of topically administered steroids: a ''mass balance'' technique, *J. Invest. Dermatol.,* 90, 29, 1988.

10. **Bucks, D. A. W., Maibach, H. I., and Guy, R. H.,** Mass balance and dose accountability in percutaneous absorption studies: development of nonocclusive application system, *Pharm. Res.,* 5, 313, 1988.

11. **Bucks, D. A. W., Maibach, H. I., and Guy, R. H.,** Occlusion does not uniformally enhance penetration *in vivo,* in *Percutaneous Absorption,* Vol. 2, Bronaugh, R. and Maibach, H., Eds., Marcel Dekker, New York, 1989, 77.

12. **Bucks, D. A. W.,** Prediction of Percutaneous Absorption, Ph.D. thesis, University of California, San Francisco, 1989.

13. **Bucks, D. A. W., Lee, M., Maibach, H. I., and Guy, R. H.,** Percutaneous penetration of *p*-substituted phenols *in vivo,* in preparation, 1990.

14. **Elias, P. M. and Brown, B. E.,** The mammalian cutaneous permeability barrier, *Lab. Invest.,* 39, 574, 1978.

15. **Elias, P. M., Cooper, E. R., Korc, A., and Brown, B.,** Percutaneous transport in relation to stratum corneum structure and lipid composition, *J. Invest. Dermatol.,* 76, 297, 1981.

16. **Elias, P.,** Epidermal lipids, barrier function, and desquamation, *J. Invest. Dermatol.,* 80, 44s, 1983.

17. **Faergemann, J., Aly, R., Wilson, D. R., and Maibach, H. I.,** Skin occlusion: effect on *Pityrosporum orbiculare,* skin permeability of carbon dioxide, pH, transepidermal water loss, and water content, *Arch. Dermatol. Res.,* 275, 383, 1983.

18. **Feldmann, R. J. and Maibach, H. I.,** Penetration of ^{14}C hydrocortisone through normal skin, *Arch. Dermatol.,* 91, 661, 1965.

19. **Feldmann, R. and Maibach, H. I.,** Percutaneous absorption of steroids in man, *J. Invest. Dermatol.,* 52, 89, 1969.

20. **Fisher, L. B. and Maibach, H. I.,** The effect of occlusive and semipermeable dressings on the mitotic activity of normal and wounded human epidermis, *Br. J. Dermatol.,* 86, 593, 1972.

21. **Garb, J.,** Nevus verrucosus unilateralis cured with podophyllin ointment, *Arch. Dermatol.,* 81, 606, 1960.

22. **Golden, G. M., Guzek, D. B., McKie, J. E., and Potts, R. O.,** Role of stratum corneum lipid fluidity in transdermal drug flux, *J. Pharm. Sci.,* 76, 25, 1987.

23. **Gordon, B. and Maibach, H. I.,** Studies on the mechanism of aluminum anhidrosis, *J. Invest. Dermatol.,* 50, 411, 1968.

24. **Guy, R. H. and Hadgraft, J.,** The prediction of plasma levels of drugs following transdermal application, *J. Control. Release,* 1, 177, 1985.

25. **Guy, R. H., Hadgraft, J., and Bucks, D. A. W.,** Transdermal drug delivery and cutaneous metabolism, *Xenobiotica,* 17, 325, 1987.

26. **Guy, R. H., Bucks, D. A. W., McMaster, J. R., Villaflor, D. A., Roskos, K. V., Hinz, R. S., and Maibach, H. I.,** Kinetics of drug absorption across human skin *in vivo,* in *Skin Pharmacokinetics,* Shroot, B. and Schaefer, H., Eds., Karger, Basel, 70, 1987.

27. **Kim, H. O., Wester, R. C., McMaster, J. R., Bucks, D. A. W., and Maibach, H. I.,** Skin absorption from patch test systems, *Contact Dermatitis,* 17, 178, 1987.

28. **Kligman, A. M.,** A biological brief on percutaneous absorption, *Drug Dev. Ind. Pharm.,* 9, 521, 1983.

29. **McKenzie, A. W.,** Percutaneous absorption of steroids, *Arch. Dermatol.,* 86, 91, 1962.

30. **McKenzie, A. W. and Stoughton, R. B.,** Method for comparing percutaneous absorption of steroids, *Arch. Dermatol.*, 86, 88, 1962.
31. **Menon, G. K., Feingold, K. R., Moser, A. H., Brown, B. E., and Elias, P. M.,** *De novo* sterologenesis in the skin. II. regulation by cutaneous barrier requirements, *J. Lipid Res.*, 26, 418, 1985.
32. **Orentreich, N., Berger, R. A., and Auerbach, R.,** Anhidrotic effects of adhesive tapes and occlusive film, *Arch. Dermatol. Res.*, 94, 709, 1966.
33. **Potts, R. O.,** Stratum corneum hydration: experimental techniques and interpretation of results, *J. Soc. Cosmet. Chem.*, 37, 9, 1986.
34. **Quisno, R. A. and Doyle, R. L.,** A new occlusive patch test system with a plastic chamber, *J. Soc. Cosmet. Chem.*, 34, 13, 1983.
35. **Schaeffer, H., Zesch, A., and Stuttgen, G., Eds.,** *Skin Permeability*, Springer-Verlag, Berlin, 1982, 541.
36. **Scholtz, J. R.,** Topical therapy of psoriasis with fluocinolone acetonide, *Arch. Dermatol.*, 84, 1029, 1961.
37. **Sulzberger, M. B. and Witten, V. H.,** Thin pliable plastic films in topical dermatological therapy, *Arch. Dermatol.*, 84, 1027, 1961.
38. **Bucks, D. A. W., et al.,** Unpublished observations.

Chapter 9

SKIN STORAGE CONDITIONS

William G. Reifenrath and Barbara W. Kemppainen

TABLE OF CONTENTS

I. EFFECTS OF STORAGE ON SPLIT-THICKNESS SKIN STRUCTURE AND FUNCTION

There has long been an interest in preserving skin for the treatment of burn injuries. Considerable literature has been developed to define and improve skin storage for grafting. Several reviews have been published.[1-5] The bulk of this literature deals with split-thickness skin or skin which contains an epidermis and some portion of the dermis. Storage conditions have been principally related to skin viability and not to effects on skin permeability or metabolism of xenobiotics. However, the information would serve as a useful baseline to those investigators wishing to store such membranes for skin penetration and disposition studies where skin viability may be an issue. The information is also important to those wishing to screen compounds for potential skin damage by their effects on skin metabolism.

A variety of techniques have been employed to store skin for the purpose of grafting, but these mainly consist of three methods: storage in a medium at 4°C; frozen storage with cryopreservative; and storage after freeze drying of skin. The quality of storage has been judged by such factors as the percent of successful "takes" of grafted skin, growth of cells in culture, or biochemical measures of the metabolic activity of skin.

Successful grafting is the ultimate objective for burn therapy. However, some investigators have cautioned against using grafting success as a measure of skin viability, as sometimes a nonviable graft serves only as a template for ingrowth of the host cells and only appears to be a "take". Indeed, the temporary covering of a wound may be in itself the objective, as it prevents fluid loss and helps to relieve pain.[6] The nude mouse has been utilized as graft recipient in tests to measure skin viability.[7] The athymic model eliminates problems of homograft rejection and the graft and donor tissues are antigenically distinct, allowing a better distinction between graft and donor skin growth.

Glucose uptake and respiratory measurements have been cited as the methods of choice to evaluate the effects of skin storage on viability. The respiratory enzymes are located in the mitochondria and are a sensitive indicator of cell damage.[3] Skin can utilize glucose by a number of alternate pathways.[8-12] After glucose enters the cells and is converted to glucose 6-phosphate, it can enter an anabolic pathway to synthesize glycogen. Alternately, glucose 6-phosphate can be used to produce energy for the cell by undergoing anaerobic glycolysis in the cytoplasm by the Embden-Myerhoff pathway to produce pyruvate. Pyruvate is principally converted to lactate by the skin, but can also enter the tricarboxylic acid cycle (via acetyl CoA) in the mitochondria to derive additional energy aerobically (respiration) and to produce CO_2. Glucose 6-phosphate can also be oxidized by the pentose phosphate pathway or hexose monophosphate shunt to produce CO_2 and energy.

By administration of radiolabeled glucose to skin slices and measuring the formation of radiolabeled CO_2, lipid and lactate, Freinkel suggested that the major portion of assimilated glucose was converted to lactic acid by the Embden-Myerhoff pathway.[13] Several investigators have reported the *in vitro* oxidation of fatty acids as an additional source of energy for the skin. The relative contributions of anaerobic glycolysis and fatty acid oxidation in the intact animal to the energy needs of the skin have been a subject of controversy. In any case, the production of CO_2 or lactate can serve as an indicator of cellular metabolism and has been used as a measure of viability for stored skin.

A. STORAGE OF SKIN AT 4°C ("REFRIGERATED STORAGE")

Cram et al.[14] stored split-thickness human skin at 4°C in saline and tissue culture media (RPMI formula 1640) in combination with low and high concentrations of penicillin and streptomycin. Viability, as judged by successful grafting to nude mice, was greatest with tissue culture media and low concentrations of antibiotics (25 units/ml penicillin G and 25 mcg/ml streptomycin). No improvement in graft success was obtained by the addition of

10% fetal calf serum. The authors stress the importance of maintaining a physiological pH, although pH alone did not explain differences in grafting success. In a subsequent study,[15] no difference in grafting success was found between RPMI-1640 media vs. Eagle's minimum essential medium (MEM) with HEPES buffer. When animal skin was stored in RPMI medium,[16] the viability of rabbit and pig skin was similar to that of human skin. Results for mouse, rat, and dog skin were significantly different from those of human skin. One should therefore be cautioned in extrapolations between species. The physical configuration of human skin stored in RPMI media was found to affect grafting success to the nude mouse.[17] For example, after 15 d storage in RPMI, greater success was obtained with skin from the outside of the roll than from the inside. The authors suggest that the optimal storage configuration for human skin be free floating, with 300 cm^2 of skin surface to 100 ml of media an appropriate ratio.

Prows and Nathan[18] used two *in vitro* methods to assess rabbit skin viability. One measured basal cell viability and the second determined the ability of skin to grow in culture. Rabbit skin growth was depressed when stored in saline at 4°C, and optimal when refrigerated in nutrient medium supplemented with fetal bovine serum in a sealed tube.

Alsbjorn et al.[19] used oxygen consumption as an index of viability in stored skin. Addition of solcoseryl, a standardized calf blood hemodialysate, to 4°C storage and freezing media had no effect during the first week of storage. Oxygen consumption was improved after the second and third weeks.

May and Wainwright[20] described the structural and metabolic degeneration of porcine skin stored in Eagle's MEM at 4°C. During the first week, a small amount of cellular debris was exfoliated. In addition, they report a significant loss of cellular enzyme activity and integrity, as indicated by leakage of aspartic and glutamic acid decarboxylases, and enzymes of the pentose shunt. Histological findings included shrinkage of the epithelium and cell loss and vacuolization in the papillary dermis. The ability of the dermis to metabolize such substrates as glucose, ornithine, and aspartic acid to CO_2 dropped to low levels during the first 10 d. Storage up to 30 d was characterized by continued cellular and vascular degeneration. Storage up to 58 d resulted in significant loss of skin structure and exfoliation of large amounts of cellular debris. For grafting purposes, the authors suggest a storage limit of one week, if physiological function is required.

In one laboratory, skin has been routinely stored in media at 4°C for skin penetration studies.[21] However, the details have not been published.

B. SUB-ZERO STORAGE ("FROZEN STORAGE")

May and De Clement[22] utilized a metabolic viability assay to determine the effects of skin storage at 4°C and cryopreservation at −196°C. The assay was based on the conversion of ^{14}C-glucose into $^{14}CO_2$ by viable cells on the dermal side of split-thickness cadaver skin. Skin was placed in Eagle's MEM/Earle's balanced salt solution with 2 m*M* fresh glutamine added directly before use.[23] The medium also contained 1 uCi of D-(U-^{14}C) glucose. After 4 h of incubation, the media was acidified with perchloric acid and the amount of radiolabeled CO_2 produced was assayed. There was no correlation between the age or sex of the skin donor and the metabolic viability. There was a strong negative correlation between the time the donor body spent in refrigerated storage and metabolic viability, decreasing to 50% of initial value after 24 h. Skin stored at 37°C in Eagle's MEM with a 5% CO_2/95% air atmosphere retained essentially all its viability after 24 h, but rapidly lost viability on continued storage, declining to 17% of initial value at 48 h. If stored at 4°C under similar conditions, skin retained 90% or more of its viability for 2 to 3 d. Thereafter, viability slowly declined to approximately 50% after 6 d of storage. Skin samples were cryopreserved at −196°C in tissue culture medium containing 10% human sera and 15% (v/v) glycerol. Skin was slowly cooled at −1°C/min to −100°C and then immersed in liquid nitrogen.

Previously, Lehr[24] had shown that maximum graft survival was obtained with slow cooling rates and rapid warming rates. Skin was warmed by immersion in a constant temperature water bath with constant warming temperatures ranging from 3 to 66°C. The maximum metabolic viability occurred over a range of 10 to 37°C. Skin viability decreased sharply above a constant warming temperature of 37°C and was completely lost at 66°C. Under carefully controlled conditions, retention of initial viability by cryopreservation at −196°C was greater than that provided by 4°C storage in tissue culture medium when stored for three or more days.

May and De Clement recommended a flat package format to aid in temperature control of skin tissue during freezing and warming.[25] May et al.[26] developed an "insulated heat sink box" for use with a −70°C refrigerator as a cost efficient alternative to a programmed controlled-rate freezer for small amounts of skin.

The use of cryoprotectants in combination with slow cooling rates have been established as beneficial in maintaining skin viability.[5] Freezing skin in the absence of cryoprotectants causes severe damage, which may be caused by the hypertonic extracellular solution produced by ice formation.[27,28] Glycerol and dimethyl sulfoxide (DMSO) have been usefully employed as cryoprotectants.[29] The cryoprotectant's effect may be related to their colligative properties in reducing osmotic dehydration.[27,28] Using reduced radiolabeled glycine incorporation into skin protein as an indicator of metabolic damage, De Loecker showed that glycerol and DMSO themselves caused damage to rat skin. A minimum exposure time was necessary to allow the protectants to penetrate cells. An exposure time of 45 min to the cryoprotectant was optimal, and longer exposures caused increased damage.[29]

Addition of ATP and inorganic phosphates, cortisol or amino acids, to the cryoprotectant media has been reported as beneficial with rat and human skin.[30-34]

Praus et al.[35] cryopreserved pig skin (−196°C) and used radiolabeled sulfate incorporation as an indicator of viability. Prior to cryopreservation, the skin surface was sterilized by washing with soap, saline, ethanolic iodine, antibiotics (chloramphenical, streptomycin, and furantoin), and saline in that order. Control skin was washed with soap and saline only. Following cyropreservation, it was found that the pretreatment with ethanolic iodine reduced radioactive sulfate uptake by approximately 20% as compared to control skin, ethanolic iodine plus antibiotics resulted in a 30% reduction, and the full treatment gave a 40% reduction. These reductions were not attributed to removal of bacteria, and the authors suggest that alternate means of skin surface preparation might improve skin resistance to freeze injury.

If cryopreserved skin is used in percutaneous absorption studies, steps would have to be taken to rule out any effect of the cryopreservation on penetration. The presence of DMSO would be of concern, as it is known to enhance skin absorption of many compounds.

One possible mechanism for the effects of sub-zero storage of skin on percutaneous penetration is the altered structure of the barrier layers of skin. Several reports have noted the structural changes induced by skin storage.[36,37,38] Fritsch and Stoughton[37] histologically examined human skin which had been stored at −17 to −22°C. The freezing resulted in reduction in cell volume and pyknosis, however the stratum corneum was intact. Kao et al.[38] incubated mouse skin in culture media at 37°C for 16 h, followed by freezing in liquid nitrogen for 5 s. The skin was thawed, incubated at 37°C in culture media, and histological examination revealed necrosis in the epidermis and follicular epithelium. Holland et al.[36] reported extensive necrosis in the epidermis and adnexal structures of mouse skin which had been stored at −20°C for one week prior to use.

C. FREEZE-DRIED SKIN STORAGE

Freeze-drying of skin permits simple storage and allows for easy shipping. However, this procedure has been reported to result in loss of viability.[39] Such skin may retain esterase activity, but this should not be confused with viability.

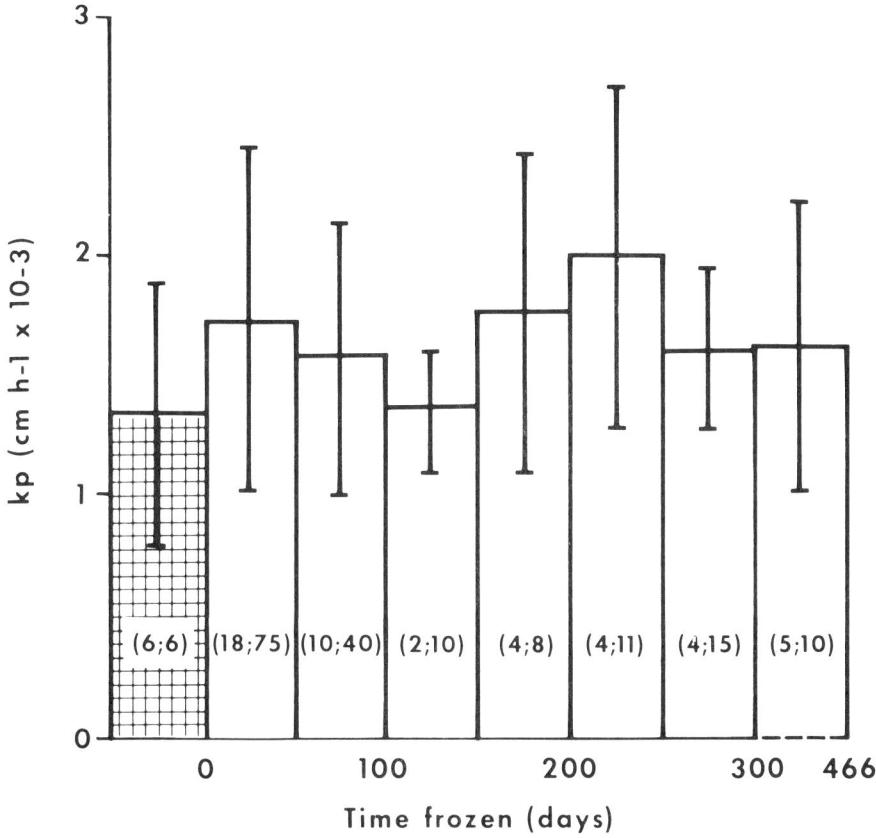

FIGURE 1. Effect of prolonged freezing on human skin permeability to water. Mean permeability coefficients (K_p) ± one standard deviation are given for 50-d intervals. Numbers in parenthesis are number of skin specimens; total number of replicates. The hatched bar represents results for skin that was not frozen. (From Harrison, S. M., Barry, B. W., and Dugard, P. H., *J. Pharm. Pharmacol.*, 36, 261, 1984.)

II. THE INFLUENCE OF STORAGE CONDITIONS ON THE *IN VITRO* PERMEABILITY PROPERTIES OF SKIN

A. ASSESSED BY PERCUTANEOUS PENETRATION OF WATER

In 1944 Burch and Winsor[40] reported that the rate of water loss through abdominal skin at various time intervals after excision was not affected by up to 27 d of frozen storage. Between successive measurements of water loss, they stored the skin samples in a refrigerator-freezer and thawed the samples slowly before the next measurement. Astley and Levine[41] obtained split-thickness (epidermis plus a portion of the dermis) of human thigh or lower leg skin from amputations. They rolled the strips of skin in sterile gauze moistened with Hanks' balanced salt solution and placed the strips in containers which were stored at −20°C. They stated that such storage did not affect the permeability of the skin to tritiated water for up to six months of storage. Harrison et al.[42] found no differences between measurements of *in vitro* percutaneous penetration of tritiated water in skin stored at −20°C for up to 466 d vs. fresh skin stored at 10°C and used within 2 to 3 d after autopsy (Figure 1). Bronaugh et al.[43] found the permeation of tritiated water through split-thickness human skin stored at 4°C until used (<24 h) or stored at −20°C (1 to 3 days) was similar. In most cases (seven out of nine comparisons) frozen (−20°C) storage of split-thickness human skin for up to 12 months did not change its permeability to water.

TABLE 1
Permeability of Cattle and Sheep
Skins to Levamisole

Animal	Skin thickness (cm)	10^6 $k_p r$ (cm$_2$ min^{-1})
Fresh		
Sheep	0.055	6.0
	0.066	7.7
	0.070	10.7
	0.071	9.2
Frozen		
Sheep	0.077	7.2
	0.077	6.5
	0.077	8.4
Fresh		
Calf	0.069	27.7
	0.071	26.8
	0.076	30.9
	0.092	33.7
	0.093	31.7
Frozen		
Calf	0.065	26.5
	0.069	25.6
	0.071	29.3
	0.082	29.5
	0.086	30.2
	0.091	33.4

Note: From a 0.85% solution in aqueous pH 8.9
buffer and water bath temperature of 37°.

From Pitman, I. H. and Downes, L. M., *J. Pharm.*
Sci., 71, 846, 1982. With permission.

It should be noted that the use of tritiated water to measure skin penetration of water is complicated by isotopic exchange and dilution with water already present in skin. It is therefore difficult to interpret such data.

B. ASSESSED BY PERCUTANEOUS PENETRATION BY XENOBIOTICS

The effects of storage on the skin permeability of xenobiotics has only recently received attention and remains a controversial point. Pitman and Downes[44] compared the penetration of levamisole through fresh cattle and sheep skin versus skin which had been stored at −30°C for 5 to 7 d and then thawed (Table 1). No significant differences were found.

In 1982 Swarbrick et al.[45] were the first to report that freezing altered the permeability of skin. The permeability of heat separated human epidermis towards a chromone acid (6,7,8,9-tetrahydro-5-hydroxy-4-oxo-10-propyl-4H-naphtho (2,3-b)-pyran-2-carboxylic acid) was increased when the tissue which had been stored frozen (−17°C), was thawed. However, the permeability of the epidermis towards the compound was unchanged when the tissue

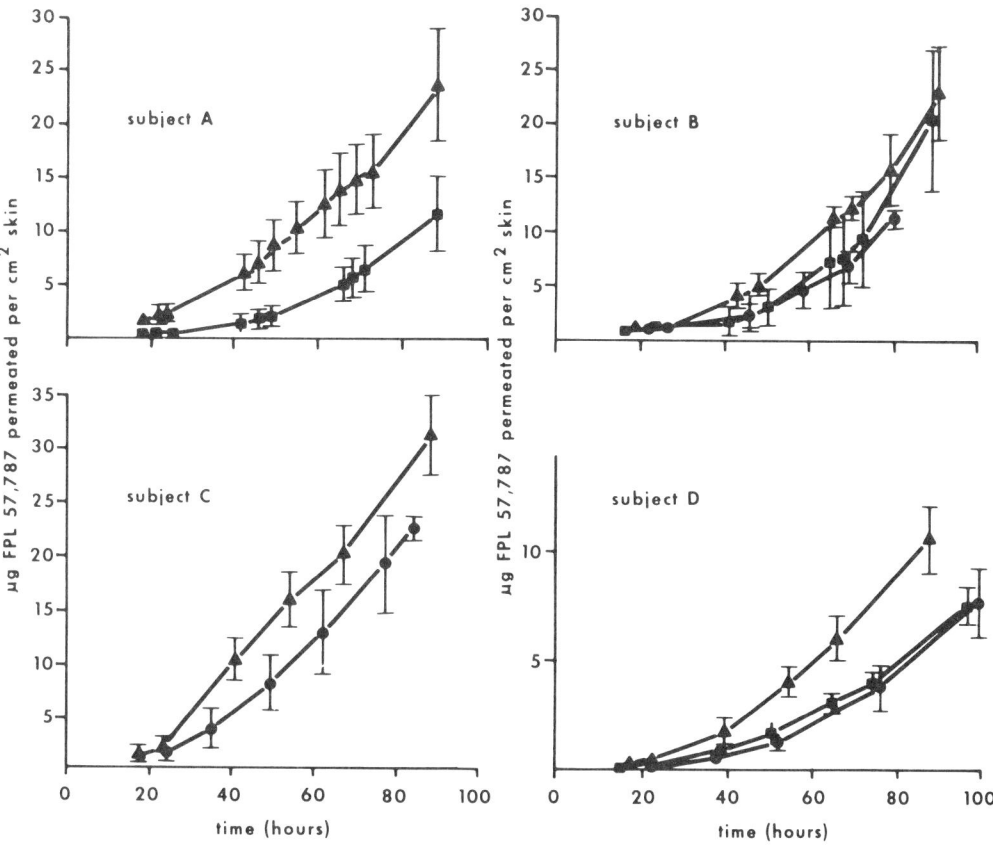

FIGURE 2. Permeation of a chromone acid (6,7,8,9-tetrahydro-4-oxo-10-propyl-4H-naptho [2,3-b] -pyran-2-carboxylic acid FPL 57, 787) through excised human skin of subjects A, B, C, and D as a function of time. Key: solid square, fresh sample; solid triangle, frozen sample; solid circle, dried sample. SD shown as vertical bars. (From Swarbrick, J., Lee, G., and Brom, J., *J. Invest. Derm.*, 78, 63, 1982. With permission.)

was stored dry as compared to fresh skin (Figure 2). This method involved drying the separated human epidermis overnight at ambient temperature in a desiccator maintained at 25% relative humidity. The dried samples were covered with aluminum foil and stored at 1°C. The deleterious effects of freezing on percutaneous barrier function were substantiated by Hawkins and Reifenrath.[46] The percutaneous penetration and evaporation of *N,N*-diethyl-m-toluamide (DEET) were determined on whole pig skin samples immediately after excision and after a freezing period of 1 to 6 weeks at $-80°C$. DEET was not metabolized by fresh skin samples under these conditions.[47] Pig skin which was held frozen for longer than 1 week was more permeable to DEET than the freshly excised skin. A positive correlation was found between mean total percentage percutaneous penetration and the number of weeks the skin was held frozen (Figure 3). As the total percutaneous penetration values of DEET increased, total evaporation values of DEET decreased. Parry et al.[48] compared the effects of storing human epidermis or isolated stratum corneum in distilled water at 4°C for up to two weeks. They found that the permeability of the epidermis increased by 36% and the permeability of the stratum corneum remained constant after 1 week of storage. The permeability of the epidermis and stratum corneum had both increased by 50% after two weeks of storage (Figure 4).

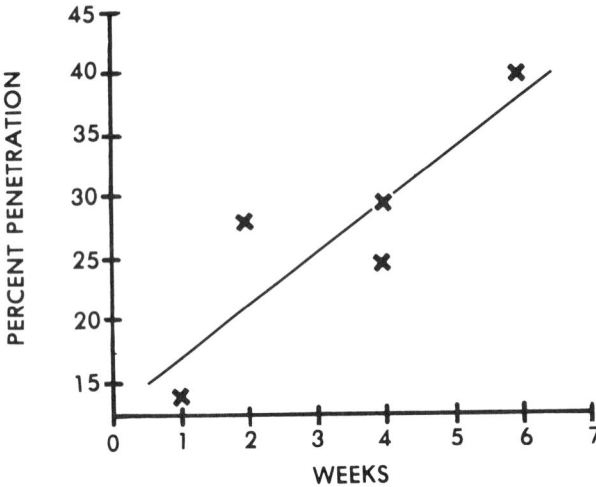

FIGURE 3.   Linear regression line and plot of mean percutaneous penetration (percentage of applied radioactive dose) of m-DEET (320 μg/cm² vs. frozen storage time of skin discs prior to use. (From Hawkins, G. S., Jr. and Reifenrath, W. G., *Fundam. Appl. Toxicol.*, 4, S133, 1984. With permission.)

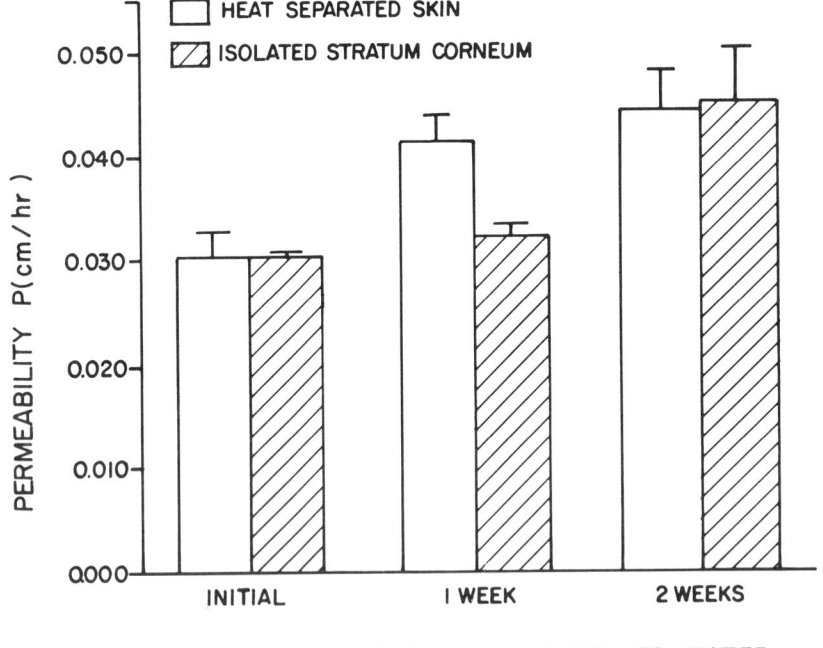

FIGURE 4.   Effect of storage time on the permeability of benzoic acid across isolated stratum corneum and heat-separated skin. (From Parry, G. E., Bunge, A. L., Silcox, G. D., Pershing, L. K., and Pershing, D. W., *Pharm. Res.*, 7, 230, 1990. With permission.)

C. CUTANEOUS METABOLISM AS MECHANISM FOR SKIN STORAGE EFFECTS ON PENETRATION

The effect of freezing on metabolic function of skin could be one of the mechanisms by which freezing alters the percutaneous penetration of xenobiotics. Holland et al.[36] compared the penetration of radiolabeled B[a]P through fresh mouse skin which had been

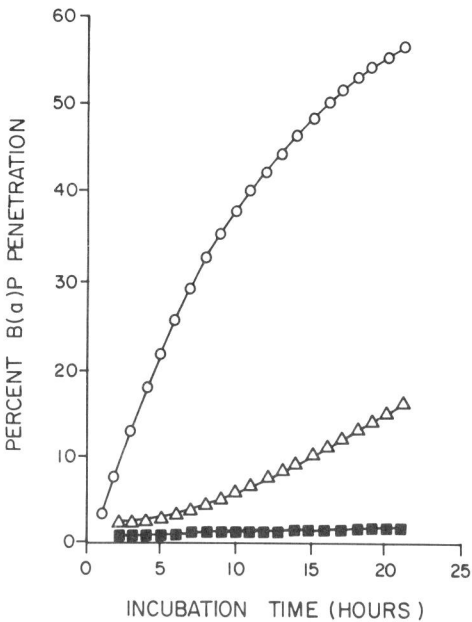

FIGURE 5. The rate of *in vitro* B[a]P penetration is compared between fresh TCDD-induced (empty circles) and acetone-treated (empty triangles) skin, vs. frozen, TCDD-induced (solid squares) skin following application of B(a)P (5000 ng/200 mm²). Pretreatment was by *in vivo* topical application 48 h prior to *in vitro* penetration experiments. Results are given as mean of five individual skin samples. (From Holland, J. M., Kao, J. Y., and Whitaker, M. J., *Toxicol. Appl. Pharmacol.*, 72, 272, 1984. With permission.)

enzymatically induced (2,3,7,8-tetrachlorodibenzo-*p*-dioxin [TCDD] treated), fresh noninduced skin, and induced skin which had been frozen on dry ice and stored at −20°C for one week. Penetration of radioactivity through the enzymatically induced skin was significantly greater than penetration through fresh skin which had not been induced, and the penetration through the noninduced fresh skin was significantly greater than through induced skin which had been frozen (Figure 5). Radiochemical analysis of the culture media which bathed the skin indicated there was a change in the relative amounts of specific metabolites generated in the enzymatically induced skin. Induction of mouse skin with TCDD resulted in a twofold reduction in the lipid soluble radiolabeled component, and a simultaneous increase in the water soluble radiolabeled component. Holland et al.[36] concluded that the metabolic status can be an important factor when determining *in vitro* skin penetration. Kemppainen et al.[49] reported that penetration of tritium-labeled T-2 toxin through excised human skin (full thickness) stored at −60°C was greater than through human skin stored at 4°C. They also found greater penetration of T-2 toxin through monkey skin stored at −60°C as compared to monkey skin stored at 4°C (Table 2). T-2 toxin is extensively metabolized during percutaneous penetration, however the metabolic profile in the receptor fluid bathing refrigerated skin was similar to that bathing skin which had been frozen. Esterases are responsible for cutaneous metabolism of T-2 toxin.[50] Esterase activity is not indicative of skin viability because cutaneous esterases are highly stable and still functional after storage of skin at −18°C followed by incubation at 37°C for 90 h.[51] The extraordinary stability of cutaneous esterases is further demonstrated by snake skin esterases which are still active for at least six months under dry storage conditions at room temperature.[52] Treatment of the snake skin at 80°C for 15 min results in loss of 90% of the esterase activity.

TABLE 2

Comparison of the Rates of Penetration of T-2 Toxin Through Human and Monkey Skin Samples Stored Under Refrigeration (4°C) or Frozen (-60°C)

Species	Refrigerated sample	Frozen sample
Rate of penetration of T-2 toxin (pg/cm²/h)*		
Human**	10.2 ± 4.0 (83)[a]	24.2 ± 4.0 (83)[b]
Monkey**	29.9 ± 12.4 (104)[b]	79.9 ± 6.3 (83)[c]
Recovery of applied dose (%)***		
Human	90.6 ± 16.7 (9)	93.4 ± 10.7 (11)
Monkey	100.4 ± 2.4 (8)	100.0 ± 3.0 (8)

* Values (Means ± 95% CL) were calculated using a general linear modelling procedure and those with different superscripts differ significantly ($p < 0.05$). Number of data points used for GLM procedure given in parentheses.

** The ranges of ambient test conditions were 23 to 25°C, 43 to 51% relative humidity for the human skin samples and 21 to 27°C, 19 to 49% relative humidity for the monkey skin samples.

*** The epidermal surfaces of human and monkey skin samples were dosed with 74 and 70 ng [³H]T-2 toxin/cm², respectively. Values are Means ± SD with the number of samples given in parenthesis.

From Kemppainen, B. W., Riley, R. T., Pace, J. G., and Hoerr, F. J., *Food Chem. Toxicol.*, 24, 221, 1986. With permission.

TABLE 3

Evaluation of Changes in the Permeability of Monkey Skin During Storage

Number of days stored	Dose penetrated after 48 h (%)
3	0.73 ± 0.49
7	1.03 ± 0.29
10	0.84 ± 0.06

Note: Values expressed as Mean ± SD (N = 3). Values are not significantly different ($p > 0.05$).

From Kemppainen, B. W., Riley, R. T., Pace, J. G., Hoerr, F. J., and Joyave, J., *Fundam. Appl. Toxicol.*, 7, 367, 1986. With permission.

Kemppainen et al.[50] determined that the permeability of monkey skin to radiolabeled T-2 toxin did not change when the skin was stored in airtight plastic bags at 4°C for up to 10 days (Table 3). Bronaugh et al.[53] assessed the extent of cutaneous metabolism of five compounds (caffeine, *p,p'*-DDT, butylated hydroxytoluene [BHT], salicylic acid, and acetyl tetramethyltetralin [AETT]) during skin penetration. Only two (AETT and BHT) of the five compounds were metabolized, and for these compounds less than 7% of the amount absorbed was metabolized. From these results Bronaugh et al.[53] concluded that for many compounds cutaneous metabolism is minimal, undetectable, or absent. They hypothesized that the dramatic effect of cutaneous metabolism on percutaneous penetration observed by Holland et al.[36] was exaggerated due to the design of *in vitro* percutaneous penetration chambers. The percutaneous penetration of lipophilic compounds can not be accurately determined by measuring accumulation of penetrant in aqueous receptor fluid which bathes the dermal surface of the skin. As the lipophilic compound penetrates into the skin, only the metabolites are sufficiently water soluble to diffuse into the aqueous receptor fluid.

III. CONCLUSION

The evidence shows that storage can affect skin permeability and measurements can vary depending on the compound and test conditions. It would be dangerous to generalize from the test results of one or a few compounds. If circumstances dictate that the skin must be stored, then the effect of storage on skin permeation should be determined for the specific compounds being tested.

REFERENCES

1. **Baxter, C., Aggarwal, S., and Diller, K. R.,** Cryopreservation of skin: a review, *Transplant. Proc.,* 17, 112, 1985.
2. **Aggarwal, S. J., Baxter, C. R., and Diller, K. R.,** Cryopreservation of skin: an assessment of current clinical applicability, *J. Burn Care Rehabil.,* 6, 469, 1985.
3. **Jensen, H. S.,** Skin viability studies *in vitro, Scand. J. Plast. Reconstr. Surg.,* 18, 55, 1984.
4. **Nathan, P. and MacMillan, B. G.,** Burn wounds: selection and preservation of skin, natural products, blood, and blood products for burn therapy, *CRC Crit. Rev. Clin. Lab. Sci.,* 7, 1, 1976.
5. **Perry, V. P.,** A review of skin preservation, *Cryobiology,* 1, 150, 1966.
6. **Pandya, N. J. and Zarem, H. A.,** The absence of vascularization in porcine skin grafts on mice, *Plast. Reconstr. Surg.,* 53, 213, 1974.
7. **Merrell, S. W., Shelby, J., Saffle, J. R., Krueger, G. G., and Navar, P. D.,** An *in vivo* test of viability for cryopreserved human skin, *Curr. Surg.,* 43, 296, 1986.
8. **Rongone, E. L.,** Skin structure, function and biochemistry, in *Dermatotoxicology,* 3rd ed., Marzulli, F. N. and Maibach, H. I., Eds., Hemisphere Publisher, Washington, D.C., 1987, 1.
9. **Hsia, S. L.,** Potentials in exploring the biochemistry of human skin, *Essays Biochem.,* 7, 1, 1971.
10. **Bickers, D. R., and Kappas, A.,** The skin as a site of chemical metabolism, *Monogr. Pharmacol. Physiol.,* 5, 295, 1980.
11. **Decker, R. H.,** Nature and regulation of energy metabolism in the epidermis, *J. Invest. Dermatol.,* 57, 351, 1971.
12. **Johnson, J. A. and Fusaro, R. M.,** The role of the skin in carbohydrate metabolism, *Adv. Metab. Disord.,* 6, 1, 1972.
13. **Freinkel, R. K.,** Metabolism of glucose-C^{14} by human skin *in vitro, J. Invest. Dermatol.,* 34, 37, 1960.
14. **Cram, A. E., Domayer, M., and Shelby, J.,** Human skin storage techniques: a study utilizing a nude mouse recipient, *J. Trauma,* 23, 924, 1983.
15. **Cram, A. E., Domayer, M. A., and Scupham, R.,** Preservation of human skin: a study of two media using the athymic (nude) mouse model, *J. Trauma,* 25, 128, 1985.
16. **Rosenquist, M. D., Cram, A. E., and Kealey, G. P.,** Skin preservation at 4°C: a species comparison, *Cryobiology,* 25, 31, 1988.
17. **Rosenquist, M. D., Cram, A. E., and Kealey, G. P.,** Short-term skin preservation at 4°C: skin storage configuration and tissue-to-volume medium ratio, *J. Burn Care Rehabil.,* 9, 52, 1988.
18. **Prows, J. and Nathan, P.,** Evaluation of storage conditions for refrigerated rabbit skin, *Cryobiology,* 17, 125, 1980.
19. **Alsbjorn, B. F., Jensen, Mk. G., and Sorensen, B.,** Effect of solcoseryl on cadaveric split-skin oxygen consumption during 4°C storage and in frozen biopsies, *Cryobiology,* 26, 119, 1989.

20. **May, S. R. and Wainwright, J. F.,** Integrated study of the structural and metabolic degeneration of skin during 4°C storage in nutrient medium, *Cryobiology,* 22, 18, 1985.

21. **Amkraut, A., Alza Corp.,** personal communication.

22. **May, S. R. and DeClement, F. A.,** Skin banking. III. Cadaveric allograft skin viability, *J. Burn Care Rehabil.,* 2, 128, 1981.

23. **Eagle, H.,** Propagation in a fluid medium of a human epidermoid carcinoma, strain KB(21811), *Proc. Soc. Exp. Biol. Med.,* 89, 362, 1955.

24. **Lehr, H. B., Berggren, R. B., Lotke, P. A. and Coriell, L. L.,** Permanent survival of preserved skin autografts, *Surgery,* 56, 742, 1964.

25. **May, S. R. and DeClement, F. A.,** Skin banking methodology: an evaluation of package format, cooling and warming rates, and storage efficiency, *Cryobiology,* 17, 33, 1980.

26. **May, S. R., Guttman, R. M., and Wainwright, J. F.,** Cryopreservation of skin using an insulated heat sink box stored at −70°C, *Cryobiology,* 22, 205, 1985.

27. **Meryman, H. T.,** The exceeding of a minimum tolerable cell volume in hypertonic suspension as a cause of freeze injury, in *The Frozen Cell,* Wolsteinholme, G. E. and O'Connor, M., Eds., Churchill Livingston, London, 1970, 51.

28. **Meryman, H. T.,** The freezing injury and its prevention in living cells, *Annu. Rev. Biophys. Bioeng.,* 3, 341, 1974.

29. **De Loecker, W., De Wever, F., Jullet, R., and Stas, M. L.,** Metabolic changes in rat skin during preservation and storage in glycerol buffer at −196°C, *Cryobiology,* 13, 24, 1976.

30. **De Loecker, W., Doms, D., Stas, M. L., and Verstreken, J.,** Metabolic changes in rat skin stored in buffer medium at −3°C, *Cryobiology,* 11, 359, 1974.

31. **De Loecker, W., Stas, M. L., De Wever, F., and Jullet, R.,** The effects of cortisol and amino acids on the metabolic activity of rat skin stored in dimethyl sulfoxide buffer at −196°C, *Cryobiology,* 15, 59, 1978.

32. **De Loecker, W., De Wever, F., Stas, M. L., and Jullet, R.,** The effects of adenosine triphosphate and inorganic phosphates on the viability of stored rat skin, *Cryobiology,* 16, 11, 1979.

33. **De Loecker, W. and De Wever, F.,** Oxidative phosphorylation in rat skin during preservation, *Cryobiology,* 16, 517, 1979.

34. **De Loecker, W., De Wever, F., and Penninckx, F.,** Metabolic changes in human skin preserved at −3 and at −196°C, *Cryobiology,* 17, 46, 1980.

35. **Praus, R., Bohm, F., and Dvorak, R.,** Skin cryopreservation. I. Incorporation of radioactive sulfate as a criterion of pigskin graft viability after freezing at −196°C in the presence of cryoprotectants, *Cryobiology,* 17, 130, 1980.

36. **Holland, J. M., Kao, J. Y., and Whitaker, M. J.,** A multisample apparatus for kinetic evaluation of skin penetration *in vitro*: The influence of viability and metabolic status of the skin, *Toxicol. Appl. Pharmacol.,* 72, 272, 1984.

37. **Fritsch, W. C. and Stoughton, R. B.,** The effect of temperature and humidity on the penetration of C^{14} acetylsalicylic acid in excised human skin, *J. Invest. Dermatol.,* 41, 307, 1963.

38. **Kao, J., Hall, J., and Holland, J. M.,** Quantitation of cutaneous toxicity: an *in vitro* approach using skin organ culture, *Toxicol. Appl. Pharmacol.,* 68, 206, 1983.

39. **May, S. R. and DeClement, F. A.,** Development of a radiometric metabolic viability testing method for human and procine skin, *Cryobiology,* 19, 362, 1982.

40. **Burch, G. E. and Winsor, T.,** Rate of insensible perspiration (diffusion of water) locally through living and through dead human skin, *Arch. Intern. Med.,* 74, 437, 1944.

41. **Astley, J. P. and Levine, M.,** Effect of dimethyl sulfoxide on permeability of human skin *in vitro, J. Pharm. Sci.,* 65, 210, 1976.

42. **Harrison, S. M., Barry, B. W., and Dugard, P. H.,** Effects of freezing on human skin permeability, *J. Pharm. Pharmacol.,* 36, 261, 1984.

43. **Bronaugh, R. L., Stewart, R. F., and Simon,** Methods for *in vitro* percutaneous absorption studies. VII. Use of excised human skin, *J. Pharm. Sci.,* 75, 1094, 1986.

44. **Pitman, I. H. and Downes, L. M.,** Cattle and sheep skin permeability: a comparison of frozen and reconstituted skin with that of fresh skin, *J. Pharm. Sci.,* 71, 846, 1982.

45. **Swarbrick, J., Lee, G., and Brom, J.,** Drug permeation through human skin. I. Effect of storage conditions on skin, *J. Invest. Dermatol.,* 78, 63, 1982.

46. **Hawkins, G. S. and Reifenrath, W. G.,** Development of an *in vitro* model for determining the fate of chemicals applied to skin, *Fund. Appl. Toxicol.,* 4, S133, 1984.

47. **Reifenrath, W. G., Hawkins, G. S., and Kurtz, M. S.,** Evaporation and skin penetration characteristics of mosquito repellent formulations, *J. Am. Mosq. Control Assoc.,* 5, 45, 1989.

48. **Parry, G. E., Bunge, A. L., Silcox, G. D., Pershing, L. K., and Pershing, D. W.,** Percutaneous absorption of benzoic acid across human skin. I. *In vitro* experiments and mathematical modeling, *Pharm. Res.,* 7, 230, 1990.

49. **Kemppainen, B. W., Riley, R. T., Pace, J. G., and Hoerr, F. J.,** Effects of skin storage conditions and concentration of applied dose on [3H]T-2 toxin penetration through excised human and monkey skin, *Fd. Chem. Toxic.,* 24, 221, 1986.

50. **Kemppainen, B. W., Riley, R. T., Pace, J. G., Hoerr, F. J., and Joyave, J.,** Evaluation of monkey skin as a model for *in vitro* percutaneous penetration and metabolism of [3H]T-2 toxin in human skin, *Fundam. Appl. Toxicol.,* 7, 367, 1986.

51. **Mollgaard, B., Hoelgaard, A., and Bundgaard, H.,** Pro-drugs as delivery systems. XXIII. Improved dermal delivery of 5-fluorouracil through human skin via *N*-acyloxymethyl pro-drug derivatives, *Int. J. Pharm.,* 12, 153, 1982.

52. **Nghiem, B. T. and Higuchi, T.,** Esterase activity in snake skin, *J. Pharm. Sci.,* 76, S53, 1987.

53. **Bronaugh, R. L., Stewart, R. F., and Storm, J. E.,** Extent of cutaneous metabolism during percutaneous absorption of xenobiotics, *Toxicol. Appl. Pharmacol.,* 99, 534, 1989.

Chapter 10

IN VITRO ABSORPTION THROUGH DAMAGED SKIN

Robert C. Scott

TABLE OF CONTENTS

I. INTRODUCTION

The factors governing the process of percutaneous absorption have been studied for many years. In the elucidation of the mechanism of absorption *in vitro* techniques have been an invaluable tool. Various studies have now confirmed that in the majority of cases the rate limiting step in the absorption of chemicals into the body through the skin is diffusion through the stratum corneum which acts as the primary barrier to the entry (and exit) of chemicals.[1] Alteration of the stratum corneum can cause changes in the permeability properties of the diffusion barrier[2] and in the majority of cases this may lead to an increased permeability. Absorption through damaged skin has been extensively studied *in vivo* in humans[3,4] and animals[5] and changes in permeability during regeneration and repair of the barrier have been quantified for a range of chemicals. Albeit in fewer studies, *in vitro* techniques have also been used to explore skin permeability changes as a result of physical alteration, chemical application, thermal injury, UV-irradiation and disease conditions. The intention of this paper is to review the utilization and contribution of *in vitro* techniques to the study of skin permeability properties in abnormal skin.

II. CHANGES IN *IN VITRO* PERMEABILITY AS A RESULT OF PHYSICAL ALTERATION

Almost 40 years ago the progressive removal of the stratum corneum by tape-stripping human skin was shown to increase *in vitro* permeability to water.[6] The results showed that removal of the first few skin strips caused little detectable change in permeability but removal of subsequent skin strips caused an exponential increase in permeability rate to over 50 times the normal value. These *in vitro* data were originally interpreted to indicate that the main permeability barrier was located near the base of the stratum corneum. On reflection[7] the data were considered more consistent with each layer of the stratum corneum contributing equally to the diffusion barrier. These data, however, clearly proved that alteration to the stratum corneum can increase permeability and that such increases can be quantified *in vitro*.

Changes in rat skin permeability as a result of mild superficial epidermal alterations have been extensively studied *in vitro*.[8] Tritiated water permeability was measured immediately after skin alteration and over the following 10 d to assess the influence of barrier regeneration. Typical of the types of physical alteration used were (A) light pressure using the blunt edge of a scalpel blade (the familiar 'Draize' abrasion); (B) light stroking with medium grade sandpaper (to mimic a grazed skin condition); and (C) tape-stripping. These types of alteration caused increased *in vitro* water permeability to different extents (Table 1). By excising the skin at various times after alteration, changes in permeability were assessed *in vitro*. After each alteration, similar patterns were observed in the regeneration of the barrier properties, a rapid partial recovery followed by a more gradual progression to normal values over a period of time.

This bi-phasic regeneration of the permeability barrier was found to be due to an initial rapid production of a temporary barrier and then at a later time by a final barrier with a fully differentiated stratum corneum.[8] The bi-phasic nature of the barrier regeneration has also been reported in *in vivo* studies. These changes can, of course, be readily followed *in vivo*[8-10] on the same subjects, which is not readily achieved using *in vitro* techniques.

Abrasion of rat skin by slight pressure with a hypodermic needle to damage the stratum corneum has also been shown to increase permeability.[11] The skin became more permeable as the number of abrasions increased. Water permeability increased threefold as a result of a single abrasion and 12.5 times above normal after four abrasions to the same skin area. In the same study tape-stripping was found to cause a 20-fold increase in the amount of water penetrating the skin over a 24-h time period compared with intact skin. This enhance-

TABLE 1

**Details of *In Vitro* Permeability
to Tritiated Water Through Altered
and Regenerating Rat Skin**

Type of alteration	A	B	C
Factor above normal permeability at			
0 h	10	25	36
24 h	2	12	30
Return to normal permeability (h)	120	190	288

TABLE 2

**Effect of Abrasion on the
In Vitro Absorption of a
Series of Penetrants Through
Human and Rat Skin**

	Factor of increase	
Penetrant	Human	Rat
Benzoic acid	2.1	2.0
Caffeine	1.6	1.6
Cortisone	3.9	4.3
Nicotinic acid	11.5	9.6
Phenol	1.7	3.7
Propylene glycol	5.3	4.0
Urea	7.1	5.1

ment is less than was reported for the rat elsewhere[8] (Table 1). However, if the maximum rate of absorption in the latter study is compared with the overall rate measured in the former study,[8] there is good agreement between the studies with approximately 30-fold increases in *in vitro* tritiated water permeability as a result of tape-stripping rat skin.

A much greater increase in permeability has been reported after tape-stripping rabbit ear skin.[12] The first tape strips (up to 5) had very little effect on the absorption of propylene glycol. However, after 10 strips the permeability had increased markedly and after 20 strips the skin was freely permeable. Twenty minutes after stripping the skin 20 times the *in vitro* absorption of propylene glycol was 300 times normal (contrast the enhancement determined for propylene glycol in Table 2). There was very little change in permeability over the next 24 h but after this time the permeability barrier was restored rapidly such that by day five the skin was only seven times more permeable. Normal barrier function was detected after 10 days (a very similar time scale to that seen after tape-stripping rat skin[8]).

The effect of abrasion on the *in vitro* permeability of human and rat skin, to a range of molecules with different physiochemical properties, has been quantified in a single study,[11] thus facilitating interchemical and interspecies comparisons. The factors of increase are summarized in Table 2. These data show that the range of increases in permeability were compound dependent and that similar factors of increase were detected in the human and the rat. This latter observation might be considered somewhat surprising as, generally, rat skin is an unreliable model for normal human skin permeability properties.[13]

No penetration of the antimicrobial agent chlorhexidine phosphanilate (CHO) was detected *in vitro* through intact human skin,[14] though it readily penetrated tape-stripped skin. Although these data do not allow the factor of increase in permeability to be quantified they provide further evidence to show that significant changes are observed in both animal and human skin *in vitro* as a result of tape-stripping.

The above shows that the permeability of animal and human skins are both affected by physical alteration and that different types of alteration cause different degrees of enhancement. Maximum enhancement in absorption is detected immediately following alteration after which permeability properties gradually return to normal. Enhancement of absorption, at least with those penetrants examined *in vitro* is similar in both human and animal skin. The degree of enhancement was found to vary greatly between twofold and several hundredfold when compared with normal skin and appears to be compound specific.

III. CHANGES IN *IN VITRO* PERMEABILITY AS A RESULT OF CHEMICAL CONTACT

The application of chemicals to the skin can result in changes to permeability due to either direct chemical interaction with the stratum corneum or indirectly as a result of a chemical affecting the differentiation process leading eventually to the production of an abnormal stratum corneum. A number of *in vitro* studies have investigated the changes in permeability resulting from chemical action.

The absorption of $CoCl_2$ has been measured[15] through intact mouse skin and mouse skin treated with acid (up to 10N HCl) or alkali (up to 10N NaOH). These treatments resulted in no significant increase in permeability (which was in marked contrast to the hundredfold enhancement reported in the same study after physical alterations). Hydrocortisone absorption, however, has been shown to be increased through hairless mouse skin as a result of 10% acetic acid or vitamin A (retinoic acid) applications.[16] The acetic acid was applied daily for 6 d and the retinoic acid for 8 d *in vivo* prior to excision of the skin and the *in vitro* determination of hydrocortisone absorption. Retinoic acid caused the largest increase in absorption but this was only two to threefold higher than normal over a 24-h absorption period. However, because of the differences in the absorption profiles through normal and treated skin, the factor of difference was up to sevenfold over the initial 12 h and even greater at earlier times. This phenomenon, of much higher enhancements being detected when short contact times are considered has been reported previously from *in vivo*[5] and *in vitro*[11] studies, yet is frequently overlooked with emphasis being placed on single time-point determinations.

Surfactants and 'accelerants' have also been shown to damage skin and induce higher *in vitro* permeability. The interaction of a series of divalent anionic surfactants with human skin and the affect on permeability was quantified using tritiated water as a marker molecule.[17] The divalent anionic surfactant phenylsulfonate CA (70% calcium dodecylbenzene sulfonate in isobutanol) caused a 40-fold increase and Empicol ML 26/F (26.5% w/v magnesium lauryl sulphate in aqueous solution) an 11-fold increase in water permeability through human skin as a result of a 12-h skin contact. In contrast, using an *in vitro* permeability technique, polyoxyethylated nonionic surfactants (from the "Synperonic" NP and "Synperonic" PE ranges) were found to have no effect on the barrier function of human skin. Further *in vitro* studies have investigated the structure of the nonionic surfactants and their ability to interact with and reduce the damage done, i.e., permeability changes by cationic surfactants to human skin. Similarly, the time course of permeability changes to ions through human skin due to damaging surfactant solutions was studied by measuring changes in electrical conductance.[18] These studies showed that anionic surfactants of 12 or more carbon atoms caused a change in conductance, indicating enhanced permeability as a result of damage, after approximately one hour. With cationic surfactants of C_{12} or above changes were detected within minutes of skin contact and prolonged contact resulted in tritiated water permeability changes of up to 100-fold. The accelerants, dimethyl sulfoxide, dimethyl formamide, and dimethyl acetonide, also caused a marked change to the permeability of human skin *in vitro*[19] as assessed by electrical conductance changes.

Studies such as those reported above, have stimulated the use of measuring electrical resistance changes *in vitro* in skin, as a result of chemical interaction with the stratum corneum, to predict potential *in vivo* corrosive chemicals.[20] An initial validation, measuring the electrical resistance change in rat epidermal slices as a result of chemical contact with 68 chemicals, has shown the *in vitro* technique to have a high sensitivity for corrosive chemicals. Water permeability and electrical resistance have been shown to be related. It can logically be assumed, therefore, that the corrosive chemicals would have caused an enhancement of skin permeability to other chemicals which diffuse through the skin by the same route as ions and water.

It appears that the application of some chemicals to skin can cause an increased *in vitro* permeability. There is sufficient evidence to indicate that the degree of enhancement can be at least as great as that following physical alteration.

IV. EFFECT OF UV-IRRADIATION ON *IN VITRO* SKIN PERMEABILITY

The effect to UV-irradiation on permeability has been investigated using both rat[11] and mouse skin.[16] In these studies different penetrants were used (hydrocortisone with the mouse and nicotinic acid with rat skin) and so comparisons are not possible between the species or penetrants. In addition, the irradiation dose levels used were different between the mouse and the rat. However, useful information can be gained from the available data. Mouse skin exhibited clinical and histological evidence of some irritation to the skin with mild erythema and some parakeratosis. The *in vitro* permeability of this skin to hydrocortisone determined immediately after excision, was twofold higher than through normal skin. Nicotinic acid absorption through the rat skin increased as a consequence of increasing doses of irradiation. The highest UV-doses resulted in cracks and eschars in rat skin. The applied nicotinic acid which was subsequently absorbed (51%) was very similar to that measured in another study through tape-stripped rat skin.[11]

Obviously, UV-irradiation, at a dose which causes inflammation and/or structural damage to skin, can affect normal barrier function. The degree of enhancement in *in vitro* permeability appears to be related to the dose. With large UV-doses the permeability can be increased to a level similar to that measured *in vitro* after physical and chemical alteration to the skin.

V. EFFECT OF THERMAL INJURY ON *IN VITRO* SKIN PERMEABILITY

The *in vitro* percutaneous absorption of water, alcohols[21], and phenol[22] has been measured through thermally damaged hairless mouse skin. When the temperature of the thermal insult was maintained at 60°C, there was very little change in skin permeability to water, methanol, ethanol, and phenol, even though deep burns were evident. The greatest effect was on butanol and octanol absorption but was no more than a threefold increase compared with normal skin. When the temperature of the thermal insult was raised above 70°C, however, larger enhancements in absorption were detected. Phenol absorption was increased by fivefold and that of methanol by fiftyfold.

Again, increases in the permeability of the skin have been detected *in vitro* and these were related to the severity of the skin damage. As seen in other studies, however, the factor of enhancement through thermally damaged skin is compound dependent.

VI. *IN VITRO* SKIN PERMEABILITY OF DISEASED SKIN

Many drug products are deliberately applied to diseased skin. Quantitation of the permeability of diseased skin, or experimentally induced "disease model" skin, has been made *in vitro*. In normal, healthy, excised, human skin, most of the applied dose of clocortolone pivalate was found in the stratum corneum[23] with very little in the epidermis and even less in the dermis. When the same dose was applied to skin with a compromised stratum corneum the distribution profile changed with the amount in the epidermis increased by 150-fold and that in the dermis by over 1000-fold. Such *in vitro* studies have been used in the development of treatment regimes for common dermatoses and the data demonstrate that the drug can reach the desired target site, i.e., the epidermis/dermis, when applied to the diseased skin, whereas absorption through adjacent intact, healthy skin will be much less.

Experimental diseased skin conditions have been induced in hairless rats by feeding them a diet deficient in essential fatty acids, which leads to the development of a scaly skin. Using *in vitro* techniques, this skin has been shown[24] to have enhanced permeability to water and hydrocortisone. As the skin condition developed, so permeability increased. Absorption of hydrocortisone was enhanced by a factor of 10 compared with normal, healthy skin, but the barrier properties of the skin were not destroyed altogether as tape-stripping led to a further increase in permeability.

These *in vitro* studies show that diseased skin still maintains some of its barrier properties which vary during the development (and regression) of the diseased state.

VII. CONCLUSIONS

Despite a few reported cases to the contrary,[25] the overwhelming weight of evidence from the literature is that abnormal skin is more permeable than normal skin and that changes in permeability can be readily and easily detected and quantified using *in vitro* techniques.

Review of the literature, however, reveals that extensive comparison between experiments is limited due to the different conditions and protocols employed. Usually, assessments have been made at a single time point and information about changes during healing after physical alteration, or during the development of abnormal skin in a disease state, has not been quantified. Many of the studies have examined physical damage and have quantified the change in permeability immediately after the insult; therefore, only maximum changes have probably been measured. The quantitation of permeability changes at a specific time after alteration, however, is possibly where the use of an *in vitro* technique is superior to the use of *in vivo* methods. In *in vivo* experiments the barrier regenerates over the time course of an experiment (typically 24 h) and this barrier regeneration can influence the permeability. The profile of changing absorption with time and healing is an important aspect of understanding the permeability properties of damaged skin. Very often the absorption through skin is much greater at early times after particularly physical damage when compared to normal skin; if permeability is compared as a result of absorption over 24 h much of this difference can be lost.[5,11] *In vitro* techniques can readily be used to quantify the early changes in absorption rates following damage to skin.

Despite the awareness that percutaneous absorption of chemicals can lead to unwanted systemic side-effects and the realization that abnormal skin is more permeable than normal skin, the assessment of abnormal skin permeability has been relatively neglected. *In vitro* techniques provide an attractive, easily controlled, economic means for the quantitative study of the absorption characteristics of both normal and abnormal skin.

REFERENCES

1. **Scheuplein, R. J. and Blank, I. H.,** Permeability of the skin, *Physiol. Rev.,* 51, 702, 1971.

2. **Pinkus, H.,** Examination of the epidermis by the strip method. II. Biometric data on regeneration of the human epidermis, *J. Invest. Dermatol.,* 19, 431, 1952.

3. **Grice, K. A., Sattar, H., and Baker, H.,** The cutaneous barrier to salts and water in psoriasis and normal skin, *Br. J. Dermatol.,* 88, 459, 1973.

4. **Spruit, D.,** Measurement of water vapour loss from very small areas of forearm skin, *J. Invest. Dermatol.,* 58, 109, 1972.

5. **Scott, R. C. and Dugard, P. H.,** A model for quantifying absorption through abnormal skin, *J. Invest. Dermatol.,* 86, 208, 1986.

6. **Blank, I. H.,** Further observations on factors which influence the water content of the stratum corneum, *J. Invest. Dermatol.,* 21, 259, 1953.

7. **Blank, I. H.,** Cutaneous barriers, *J. Invest. Dermatol.,* 45, 249, 1965.

8. **Scott, R. C., Dugard, P. H., and Doss, A. W.,** Permeability of abnormal rat skin, *J. Invest. Dermatol.,* 86, 201, 1986.

9. **Matoltsy, A. G., Schragger, A., and Matoltsy, M. N.,** Observations on regeneration of the skin barrier, *J. Invest. Dermatol.,* 38, 5, 251, 1962.

10. **Spruit, D. and Malten, K. E.,** The regeneration of the water vapour loss of heavily damaged skin, *Dermatologica,* 132, 115, 1966.

11. **Bronaugh, R. L. and Stewart, R. F.,** Method for *in vitro* percutaneous absorption studies. V. Permeation through damaged skin, *J. Pharm. Sci.,* 74, 1062, 1985.

12. **Komatsu, H. and Suzuki, M.,** Studies on the regeneration of the skin barrier and the changes in ^{32}P incorporation into the epidermis after stripping, *Br. J. Dermatol.,* 106, 551, 1982.

13. **Scott, R. C., Walker, M., and Dugard, P. H.,** A comparison of the *in vitro* permeability properties of human and some laboratory animal skins, *Int. J. Cosmet. Sci.,* 8, 189, 1987.

14. **Wang, J. C. T., Williams, R. R., Wang, L., and Loder, J.,** *In vitro* skin permeation and bioassay of chlorhexidine phosphanilate, a new antimicrobial agent, to be published in Pharm. Res.

15. **Kusama, T., Itoh, S., and Yoshizawa, Y.,** Absorption of radionuclides through wounded skin, *Health Phys.,* 51, 138, 1986.

16. **Solomon, A. E. and Lowe, N. J.,** Percutaneous absorption in experimental epidermal disease, *Br. J. Dermatol.,* 100, 717, 1979.

17. **Eagle, S. C., Barry, B. W., and Scott, R. C.,** Inhibition of the damaging activity of divalent anionic surfactants on human skin, in *Prediction of Percutaneous Penetration—Methods, Measurements, Modelling,* Scott, R. C., Guy, R. H., and Hadgraft, J., Eds., IBC Technical Services, London, 1990, 417.

18. **Dugard, P. H. and Scheuplein, R. J.,** Effects of ionic surfactants on the permeability of human epidermis: an electometric study, *J. Invest. Dermatol.,* 60, 263, 1973.

19. **Allenby, A. C., Creasey, N. H., Edington, J. A. G., Fletcher, J. A., and Schock, C.,** Mechanism of action of accelerants on skin penetration, *Br. J. Dermatol.,* 81, 47, 1969.

20. **Oliver, G. J. A., Pemberton, M. A., and Rhodes, C.,** An *in vitro* model for identifying skin-corrosive chemicals. I. Initial validation, *Toxicology In Vitro,* 2, 7, 1988.

21. **Behl, C. R., Flynn, G. L., Kurihara, T., Smith, W., Gatmaitan, O., Higuchi, W. I., Ho, N. F. H., and Pierson, C. L.,** Permeability of thermally damaged skin. I. Immediate influences of 60°C scalding on hairless mouse skin, *J. Invest. Dermatol.,* 75, 340, 1980.

22. **Behl, C. R., Linn, E. E., Flynn, G. L., Pierson, C. L., Higuchi, W. I., and Ho, N. F. H.,** Permeation of skin eschar by antiseptics. I. Baseline studies with phenol, *J. Pharm. Sci.,* 72, 391, 1983.

23. **Stuttgen, G.,** Drug absorption by intact and damaged skin, in *Dermal and Transdermal Absorption,* Wissenschaftliche Verglagsgesellschaft, Stuttgart, Germany, 1982, 27.

24. **Lambrey, B., Schalla, W., Kail, N., Daniel, M. H., Blachon, F., and Schaefer, H.,** Cutaneous permeation of hydrocortisone in essential fatty acid deficient hairless rats. Importance of the horny layer barrier, *Europ. Congr. Bioph. Biopharmacokinetics,* 1, 331, 1987.

25. **Degen, P. H., Moppert, J., Schmid, K., and Weirich, E. G.,** Percutaneous absorption of clioquinol (Vioform), *Dermatologia,* 159, 259, 1979.

Chapter 11

LOCALIZATION OF DRUGS IN THE SKIN

Alain Rolland

TABLE OF CONTENTS

I. INTRODUCTION

Over the past decades, localization of drugs within the skin, after topical application of dermal or transdermal therapeutic systems, has stimulated an increasing interest. When a dermatological disease is treated locally, i.e., via the topical route, the drug is directly applied to the target organ. As a consequence, evaluation of drug bioavailability by determination of drug concentration in plasma and in various organs is neither sufficient nor representative after topical application of a drug. Thus, apart from percutaneous systemic therapy with transdermal drug delivery systems, classical pharmacokinetics can only give information concerning side effects and systemic toxicity when a drug has been applied topically. Therefore, after topical application of a drug, its bioavailability in the target organ, the skin, has to be evaluated and this is of major importance for a better understanding of the pharmacological effects. In the past few years, a more precise knowledge of the skin physiology and an evolution of the technology have allowed a more accurate localization and quantification of various drugs in the different skin layers.

In vitro diffusion experiments represent a valuable approach to *in vivo* percutaneous absorption studies and a convenient means for evaluating the permeation characteristics of drugs through human and animal skin and their disposition after topical application. The main objective of the *in vitro* experiments is to explore the intrinsic penetration of drugs through the skin, with regard to dermal permeation and percutaneous absorption. For new drugs with unknown toxicity, *in vitro* methodology may be an early way of investigating percutaneous absorption with human skin, since human skin is unique in its structure, e.g., stratum corneum thickness and hair follicle density. Unfortunately, supply of human skin samples is often limited and *in vitro* studies have to be performed with skin from various animal species, bearing in mind that this substitute can produce deviations from human permeation characteristics. Finally, compared to *in vivo* methods, *in vitro* experiments are generally less time-consuming, ethically acceptable, and easier to carry out on a large scale for evaluating, for instance, variables such as the effect of temperature, vehicle or drug delivery system, moisture, pretreatment, dose, etc.

II. *IN VITRO* METHODOLOGY

A. DIFFUSION CELLS

In many of the early percutaneous absorption studies, the penetration of drugs or chemicals through the skin was followed using a two-chambered diffusion cell.[1-3] In that particular procedure, the drug was applied in solution on one side of the membrane and its rate of permeation was obtained by analyzing the drug concentration in the same solvent (generally water or saline) on the other side; homogeneity of both solutions (donor and receptor) was obtained with stirring devices on both sides.

One-chambered cells are nowadays more commonly used, so that the tested drug can be applied to the skin in any type of formulation (ointment, emulsion, gel, lotion) and so that the skin surface is maintained at ambient conditions and is thus not excessively hydrated as in the two-chamber procedure. The static Franz cell[4,5] has been widely used and many similar cells have been designed on this model.[6,7]

Flow-through diffusion cells[7-9] may promote the solubility of hydrophobic compounds in the receptor phase since the receptor fluid is perfused below the surface of the skin, mimicking the effect of blood flow through the skin. Automatic sampling can be employed and sink conditions can be easily maintained with such a technique.

Independently of the cell type, human or animal skin prepared according to defined procedures is mounted in the diffusion cell with conditions which approximate the physiological situation. The underside of the skin is in contact with the receptor fluid (generally physiological saline), the temperature being constant, generally 32°C at the skin surface.[10] An appropriate amount of drug formulation (finite or infinite dose), generally containing a defined amount of radiolabeled drug, is applied to a given surface area, and after a certain period of time the drug concentration is analyzed in the receptor phase and in the different skin layers.

B. SKIN LAYERS AS *IN VITRO* DIFFUSION MEMBRANES

Most of the *in vitro* tests for the localization of drugs in the skin are conducted with human *in vivo* studies in mind; excised human skin is therefore obviously the best choice for *in vitro* experiments. However, its limited supply precludes the sole use of this membrane. Animal skin can serve as a substitute to human skin if the absolute differences in the drug permeation can be accepted.[11]

Human skin samples are generally obtained from surgical material (mastectomies, amputations, etc.) or from cadavers. To date, no clear changes in barrier function have been reported for postmortem skin samples.[12] However, an important concern is to verify, particularly with cadaver skin, that the skin barrier function is maintained. Several causes may alter skin integrity: disease or trauma of the donor; alteration of skin between death and sample harvesting; or alteration during skin preparation and storage. In addition to skin barrier function, biochemical viability and metabolic capacities of the tissue may be of major importance and have to be verified, particularly for drugs that may be metabolized within the skin.

When trying to substitute human skin, the choice of the animal skin may depend on the tested drug. Thus, investigators[13-16] reviewed comparative permeability data of animal skin relative to human skin for different compounds. The results were highly dependent on both skin origin and drug structure.

1. Full-Thickness Skin

Investigations of drug penetration *in vitro* with total excised skin should always be discussed with respect to possible drug accumulation in the dermis and subcutaneous fat, since there is a lack of vascular transport. Therefore, the skin must be first freed of subcutaneous fat, in which lipophilic drugs particularly tend otherwise to accumulate. Special cautions have to be taken in removing the fat in order to avoid any fundamental damage to the dermis. However, if full-thickness skin is to be used in the diffusion cell assay, the thick dermal tissue ($= 2$ to 3 mm in humans, rats) can represent a depot especially for lipophilic drugs.[14]

2. Horny Layer and Epidermis

In vivo, compounds that are absorbed through the skin are taken up by blood vessels directly beneath the epidermis; compounds are consequently not required to penetrate the full thickness of the skin before entering the vasculature system. Accordingly, for *in vitro* studies, the epidermis, including the horny layer, can be separated from the underlying dermis using different techniques.[12,17,18] Heat separation (e.g., at 60°C for 2 min in water) has been used for human and animal skin,[6,13,19,20] and chemical separation[21,22,23] (e.g., by skin immersion in $2M$ aqueous sodium bromide solution) has been applied to fresh human abdominal skin[21] and to rat and human cadaver skin[22,23] in order to prepare epidermal sheets. Nevertheless, this technique cannot be applied to haired animal skin, the hair shafts leaving holes in the separated epidermis.

Another means of preparing split-thickness skin for diffusion studies is to use a dermatome. A layer thickness of 350 μm can be obtained, for example, from rat skin, corresponding to the whole epidermis and upper papillary dermis.[24] A similar technique can also be applied to human skin; dermatome sections, 200 μm thick, have been prepared from human abdominal skin.[15,25] When the animal skin is relatively thin (≤ 1 mm, e.g., hairless mouse, rabbit, etc.), the sectioning of the skin is not necessary.

3. Isolated Stratum Corneum

Percutaneous absorption measurements can also be performed using diffusion cells containing isolated stratum corneum obtained from human cadaver abdominal skin.[19] However, it must be remembered that the stratum corneum is quite difficult to handle and tends to undergo marked swelling[26,27] and can hence present modified physical properties. Moreover, isolated stratum corneum used as a diffusion membrane for *in vitro* assays cannot provide any accurate information regarding the diffusion of a drug into the other underlying skin layers. Following any chemical or physical treatment for reducing skin thickness, the barrier function of the skin sheets should be controlled using a test molecule of known penetration, such as water, cortisone, or urea.[23-25,28-30]

C. SKIN STORAGE

Since human skin supply is generally limited, possibilities of skin storage are of major concern. Skin pieces can be stored at 4°C for 10 h[27] and for weeks and even months if deep frozen and protected from drying out.[12,28,31,32] The *in vitro* penetration kinetics of n-alkanols through previously frozen or nonfrozen rat skin were not significantly different.[32] Statistically, no difference was observed in the lidocaine flux between fresh and frozen (up to 2 months) hairless mouse skin and between fresh and frozen heat-separated epidermis samples from the same human subject.[33] In another study using cadaver skin stored in either frozen or dried state, it was found that skin samples that had been frozen were more permeable than fresh skin to the investigated chromone acid. However, with dried skin, when appropriately rehydrated before use, rates of drug permeation were found to be similar to those obtained with fresh skin.[34] Therefore, precise conditions of freezing techniques have to be established so that the skin viability and barrier function will not be altered during storage.

The use of frozen "nonviable" skin for the *in vitro* estimation of drug permeation may provide information on the diffusional aspect of penetration, whereas the use of metabolically viable and structurally intact skin may give results depending on both diffusion and metabolism. The relative prevalence of each process will depend on the physico-chemical properties of the drug and on its metabolism within the skin.

For highly lipophilic compounds, such as benzo[a] pyrene (B[a]P), diffusion may be controlled by the epidermis and dermis to a greater extent than by the stratum corneum, and the cutaneous metabolism may become the rate-limiting process in drug penetration.[35]

III. RECOVERY OF DRUG MOLECULES *IN VITRO*

A. OPTIMIZATION OF THE RECEPTOR PHASE COMPOSITION

One of the most common ways of assessing drug permeation *in vitro* is to measure drug concentration in the receptor fluid. Typically, *in vivo* conditions, i.e., blood flow through the skin, are represented *in vitro* by static or circulating physiological saline. However, if normal saline or physiological buffer solution (pH 7.4) is used as the receptor solution, lipophilic drugs may not partition into the aqueous medium and remain in the skin. For instance, Cole et al.[36] showed that the penetration of cyclosporin A through full-thickness cadaver skin over a 48-h period was very low (< 0.3%) when using normal saline as the receptor phase. The thick dermal layer can, indeed, impose an artificial barrier to the diffusion of water-insoluble drugs and an aqueous receptor fluid can be unfavorable for the drug partition. Therefore, many investigators have studied the effect of the receptor fluid composition on the permeation of hydrophobic molecules.[12,37]

Bronaugh and Stewart[24] showed that the receptor fluid composition dramatically affects the apparent absorption of cinnamyl anthranilate and acetyl ethyl tetramethyl tetralin. Simply replacing normal saline by a 6% aqueous solution of polyethylene glycol 20 oleyl ether (Volpo 20) gave a threefold and a tenfold increase in the *in vitro* absorption of cinnamyl anthranilate and acetyl ethyl tetramethyl tetralin, respectively. The permeation of standard molecules such as cortisone, water, and urea was not affected by the use of 6% Volpo 20 as the receptor fluid, which means that the integrity of the skin barrier was apparently not modified by the surfactant solution.

In another *in vitro* percutaneous absorption/metabolism study, Collier et al.[38] investigated the effect of various receptor phases on the viability of dermatomed skin in a flow-through cell. Skin viability was examined through aerobic glucose utilization, maintenance of oestradiol and testosterone metabolism for 24 h, and through the histological aspect of skin. Phosphate-buffered saline did not maintain skin viability for 24 h, as shown by decrease of glucose utilization and steroid metabolism. However, HEPES-buffered Hanks' balanced salt solution was as potent as minimal essential medium (MEM) in maintaining skin viability and metabolic activity for 24 h. Addition of fetal bovine serum to the media was shown to hinder the extraction of steroids and metabolites from the receptor fluids, thus masking part or whole of the metabolism due to protein binding of steroids. The choice of a receptor fluid is therefore of major importance for *in vitro* absorption studies evaluating metabolism of drugs within the skin, since skin viability and metabolic activity have to be maintained throughout the experiments.

B. DRUG SAMPLING IN THE RECEPTOR PHASE
1. Static Diffusion Cells
Typically, small aliquots are collected from static diffusion cells and the volume of the receptor phase ($\simeq$ 2 to 10 ml) has to be replenished after sampling. Owing to the replenishment of the aliquots, continual dilution of the receptor fluid occurs, especially when larger samples are taken; this must be considered in the final calculations. In addition, the receptor compartment is generally made as small as possible and boundary layers may sometimes appear when stirring of the receptor phase is not optimized. If inadequate stirring occurs in the receptor phase, stagnant or boundary layers are obtained and they can affect the homogeneity of the receptor fluid; in these conditions, the initial aliquots may not be representative of the total receptor phase. A simple method has been proposed by Gummer[39] to establish if stirring is homogeneous within the receptor phase. A crystal of potassium permanganate is added to the receptor fluid and if the receptor compartment has not turned homogeneously mauve within 30 s, boundary layers can be observed in the receptor compartment,[7] and the stirring is judged as being inadequate.

2. Flow-Through Diffusion Cells
The main advantage of flow-through cell designs is that continuous sampling can be used. However, since a small receptor compartment is mainly used, it is essential to maintain a sufficient flow rate in order to ensure sink conditions throughout the experiment. Although variation of the flow rate between 2 and 7.5 ml/h has been shown to induce no significant modification of B[a]P penetration through mouse skin,[35] one cannot conclude that the same results will be systematically obtained in other *in vitro* experiments. One parameter that should be particularly considered is the ratio between flow rates and the volume of the receptor compartment.

Care must also be taken in controlling that no contamination of the equipment can occur, especially when the drug molecule may adsorb onto parts of the diffusion cell, e.g., interaction with rubber, glass, etc. Thus preliminary experiments should be performed to verify potential compound/equipment interactions, especially when nonradioactive compounds are under investigation.

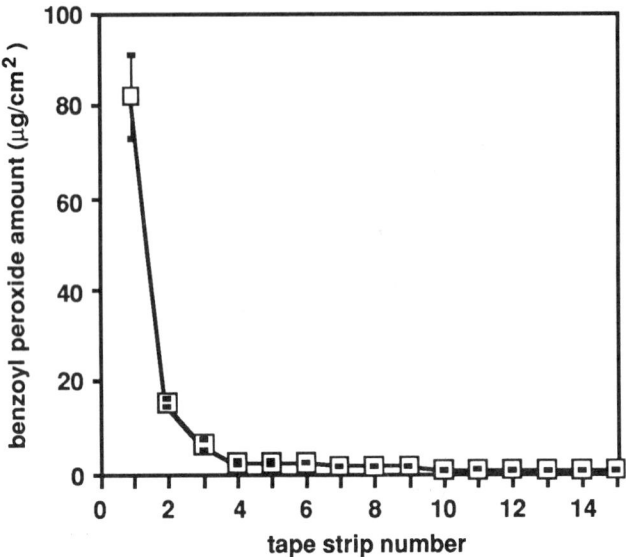

FIGURE 1.   Benzoyl peroxide amount in stratum corneum removed by tape stripping after *in vitro* application on human skin of 5% benzoyl peroxide in an aqueous gel for 8 h under occlusion. (Drawn from Leclercq, M., Ph.D. thesis, University of Paris-Sud, France, 1981. With permission.)

C. DRUG CONCENTRATION IN THE RECEPTOR FLUID

The most common technique for studying *in vitro* drug permeation through the skin is to measure the amount of radioactivity in the receptor fluid after application of a formulation containing a defined amount of a drug generally radiolabeled with tritium or carbon-14. Radioactive labeling is very convenient since it is quite easy to use, assuming metabolic pathways of the tested drug are well known, and it also allows a simple calculation of the mass balance. However, when radioactive labeling of the compound is not possible or when *in vitro* drug metabolism has to be investigated, other analytical procedures such as high performance liquid chromatography (HPLC),[3,34] high performance thin layer chromatography (HPTLC),[35,40] adsorption differential pulse voltammetry (ADPV),[41] etc., can be used.

D. DISTRIBUTION OF DRUGS WITHIN THE SKIN

In addition to the quantification of drugs in the receptor compartment, it is of main interest, particularly for dermal therapeutic systems, to follow the kinetics of drug distribution within the skin. Although the evaluation of drug concentration in the receptor phase is of major importance for transdermal therapeutic systems since *in vivo* drug activity will be correlated with percutaneous absorption, for dermal formulations the drug localization within the skin layers is essential, while drug concentration in the receptor fluid is of interest for estimating potential systemic side-effects that may occur *in vivo*.

There are different techniques for identifying and quantifying drug distribution within the skin, such as tape stripping, skin sectioning, microscopy, and autoradiography.

1. Stripping Methodology

The stripping technique can be used to measure drug concentration in the stratum corneum, the outermost skin layer, by repetitive application of adhesive tape to the skin sample (generally 10 to 15 times). The amount of drug is then measured in each single strip, and a gradient is generally observed within the horny layer. The amount of penetrating substance decreases with increasing depth (see Figure 1) after *in vitro* application of 5% benzoyl peroxide in an aqueous gel on human skin for 8 h under occlusion.[42] The amount

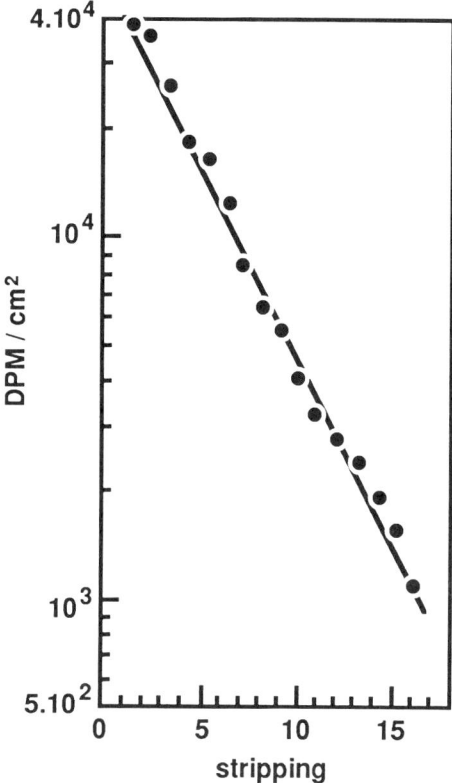

FIGURE 2. Penetration of 0.1% tritium-labeled 4-chlorotestosterone acetate in vaseline into the horny layer of human skin *in vitro*. (Reproduced from Schaefer, H., Stüttgen, G., Zesch, A., Schalla, W., and Gazith, J., *Curr. Probl. Dermatol.*, 7, 80, 1978. With permission.)

of drug decreases rapidly within the first strips, with almost 70 and 85% of the total amount in the horny layer recovered in the first strip and in the first three strips, respectively. Similarly, Schaefer et al.[43] observed a logarithmic decrease of drug amount as a function of depth of the horny layer (Figure 2).

The stripping technique widely used *in vivo* presents some drawbacks when used for *in vitro* tests. No precise information is generally given regarding the amount of tissue removed with each tape strip, although it is well established that it decreases with each consecutive tape strip.[18,44] The uppermost cell layers of the stratum corneum are readily removed because they stick correctly to the adhesive tape. The lower layers adhere less to the tape, owing to the interstitial lipid and water content, so that less tissue will consequently be removed. King et al.[45] also showed a significant drop in the number of corneocytes released after stripping skin from the surface downward.

Recent studies suggest that the stratum corneum may be considered as a homogeneous penetration barrier, the distribution of drugs throughout the stratum corneum being rather uniform. Pitts and Swarbrick[46] studied the distribution of ³H-clemastine in human skin slabs *in vitro* and they showed that the amount of drug recovered in the successive tape strips was directly related to the mass of stratum corneum removed (Figure 3). Therefore, although a gradient of drug amount was observed within the horny layer, when the amount of ³H-clemastine removed by tape stripping was related to the weight of stratum corneum in each tape strip, the drug concentration was found to be uniform through the stratum corneum (Figure 4). In another study, following *in vitro* application of an aqueous solution of nicotinic acid to intact excised guinea pig skin,[47] radioactivity in the tape strips of the stratum corneum

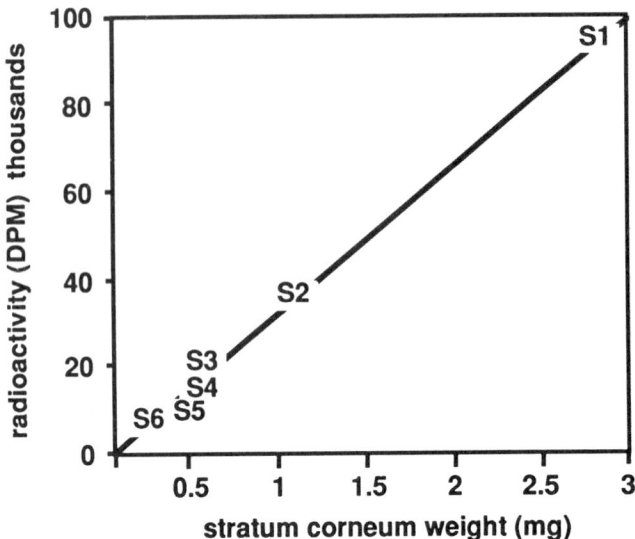

FIGURE 3.   Drug radioactivity — stratum corneum weight profile after application of [^{3}H]-clemastine in DMSO to a rat skin slab. The slab was incubated over Drierite at 32°C for 3 h and tape stripped six times (S1 to S6). (Drawn from Pitts, J. and Swarbrick, J., *Pharm. Res.,* 6 (Suppl.), 174, 1989. With permission.)

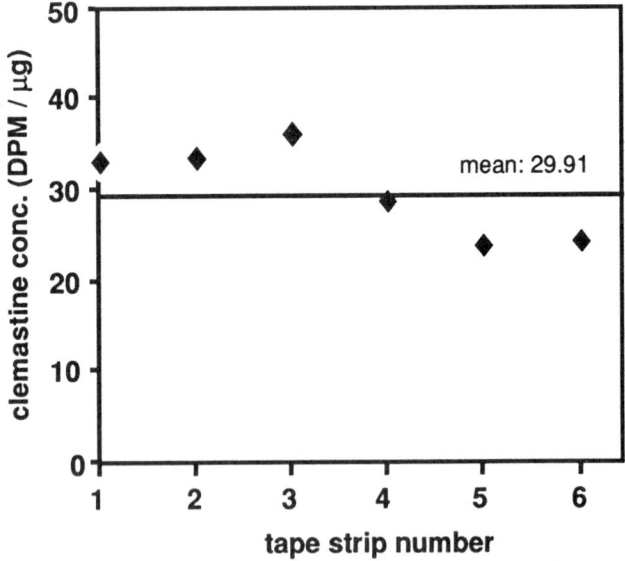

FIGURE 4.   Drug concentration in stratum corneum recovered by successive tape stripping after application of [^{3}H]-clemastine in DMSO to a rat skin slab incubated over Drierite at 32°C for 3 h. (Drawn from Pitts, J. and Swarbrick, J., *Pharm. Res.,* 6 (Suppl.), 174, 1989. With permission.)

decreased dramatically within the first strips after 24 h (Figure 5). Nevertheless, whether drug concentration within the stratum corneum layers follows the same pattern as drug amount is not obvious. In fact, if one can assume a similarity between *in vitro* and *in vivo* stripping in terms of weight of tissue sticking to the adhesive tape, the dramatic decrease of nicotinic acid radioactivity in the first strippings *in vitro* might correspond to the drop of stratum corneum weight in each consecutive tape strip. From these data the concentration

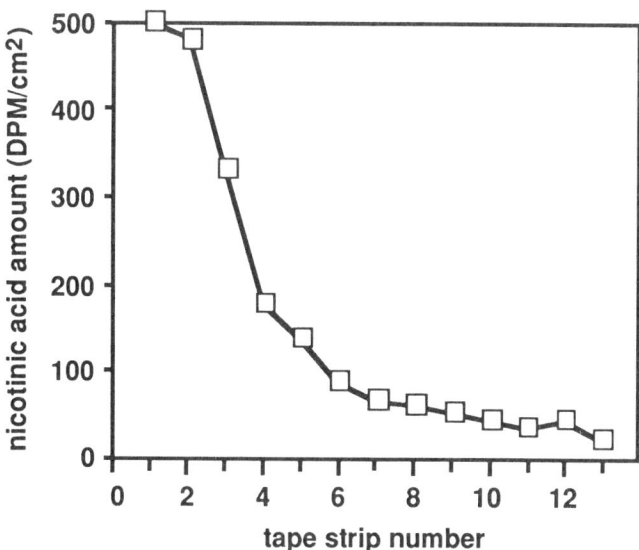

FIGURE 5. Drug radioactivity in each consecutive tape strip after application of an aqueous solution of [¹⁴C]-nicotinic acid to intact excised guinea pig abdominal skin for 24 h. (Redrawn from Osamura, H., Jimbo, Y., and Ishihara, M., *J. Dermatol.*, 11, 471, 1984. With permission.)

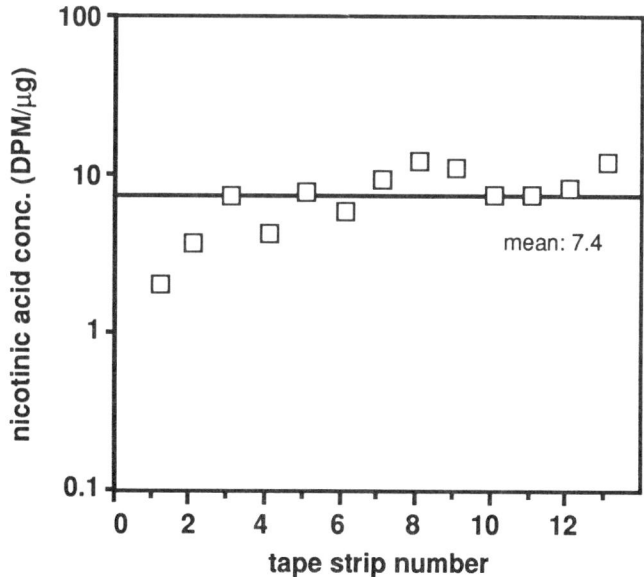

FIGURE 6. Drug concentration in stratum corneum recovered by successive tape stripping, after application of an aqueous solution of [¹⁴C]-nicotinic acid to intact excised guinea pig abdominal skin for 24 h. (Drawn from Osamura, H., Jimbo, Y., and Ishihara, M., *J. Dermatol.*, 11, 471, 1984. With permission.)

of nicotinic acid in each tape strip can be calculated and as shown in Figure 6, drug concentration is almost constant in the stratum corneum layers. The first two lower concentrations could be explained by the extensive skin surface washing using a mixture of methanol/toluene/dioxane (1:1:2) in the procedure.

Stratum corneum should therefore be regarded as a uniform skin layer with regard to drug concentration and the amount of tissue per tape strip should systematically be evaluated

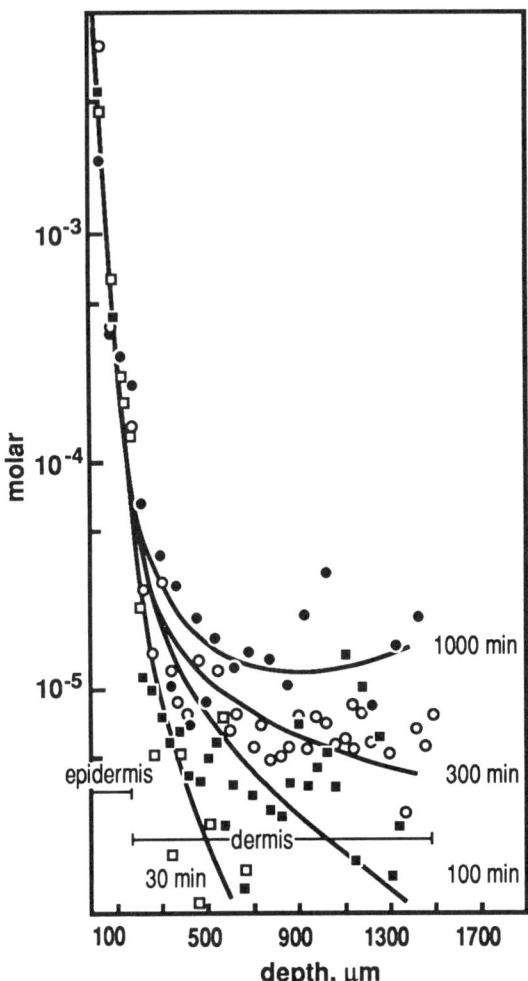

FIGURE 7.   Distribution of triphenylstibine sulfide in the skin as a function of penetration time: □, 30 min; ■, 100 min; ○, 300 min; ●, 1000 min. (Reproduced from Schaefer, H., Zesch, A., and Stüttgen, G., Eds., *Skin Permeability*, Springer Verlag, Berlin, 1982, 541. With permission.)

for a precise determination of drug concentration within the stratum corneum. In addition, it is quite obvious that the efficiency of tape stripping for removing the horny layer will highly depend on the duration of the *in vitro* experiments. After at least 120-h assays, stratum corneum was shown, for instance, to peel readily from the dermis, with some epidermis adhering both to the stratum corneum and to the dermis[48] and this phenomenon may even occur after much shorter periods of time (24 h).

2. Skin Sectioning

Skin sections, parallel to the skin surface, can be used for localizing drugs within the different skin layers.[49] Generally, frozen skin is sectioned parallel to the skin surface using a cryomicrotome. The thickness of the skin sections will be highly dependent on the species and skin site.[50] With this technique, reproducible section thickness can be obtained so that the amount of tissue removed is less variable. The distribution of triphenylstibine sulfide in the skin was thus investigated *in vitro* by Schaefer et al.[44] and an accumulation of the compound in the dermis was observed as a function of exposure time of the preparation (Figure 7). The gradient between the superficial and deeper layer of the skin decreases with

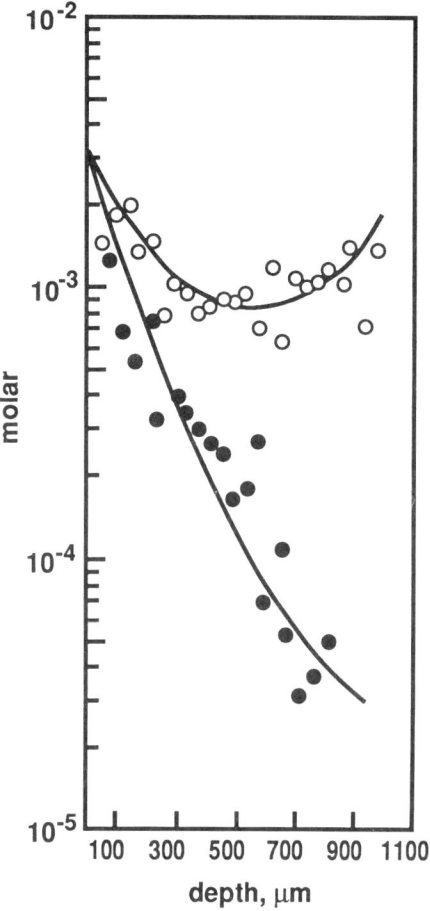

FIGURE 8. Molar distribution of hydrocortisone in human skin *in vitro* (○) and *in vivo* (●); 1% hydrocortisone in polyethylene glycol ointment; 1000-min penetration period. (Reproduced from Zesch, A. and Schaefer, H., *Arch. Dermatol. Res.*, 252, 245, 1975. With permission.)

time, since in *in vitro* experiments accumulation of the substance in the dermis may be related to its poor solubility in the aqueous receptor phase. Therefore, the difference between *in vitro* and *in vivo* experiments, particularly when comparing the distribution profile of certain drugs in the dermis with time, may be partly explained by the lack of a "resorption-like" process in *in vitro* assays. As shown for instance in Figure 8, the molar distribution of hydrocortisone in human skin, 1000 min after application of 1% hydrocortisone in polyethylene glycol ointments, was very similar in the horny layer and epidermis under *in vitro* and *in vivo* conditions; whereas drug accumulation within the dermis was only observed *in vitro*.[51] Drug enrichment of the deeper lying layers *in vitro* may result from a lack of vascular transport and/or drug preferential localization in the dermis.

However, skin sectioning cannot take into account the variations in the dermo-epidermal junction, since it depends on the animal or human skin types (age, sex, test site, etc.) and this method does not give any precise information regarding a preferential penetration route, e.g., trans-follicular vs. trans-epidermal. Therefore, other techniques such as microscopy and histoautoradiography are often combined with skin sectioning in order to get more information on drug penetration.

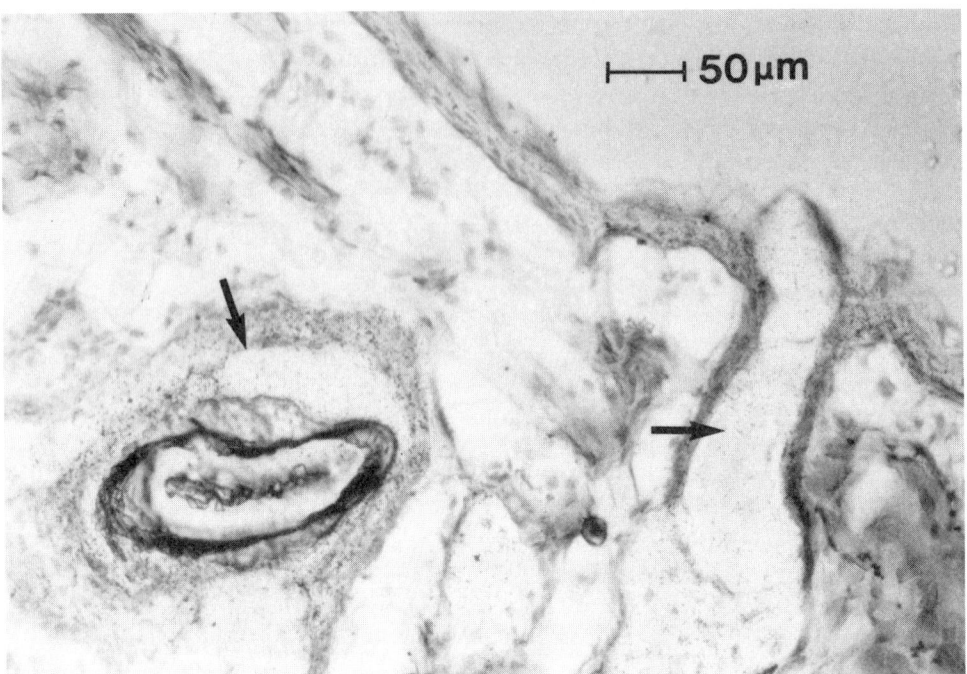

FIGURE 9.   Follicular localization of CD 271 radioactivity after application of a lotion containing 0.3% [^{14}C]-CD 271 to hairless rats for 6 h under occlusion. (Reproduced from Jamoulle, J. C., Lamaud, E., Grandjean, L., Shroot, B., and Schaefer, H., in *Pharmacology of Retinoids in the Skin, Vol. 3*, Reichert, U. and Shroot, B., Eds., S. Karger, Basel, 1989, 198. With permission.)

3. Histoautoradiography

Histoautoradiography has been used for visualizing drug distribution within the skin. Generally, vertical sections of skin are placed on photographic emulsions in the dark, and after an exposure time proportional to the radiolabeled drug concentration in the sections (from weeks up to months), sections are analyzed by light or electron microscopy. Direction of cutting is obviously crucial in this procedure: if the most intensive labeling after topical application is found in the stratum corneum, sections will have to be taken from the lower skin part to the skin surface after eventual stripping of the stratum corneum and the knife be cleaned after each section to avoid artifacts. This technique does not allow quantitative estimation of drug concentration in the different skin layers, but it may distinguish between various routes of penetration. After a 6-h application to hairless rats of a single dose of a lotion containing 0.3% [^{14}C]-CD 271, a new naphthoic acid derivative for topical acne treatment (CIRD GALDERMA, France), histoautoradiographic studies after stratum corneum tape stripping showed, for instance, that drug radioactivity was mainly located in the hair follicles and sebaceous glands (Figure 9).[52]

However, this type of technique is often hampered by difficulties encountered in localizing diffusible drugs, since during sample preparation for light or electron microscopic examination extraction or displacement of the tested substance may occur. Extraction of diffusible compounds during sample preparation can be circumvented by using cryotechniques instead of chemical precipitation.[53,54] ''Dry'' sample techniques seem to be a rational approach.[21] Steroid hormone penetration through excised human stratum corneum was thus studied by combining ''dry'' sample preparation and electron microscope autoradiography.[21] The ultrastructure of the stratum corneum was preserved without significant drug extraction during sample preparation and without problems related to detection limit.

4. Microscopy

Other techniques have been used for visualizing drug distribution in skin. The permeability pathways of n-butanol in mammalian stratum corneum were observed by *in situ* precipitation with osmium vapor associated with electron microscopy and X-ray microanalysis.[53] The route of penetration of mercuric chloride, as a marker, through excised human skin and the effect of penetration enhancer were investigated using *in situ* precipitation technique coupled to electron microscopy.[55] Several studies also used X-ray microanalysis with precipitation or cryotechniques for localizing native electrolytes within the epidermis or for the localization of topically applied formulations.[54,56,57]

For studying *in vitro* the involvement of skin appendages (such as hair follicles and adjacent sebaceous glands together with sweat glands) in the skin absorption of B[a]P, cryosections of skin samples were prepared 1 h after *in vitro* application of the compound to mouse skin, and were examined by fluorescence microscopy.[58] This technique showed an intense fluorescence deep within the nonfluorescing dermis of skin from the haired mice, these fluorescent areas being correlated with follicular ducts and sebaceous glands. Conversely, no fluorescence was detected in the dermis of skin from hairless mice. This analysis of *in vitro* experiments by fluorescence microscopy pointed out the important part played by the transfollicular pathway in skin penetration of a compound such as B[a]P.

The relationship between microsphere size and penetration in stratum corneum and/or hair follicles was also investigated using fluorescent polystyrene microspheres.[59] After application to hairless rat and human skin, the microspheres were selectively targeted to skin appendages as a function of their size. Microparticles with a mean size of at least 10 μm did not penetrate the skin, while microspheres with a diameter less than 3 μm were localized in skin appendages and stratum corneum. However, 7 μm microspheres were mainly observed deep down the follicular ducts and rarely in the stratum corneum (Figure 10). The influence of various parameters (formulation lipophilicity, massage time, etc.) on the passive targeting of calibrated microspheres to hair follicles was also investigated by fluorescence microscopy.

Confocal microscopy is a new powerful technique that will permit one to investigate accurately the different penetration pathways of drugs through the skin layers. Laser scanning confocal fluorescence microscopy has already been used to visualize the passive and iontophoretic permeation pathways of hydrophilic, lipophilic, and polar compounds through the stratum corneum of hairless mouse skin.[60] The distribution of lamellar bodies within the mouse stratum granulosum, followed by their deposition into the intercellular spaces at the stratum granulosum — stratum corneum interface, was also clearly visualized by laser-scanning confocal microscopy.[61] This new technique offers a convenient way of studying the dynamics of lamellar body secretion under normal and experimentally perturbed conditions.

In vitro penetration of nonionic surfactant vesicles (niosomes) through dermatomed human skin was also investigated by freeze fracture electron microscopy.[62] Electron micrographs tended to show structural changes in the stratum corneum when niosomes were applied to human skin *in vitro* over a 48-h period.

IV. FACTORS INFLUENCING *IN VITRO* CUTANEOUS PERMEATION OF DRUGS

In vitro permeation and/or percutaneous absorption of drugs depend(s) on several parameters which can be evaluated and compared using the diffusion cell technique. Apart from experimental parameters (skin origin, storage and preparation, temperature, composition of the receptor phase, applied dose of formulation, occlusion, and diffusion cell type) the main factors investigated in *in vitro* penetration studies are the effects of the physico-chemical characteristics of the drugs and the drug delivery systems.

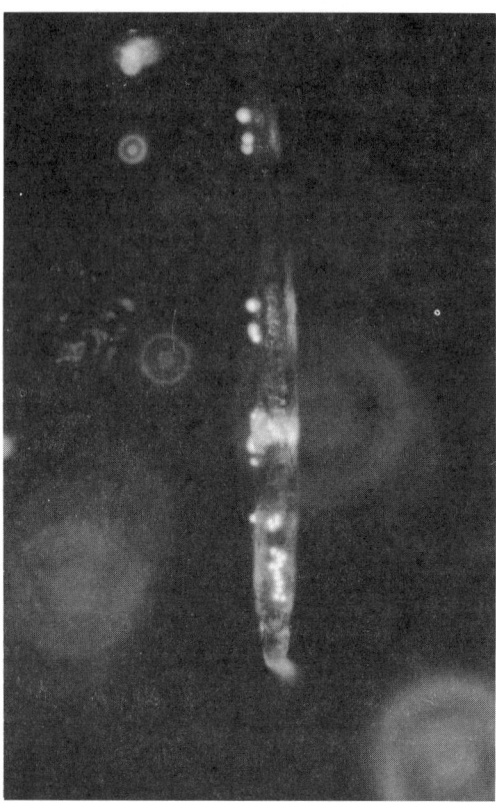

FIGURE 10.   Follicular biopsy observed by fluorescence microscopy after application of fluorescent 7 μm microspheres to human skin. (Reproduced from Schaefer, H., Watts, F., Brod, J., and Illel, B., in *Prediction of Percutaneous Penetration, Methods, Measurements, Modelling,* Scott, R. C., Guy, R. H., and Hadgraft, J., Eds., IBC Technical Services, London, 1990, 163. With permission.)

A. DRUG LIPOPHILICITY

For a series of phenolic compounds[63] and steroids,[64] the permeability coefficient was correlated with their lipophilic character as assessed by their octanol/water partition coefficients.

In vitro drug penetration across hairless mouse skin also showed marked differences between nicotinate esters applied in acetone.[65] Thus, the more polar esters (methyl and ethyl nicotinates) were readily available within the skin; whereas the lipophilic molecules (benzyl and hexyl nicotinates) did not appear to be delivered directly into the skin to any appreciable extent as assessed by radioactivity measurements in the receptor phase. These results are highly dependent on the physico-chemical properties of the nicotinate esters, and their lipid solubility is one key parameter for interpreting differences in their *in vitro* percutaneous absorption characteristics.

The *in vitro* permeabilities of six narcotic analgesics through human cadaver skin were compared by Roy and Flynn.[66] A relationship was established between the *in vitro* permeation rates of the narcotic analgesics through human skin and their physicochemical properties. Thus, the relative permeability coefficients of the three investigated phenylpiperidine analogs (meperidine, fentanyl, and sufentanil) were much higher than the permeability coefficients of three opioid alkaloids (morphine, codeine, and hydromorphone). The higher permeability coefficients of the phenylpiperidine analogs were explained by their much higher lipophilicity as reflected by their high octanol per water partition coefficients (K-octanol/water > 40).

B. FORMULATION COMPOSITION

The effect of the formulations on the permeation rate of various compounds has been extensively investigated over the past years using *in vitro* procedures.[6,44,67,68] One means of altering permeation rate is to modify a formulation by incorporating solvents[69] which may have predominantly two effects:

1. Raise the thermodynamic activity of the drug in the formulation, thereby promoting the interfacial drug transfer into the stratum corneum.
2. Modify skin barrier by using solvents that penetrate and alter the stratum corneum.

The influence of 21 different solvents on the *in vitro* cutaneous permeation of estradiol was investigated using excised human skin mounted in open diffusion cell.[70] Significant variations were found in the steady-state flux and lag time with the different tested solvents. Since the same amount of drug and solvent were applied in the experiments, the solubility of estradiol in the different vehicles was determined. The thermodynamic activity of the drug is indeed correlated to its degree of saturation in the solvent. However, no correlation was shown between drug solubility in the solvent and steady-state flux. For example, the solubility of estradiol in ethylene glycol, ethyl oleate, and diethanolamine was similar ($\simeq$ 16 mg/g at 22°C) but skin permeation rates were quite different; the steady-state flux with ethylene glycol was about 9-fold and 31-fold higher than with ethyl oleate and diethanolamine, respectively. It seems that drug permeability is influenced in this case by an effect of the solvent on the barrier function rather than by the thermodynamic activity of the drug in the solvent.

Another means of modifying permeation rate is to add penetration enhancers to a formulation. Although nonionic surfactants may reduce the thermodynamic activity of a drug and thus retard penetration, they may also act as penetration enhancers by entering the skin and changing its permeability characteristics. Sarpotdar and Zatz[71] demonstrated that the two nonionic surfactants, polysorbate 20 and polysorbate 60, increased the steady-state flux of lidocaine in the presence of propylene glycol *in vitro*, using excised hairless mouse skin as the membrane. In infinite dose experiments, the effect of both surfactants on lidocaine penetration was more pronounced as the propylene glycol concentration increased. Similar results were obtained under finite dose conditions; the skin permeation rate of lidocaine being enhanced in presence of both nonionic surfactants and propylene glycol. *In vitro* penetration of ibuprofen and flurbiprofen through human skin was also investigated.[48] N-methyl-2-pyrrolidone was shown to act as a penetration enhancer, significantly increasing the penetration flux of both phenylalkanoic acids, ibuprofen and flurbiprofen. N-methyl-2-pyrrolidone increased the dissolution of the drugs on the skin, and since it permeated the skin readily, it also promoted the passage of drugs into the skin. The main role of N-methyl-2-pyrrolidone was, however, of penetration enhancer-type; i.e., by modifying the diffusional resistance of the skin it increased the diffusion coefficient of both drugs. The penetration of mercuric chloride through human skin *in vitro* was also shown to be enhanced by N-alkyl-azacycloheptan-2-ones (alkyl = hexyl, dodecyl, hexadecyl). Hexyl- and hexadecyl-azacycloheptan-2-one exhibited the same activity *in vitro*, resulting in accumulation of mercury inside apical corneocytes and in its predominance between medial and proximal corneocytes; whereas azone (1-dodecyl-azacycloheptan-2-one) proved to particularly enhance the intercellular presence of mercury throughout the stratum corneum.[55]

In addition to the previous parameters modifying the formulation, and consequently drug permeation, new drug delivery systems such as drug carriers have been proposed in order to modulate either drug penetration and/or release.

For example, liposomal transdermal delivery systems have been designed for modulating the release of progesterone and consequently controlling *in vitro* drug percutaneous absorption through hairless mouse skin.[72,73] The rate of drug appearance in the receptor compartment after application of a standard progesterone patch was skin-controlled, the skin acting as a rate limiting membrane and the patch as a drug reservoir. However, association of progesterone with multilamellar liposomes (phosphatidylcholine [PC] or dipalmitoylphosphatidylcholine [DPPC]) in the delivery system slowed the transdermal passage by about one half of the value obtained with the standard patch, and by more than one order of magnitude for the PC and DPPC liposomes, respectively. The liposomal delivery systems were capable of controlling *in vitro* progesterone absorption through the skin.

Particulate carriers such as liposomes, have also been investigated *in vivo*[74-76] and *in vitro* for the topical delivery of drugs. In a study using a two-compartment diffusion cell, liposomes were shown not to enhance the penetration of ciclosporin through human skin. Even after a 45-h experiment, with either neutral or negative liposomes, no ciclosporin was detected by HPLC in the receptor compartment.[3] However, encapsulation of hydrocortisone into liposomes dramatically increased its concentration into the epidermis and dermis, as compared to classical hydrocortisone ointments, and drug amounts remained almost constant in the different skin layers over a 5-h period.[77] Liposomes could therefore find new development in corticotherapy where they could lead to a significant improvement of the therapeutic efficacy of drugs.

V. CONCLUDING REMARKS

In vitro experimentation has been shown to be of great interest for investigating either localization and distribution of drugs within the skin, or penetration kinetics and pathways after topical application of dermal or transdermal therapeutic systems. However, since *in vivo* conditions will never be completely reproduced *in vitro*, care should be exercised when extrapolating from *in vitro* experiments for predicting the *in vivo* behavior of a topically applied drug. A recent FDA and AAPS report,[78] in particular regarding *in vitro* percutaneous penetration studies, proposed that *in vitro* experiments could be used in pharmaceutical development of topical dosage forms:

1.   To investigate drug release from topical formulations and drug permeation rate
2.   As a quality control assay
3.   To ensure the drug delivery characteristics on a lot to lot basis
4.   To compare reformulations of topical products on which the bioavailability has already been defined

It is noteworthy that the guidelines for *in vitro* assays include its potential use as a predictive test, that is to say, it could be an alternative to *in vivo* experiments for assessing drug bioavailability when only minor changes in a topical formulation are made. However, this particular orientation of *in vitro* tests must be considered with caution: firstly, because large quantities of well-defined and comparable skin samples are required to carry out tests of sufficient precision; and secondly, it is now well known that for other routes of administration, e.g., the oral route, *in vitro* tests (i.e., dissolution test) cannot always be predictive of drug bioavailability, even when only minor modifications are made in formulations.[79,80]

In vitro penetration studies should mainly be devoted to selection of topical formulations and to optimization of drug concentration in the selected drug delivery system.

As well as trying to reproduce *in vivo* conditions and to find animal alternatives to human skin for *in vitro* experiments, investigators are designing *in vitro* procedures in such a way that drug penetration will only be related to the physico-chemical properties of the drug and to the drug delivery system, so that accurate comparisons between different drugs and drug formulations can be made. *In vitro* experimentation also represents a precious tool for studying diffusion processes within the different skin layers, barrier properties of the skin, penetration pathways, etc. As an alternative to animal or human skin, artificial membranes or reconstructed skin will prove to be valuable tools in an early screening of new topical formulations.[81] However, whether *in vitro* — *in vivo* correlations will systematically exist, still remains another debate.

ACKNOWLEDGMENTS

We would like to thank Mrs. Jeannette Foulon for typing the manuscript, Mrs. Nathalie Wagner for her assistance, and Mrs. Dominique Poisson for art work. We are also indebted to Pr. H. Schaefer, Dr. B. Shroot, and to colleagues of the CIRD GALDERMA for valuable discussion and comments.

REFERENCES

1. **Blank, I. H. and Scheuplein, R. J.,** Transport into and within the skin, *Br. J. Dermatol.,* 81, 4, 1969.
2. **Valia, K. H. and Chien, Y. W.,** Long-term skin permeation kinetics of estradiol. II. Kinetics of skin uptake, binding, and metabolism, *Drug Dev. Ind. Pharm.,* 10, 991, 1984.
3. **Hermann, R. C., Taylor, R. S., Ellis, C. N., Williams, N. A., Weiner, N. D., Flynn, G. L., Annesley, T. M., and Voorhees, J. J.,** Topical ciclosporin for psoriasis: *in vitro* skin penetration and clinical study, *Skin Pharmacol.,* 1, 246, 1988.
4. **Franz, T. J.,** Percutaneous absorption on the relevance of *in vitro* data, *J. Invest. Dermatol.,* 64, 190, 1975.
5. **Franz, T. J.,** The finite dose technique as a valid *in vitro* model for the study of percutaneous absorption in man, *Curr. Probl. Dermatol.,* 7, 58, 1978.
6. **Bronaugh, R. L., Congdon, E. R., and Scheuplein, R. J.,** The effect of cosmetic vehicles on the penetration of *N*-nitrosodiethanolamine through excised human skin, *J. Invest. Dermatol.,* 76, 94, 1981.
7. **Gummer, C. L., Hinz, R. S., and Maibach, H. I.,** The skin penetration cell: a design uptake, *Int. J. Pharm.,* 40, 101, 1987.
8. **Bronaugh, R. L. and Stewart, R. F.,** Methods for *in vitro* percutaneous absorption studies. IV. The flow-through diffusion cell, *J. Pharm. Sci.,* 74, 64, 1985.
9. **Okamoto, H., Yamashita, F., Saito, K., and Hashida, M.,** Analysis of drug penetration through the skin by the two-layer skin model, *Pharm. Res.,* 6, 931, 1989.
10. **Wester, R. C. and Maibach, H. I.,** *In vitro* testing of topical pharmaceutical formulations, in *Percutaneous Absorption: Mechanisms, Methodology, Drug Delivery,* 2nd ed., Bronaugh, R. L. and Maibach, H. I., Eds., Marcel Dekker, New York, 1989, 653.
11. **Lambert, W. J., Higuchi, W. I., Knutson, K., and Krill, S. L.,** Effects of long-term hydration leading to the development of polar channels in hairless mouse stratum corneum, *J. Pharm. Sci.,* 78, 925, 1989.
12. **Bronaugh, R. L.,** Determination of percutaneous absorption by *in vitro* techniques, in *Percutaneous Absorption: Mechanisms, Methodology, Drug Delivery,* 2nd ed., Bronaugh, R. L. and Maibach, H. I., Eds., Marcel Dekker, New York, 1989, 239.
13. **Foreman, M. I., Clanachan, I., and Kelly, I. P.,** Diffusion barriers in skin—a new method of comparison, *Br. J. Dermatol.,* 108, 549, 1983.
14. **Bronaugh, R. L. and Maibach, H. I.,** *In vitro* models for human percutaneous absorption, in *Models in Dermatology,* Maibach, H. I. and Lowe, N. J., Eds., S. Karger, Basel, 1985, 178.
15. **Bond, J. R. and Barry, B. W.,** Limitations of hairless mouse skin as a model for *in vitro* permeation studies through human skin: hydration damage, *J. Invest. Dermatol.,* 90, 486, 1988.
16. **Everson, A., Van der Merwe, E., and Ackermann, C.,** Permeation of tretinoin through nude mouse and human skin *in vitro* and histological changes during permeation as observed with transmission electron microscopy, in *Pharmacology of Retinoids in the Skin,* Reichert, U. and Shroot, B., Eds., S. Karger, Basel, 1989, 195.
17. **Lee, G. and Parlicharla, P.,** An examination of excised skin tissues used for *in vitro* membrane permeation studies, *Pharm. Res.,* 3, 356, 1986.

18. **Schalla, W., Jamoulle, J. C., and Schaefer, H.,** Localization of compounds in different skin layers and its use as an indicator of percutaneous absorption, in *Percutaneous Absorption: Mechanisms, Methodology, Drug Delivery,* 2nd ed., Bronaugh, R. L. and Maibach, H. I., Eds., Marcel Dekker, New York, 1989, 283.

19. **Kligman, A. M. and Christophers, E.,** Preparation of isolated sheets of human stratum corneum, *Arch. Dermatol.,* 88, 702, 1963.

20. **Blank, I. H. and McAuliffe, D. J.,** Penetration of benzene through human skin, *J. Invest. Dermatol.,* 85, 522, 1985.

21. **De Hann, F. H. N., Boddé, H. E., de Bruijn, W. C., Ginsel, L. A., and Junjinger, H. E.,** Visualizing drug transport across stratum corneum: cryotechniques, vapour fixation, autoradiography, *Int. J. Pharm.,* 56, 75, 1989.

22. **Baker, K. W. and Habowsky, J. E. J.,** EDTA separation and ATPase Langerhans cell staining in the mouse epidermis, *J. Invest. Dermatol.,* 80, 104, 1983.

23. **Scott, R. C., Walker, M., and Dugard, P. H.,** *In vitro* percutaneous absorption experiments: a technique for the production of intact epidermal membranes from rat skin, *J. Soc. Cosmet. Chem.,* 37, 35, 1986.

24. **Bronaugh, R. L. and Stewart, R. F.,** Methods for *in vitro* percutaneous absorption studies. III. Hydrophobic compounds, *J. Pharm. Sci.,* 73, 1255, 1984.

25. **Bronaugh, R. L. and Stewart, R. F.,** Methods for *in vitro* percutaneous absorption studies. VI. Preparation of the barrier layer, *J. Pharm. Sci.,* 75, 487, 1986.

26. **Scheuplein, R. J. and Blank, I. H.,** Mechanism of percutaneous absorption. IV. Penetration of nonelectrolytes (alcohols) from aqueous solutions and pure liquids, *J. Invest. Dermatol.,* 60, 286, 1973.

27. **Zesch, A.,** Methods for evaluation of drug concentration in human skin, in *Dermal and Transdermal Absorption,* Brandau, R. and Lippold, B. H., Eds., Wissenschaftliche Verlagsgesellschaft mbH, Stuttgart, Germany, 1982, 116.

28. **Bronaugh, R. L., Stewart, R. F., and Simon, M.,** Methods for *in vitro* percutaneous absorption studies. VII. Use of excised human skin, *J. Pharm. Sci.,* 75, 1094, 1986.

29. **Van der Merwe, E., Ackermann, C., and Van Wyk, C. J.,** Factors affecting the permeability of urea and water through nude mouse skin *in vitro.* I. Temperature and time of hydration, *Int. J. Pharm.,* 44, 71, 1988.

30. **Bronaugh, R. L. and Stewart, R. F.,** Methods for *in vitro* percutaneous absorption studies. V. Permeation through damaged skin, *J. Pharm. Sci.,* 74, 1062, 1985.

31. **Harrison, S. M., Barry, B. W., and Dugard, P. H.,** Effects of freezing on human skin permeability, *J. Pharm. Pharmacol.,* 36, 261, 1984.

32. **DelTerzo, S., Behl, C. R., Nash, R. A., Bellantone, N. H., and Malick, A. W.,** Evaluation of the nude rat as a model: effects of short-term freezing and alkyl chain length on the permeabilities of n-alkanols and water, *J. Soc. Cosmet. Chem.,* 37, 297, 1986.

33. **Kushla, G. P. and Zatz, J. L.,** Lidocaine penetration through human and hairless mouse skin *in vitro, J. Soc. Cosmet. Chem.,* 40, 41, 1989.

34. **Swarbrick, J., Lee, G., and Brom, J.,** Drug permeation through human skin. I. Effects of storage conditions of skin, *J. Invest. Dermatol.,* 78, 63, 1982.

35. **Holland, J. M., Kao, J. Y., and Whitaker, M. J.,** A multisample apparatus for kinetic evaluation of skin penetration *in vitro*: the influence of viability and metabolic status of the skin, *Toxicol. Appl. Pharmacol.,* 72, 272, 1984.

36. **Cole, G. W., Shimonaye, S., and Goodman, M.,** The effect of topical ciclosporin A on the elicitation phase of allergic contact dermatitis, *Contact Dermatitis,* 19, 129, 1988.

37. **Riley, R. T. and Kemppainen, B. W.,** Receptor fluid penetrant interactions and the *in vitro* cutaneous penetration of chemicals, in *Percutaneous Absorption: Mechanisms, Methodology, Drug Delivery,* Bronaugh, R. L. and Maibach, H. I., Eds., Marcel Dekker, New York, 1985, 387.

38. **Collier, S. W., Sheikh, N. M., Sakr, A., Lichtin, J. L., Stewart, R. F., and Bronaugh, R. L.,** Maintenance of skin viability during *in vitro* percutaneous absorption/metabolism studies, *Toxicologist,* 8, 125, 1988.

39. **Gummer, C. L.,** The *in vitro* evaluation of transdermal delivery, in *Transdermal Drug Delivery, Developmental Issues and Research Initiatives,* Hadgraft, J. and Guy, R. H., Eds., Marcel Dekker, New York, 1989, 177.

40. **Kao, J. and Hall, J.,** Skin absorption and cutaneous first pass metabolism of topical steroids: *in vitro* studies with mouse skin in organ culture, *J. Pharmacol. Exp. Ther.,* 241, 482, 1987.

41. **Fullerton, A., Andersen, J. R., Hoelgaard, A., and Menné, T.,** Permeation of nickel salts through human skin *in vitro, Contact Dermatitis,* 15, 173, 1986.

42. **Leclercq, M.,** Percutaneous absorption of benzoyl peroxide and benzoic acid, Ph.D. thesis, No. 31, University of Paris-Sud, France, 1981.

43. **Schaefer, H., Stüttgen, G., Zesch, A., Schalla, W., and Gazith, J.,** Quantitative determination of percutaneous absorption of radiolabeled drugs *in vitro* and *in vivo* by human skin, *Curr. Probl. Dermatol.,* 7, 80, 1978.

44. **Schaefer, H., Zesch, A., and Stüttgen, G., Eds.**, *Skin Permeability*, Springer Verlag, Berlin, 1982, 541.

45. **King, C. S., Barton, S. P., Nicholls, S., and Marks, R.**, The change in properties of the stratum corneum as a function of depth, *Br. J. Dermatol.*, 100, 165, 1979.

46. **Pitts, J. and Swarbrick, J.**, An *in vitro* skin slab model for percutaneous absorption studies, *Pharm. Res.*, 6 (Suppl.), 174, 1989.

47. **Osamura, H., Jimbo, Y., and Ishihara, M.**, Skin penetration of nicotinic acid, methyl nicotinate and butyl nicotinate in the guinea pig, *J. Dermatol.*, 11, 471, 1984.

48. **Akhter, S. A. and Barry, B. W.**, Absorption through human skin of ibuprofen and flurbiprofen: effect of dose variation, deposited drug films, occlusion and the penetration enhancer *N*-methyl-2-pyrrolidone, *J. Pharm. Pharmacol.*, 37, 27, 1985.

49. **Schaefer, H. and Lamaud, E.**, Standardization of experimental models, in *Skin Pharmacokinetics*, Shroot, B. and Schaefer, H., Eds., S. Karger, Basel, 1987, 77.

50. **Bronaugh, R. L. and Maibach, H. I.**, *In vitro* percutaneous absorption, in *Models in Dermatology*, Maibach, H. I. and Lowe, N. J., Eds., S. Karger, Basel, 1985, 121.

51. **Zesch, A. and Schaefer, H.**, Penetration of radioactive hydrocortisone in human skin from various ointment bases. II. *In vivo* experiments, *Arch. Dermatol. Res.*, 252, 245, 1975.

52. **Jamoulle, J. C., Lamaud, E., Grandjean, L., Shroot, B., and Schaefer, H.**, Follicular penetration, distribution and migration of CD 271, a new naphthoic acid derivative for topical acne treatment, in *Pharmacology of Retinoids in the Skin, Vol. 3*, Reichert, U. and Shroot, B., Eds., S. Karger, Basel, 1989, 198.

53. **Nemanic, M. K. and Elias, P. M.**, *In situ* precipitation: a novel cytochemical technique for visualization of permeability pathways in mammalian stratum corneum, *J. Histochem. Cytochem.*, 28, 573, 1980.

54. **Sharata, H. H. and Burnette, R. R.**, Effect of dipolar aprotic permeability enhancers on the basal stratum corneum, *J. Pharm. Sci.*, 77, 27, 1988.

55. **Boddé, H. E., Kruithof, M. A. M., Brussee, J., and Koerten, H. K.**, Visualisation of normal and enhanced $HgCl_2$ transport through human skin *in vitro*, *Intern. Symp. Control. Rel. Bioact. Mater.*, 16, 37, 1989.

56. **Forslind, B.**, Clinical applications of scanning electron microscopy and X-ray microanalysis in dermatology, *Scanning Electron Micros.*, 183, 1984.

57. **Grundin, T. G., Roomans, G. M., Forslind, B., Lindberg, M., and Werner, Y.**, X-ray microanalysis of psoriatic skin, *J. Invest. Dermatol.*, 85, 378, 1985.

58. **Kao, J., Hall, J., and Helman, G.**, *In vitro* percutaneous absorption in mouse skin: influence of skin appendages, *Toxicol. Appl. Pharmacol.*, 94, 93, 1988.

59. **Schaefer, H., Watts, F., Brod, J., and Illel, B.**, Follicular Penetration, in *Prediction of Percutaneous Penetration, Methods, Measurements, Modelling*, Scott, R. C., Guy, R. H., and Hadgraft, J., Eds., IBC Technical Services, London, 1990, 163.

60. **Cullander, C. and Guy, R. H.**, The transdermal pathways of hydrophilic, lipophilic, and polar substances can be imaged with confocal microscopy, *Clin. Res.*, 38, 223A, 1990.

61. **Cullander, C., Menon, G. K., Guy, R. H., and Elias, P. M.**, *In situ* visualization of the lamellar body secretory system by confocal microscopy, *J. Invest. Dermatol.*, 94, 517, 1990.

62. **Hofland, H. E. J., Bouwstra, J. A., Boddé, H. E., Spies, F., and Junginger, H. E.**, *In vitro* interactions between non-ionic surfactant vesicles and human skin, *Skin Pharmacol.*, 2, 117, 1989.

63. **Roberts, M. S., Anderson, R. A., and Swarbrick, J.**, Permeability of human epidermis to phenolic compounds, *J. Pharm. Pharmac.*, 29, 677, 1977.

64. **Grasso, P. and Lansdown, A. B. G.**, Methods of measuring, and factors affecting, percutaneous absorption, *J. Soc. Cosmet. Chem.*, 23, 481, 1972.

65. **Guy, R. H., Carlström, E. M., Bucks, D. A. W., Hinz, R. S., and Maibach, H. I.**, Percutaneous penetration of nicotinates: *in vivo* and *in vitro* measurements, *J. Pharm. Sci.*, 75, 968, 1986.

66. **Roy, S. D. and Flynn, G. L.**, Transdermal delivery of narcotic analgesics: comparative permeabilities of narcotic analgesics through human cadaver skin, *Pharm. Res.*, 6, 825, 1989.

67. **Jimbo, Y., Ishihara, M., Osamura, H., Takano, M., and Ohara, M.**, Influence of vehicles on penetration through human epidermis of benzyl alcohol, isoeugenol and methyl isoeugenol, *J. Dermatol.*, 10, 241, 1983.

68. **Stüttgen, G. and Bauer, E.**, Penetration and permeation into human skin of fusidic acid in different galenical formulations, Arzneimittelforsch./ *Drug Res.*, 38, 730, 1988.

69. **Twist, J. N. and Zatz, J. L.**, The effect of solvents on solute penetration through fuzzy rat skin *in vitro*, *J. Soc. Cosmet. Chem.*, 40, 231, 1989.

70. **Møllgaard, B. and Hoelgaard, A.**, Permeation of estradiol through the skin—effect of vehicles, *Int. J. Pharm.*, 15, 185, 1983.

71. **Sarpotdar, P. P. and Zatz, J. L.**, Evaluation of penetration enhancement of lidocaine by nonionic surfactants through hairless mouse skin *in vitro*, *J. Pharm. Sci.*, 75, 176, 1986.

72. **Knepp, V. M., Hinz, R. S., Szoka, F. C., and Guy, R. H.,** Controlled drug release from a novel liposomal delivery system. I. Investigation of transdermal potential, *J. Controlled Release,* 5, 211, 1988.
73. **Knepp, V. M., Szoka, F. C., and Guy, R. H.,** Controlled drug release from a novel liposomal delivery system. II. Transdermal delivery characteristics, *J. Controlled Release,* 12, 25, 1990.
74. **Mezei, M. and Gulasekharam, V.,** Liposomes—A selective drug delivery system for the topical route of administration. I. Lotion dosage form, *Life Sci.,* 26, 1473, 1980.
75. **Krowczynski, L. and Stozek, T.,** Liposomen als Wirkstoffträger in der percutanen Therapie, *Pharmazie,* 39, 627, 1984.
76. **Weiner, N., Williams, N., Birch, G., Ramachandran, C., Shipman, C., and Flynn, G.,** Topical delivery of liposomally encapsulated interferon evaluated in a cutaneous herpes guinea pig model, *Antimicrob. Agents Chemother.,* 33, 1217, 1989.
77. **Wohlrab, W. and Lasch, J.,** Penetration kinetics of liposomal hydrocortisone in human skin, *Dermatologica,* 174, 18, 1987.
78. **Skelly, J. P., Shah, V. P., Maibach, H. I., Guy, R. H., Wester, R. C., Flynn, G., and Yacobi, A.,** FDA and AAPS report of the workshop on principles and practices of *in vitro* percutaneous penetration studies: relevance to bioavailability and bioequivalence, *Pharm. Res.,* 4, 265, 1987.
79. **Rolland, A., Gibassier, D., Chemtob, C., and Le Verge, R.,** Etude comparative de la disponibilité *"in vitro"* de la vincamine à partir de formes "à libération modifiée", I, *J. Pharm. Belg.,* 39, 93, 1984.
80. **Rolland, A., Gibassier, D., Chemtob, C., and Le Verge, R.,** Etude de la biodisponibilité de la vincamine à partir de comprimés à matrice hydrophile, II, *J. Pharm. Belg.,* 39, 136, 1984.
81. **Regnier, M., Darmon, M., and Schaefer, H.,** Human epidermis reconstructed *in vitro* as a model for studying epidermal permeability: effects of modulators of keratinocyte differentiation, *J. Invest. Dermatol.,* 95, 485, 1990.

Chapter 12

EXPRESSION OF ABSORPTION DATA

Steven W. Collier and Robert L. Bronaugh

TABLE OF CONTENTS

I. INTRODUCTION

The absorption of chemicals percutaneously is studied by investigators from divergent backgrounds. The percutaneous absorption literature unites the fields of toxicology, pharmacology, pharmaceutics, and medicine and brings with it similar terminology and techniques for its study and description. As with all experimental results, there is a need to measure and express percutaneous absorption data in a manner which is relevant, concise, and unambiguous. The purpose of this chapter is to discuss this need in the context of the current percutaneous absorption literature and the application of reported results by its readers.

II. DOSAGE

The absorption process begins with the application of the test material to the skin. The conditions of its application need to be explicitly stated at the beginning of the description of the experimental procedure in a manuscript. Percutaneous absorption results reported in published abstracts are useful only when dosing conditions are given. The dosage expressed for a percutaneous absorption study may be a concentration in a vehicle if the study is conducted under infinite dosing (or steady state) conditions. More commonly, finite dosing techniques are employed with dosage expressed as a surface density (mass per unit area) of the vehicle or the neat substance. If the test material is applied in a nonvolatile vehicle, a concentration of the compound in the vehicle should be given. The dose should be in solution (not a suspension) when it is applied unless it is a suspension under conditions of use. Otherwise, the percentage of the applied dose which is absorbed will be underestimated. The selection of the dose or range of dosages in a percutaneous absorption study depends on the intended use or relevance of the data. If the intention of the study is extrapolation to human exposure conditions for drug development or toxicological risk assessment, the dosages selected must be applicable.

The use of radiolabeled test materials in *in vitro* absorption studies makes for sensitive and unambiguous measurements of the test material in the receptor fluid and skin. Studies conducted with radiolabeled test compounds frequently use applied surface densities in the low $\mu g/cm^2$ range. The relationship between dosage and the fraction of the dose absorbed is rather linear over a finite range. For many compounds this linear concentration/absorption response is in the low $\mu g/cm^2$ range. For other compounds, a larger linear range has been found. Predictions of high exposures cannot necessarily be interpreted from much lower dose absorption values. Similarly, the thermodynamic activity of the test material in the vehicle in which the dosage is applied will uniquely affect the absorption of the compound. These variables suggest the need for an adequate forethought in the design of the study and an explicit description of dosing conditions in the reporting of the data.

A. SELECTION OF AN APPROPRIATE DOSE

The dose used in a percutaneous absorption study may be predetermined from an analytical investigation of exposure conditions, such as the concentration of an organic compound in bathing water, the average rate of aerial application of a pesticide, or the average usage of cosmetic containing a toxic contaminant. For other types of studies, the dermal dose to test must be extrapolated from the pharmacodynamic relationship between effective dose and biological response. Examples of these types of data might be the average blood level required for a transdermally delivered drug to be effective or the systemic dose in an animal toxicological study where no effect is observed. The extrapolation of the dose-effect relationship from the animal model to humans is done by a dose conversion. The principles of dose conversion and the application of suitable safety factors for human exposure are highly

TABLE 1
General Factors Used in Dose Extrapolation

Parameter	Human factors
Mass of standard humans	
Man	70 kg
Woman	60 kg
Child	20 kg
Skin surface area	
Totally exposed (man 180 cm tall)	1.8 m^2
Clothed with short-sleeved shirt, pants, shoes	0.3 m^2
Clothed with long-sleeved shirt, pants, shoes, gloves	0.1 m^2

	Male	Female
Weight and thickness of skin for reference adult		
Total skin weight (g)	2,600	1,790
Epidermis (g)	100	90
Dermis (g)	2,500	1,700
Hypodermis (g)	7,500	13,000
Total skin thickness (μm)	1,300	1,300
Epidermis (μm)	50	50
Dermis (μm)	1,250	1,250
Hypodermis (μm)	3,750	6,600

Parameter	Human factors
Factors used in extrapolation	
Weight	
Rat	0.35 kg
Mouse	0.03 kg

Data from *Report of the Task Force on Reference Man.*, ICRP Publ. No. 23, Pergamon, Elmsford, New York, 1975.

dependent on the test material and the type of study from which the data were obtained (i.e., chronic or subchronic). The rationale for their selection has been described by Kokoski et al.[1] The Food and Drug Administration (FDA) accepts equivalent dose calculations based on a dose per unit body mass basis (i.e., mg/kg) whereas the Environmental Protection Agency (EPA) scales this to a 2/3 exponent to give a dose per unit body surface area.[2] A scaling exponent of 3/4 has been suggested to reflect a conversion for basal metabolic rate.[3,4] These conversions are based on standard parameters for humans and animals such as those described by the International Commission on Radiological Protection.[5] Similarly, dosages based on exposure conditions anticipated for humans can be calculated from standard parameters and scaled for animal testing. A few such standard parameters related to dermal exposure and percutaneous absorption are listed in Table 1.

For establishing the acceptable daily intake (ADI) for noncarcinogens in humans, the no observed effect level (NOEL) in the animal model from a systemic dose is scaled to humans and multiplied by a safety factor. The safety factor multiplier is usually in the range[6] from 10 to 1000 with a factor of 100 commonly used by the FDA when chronic toxicological data is available. The permeability of human skin is usually less than that of animal skin. In toxicological testing, this difference might contribute to the overall safety factor if animal percutaneous absorption data is substituted for human data. However, if human percutaneous studies are planned and animal dose-response data and absorption data are determined, a nominal dosage range for testing on human skin can be calculated. Currently, there is no standard method of scaling doses among regulatory agencies. The exact manner in which these calculations are performed will depend on test material and the manner in which humans are exposed to it. The ADI for a noncarcinogen using the dose per unit body mass paradigm is calculated as shown below:

$$ADI = \frac{\text{Test Species NOEL} \times \text{Standard adult body mass (70 kg)}}{\text{Safety Factor}}$$

where the safety factor is an acceptable safety factor such as 100. For example: a toxicant was determined to have a NOEL in rats (average body weight of 0.35 kg) of 0.5 mg/kg/d. What would be an appropriate range of doses to test in a percutaneous absorption study at a human ADI? The average adult man has a body mass of 70 kg (Table 1). Based on a dose per unit body weight calculation with a safety factor of 100, the ADI can be derived.

$$ADI = \frac{0.5 \text{ mg/kg/day} \times 70 \text{ kg}}{100} = 0.35 \text{ mg/day}$$

If we assume that all of the applied dose will be absorbed in 24 h, the skin surface density at the ADI can be calculated. If it is assumed that the exposure is encountered by an unprotected man with an exposed skin surface area of 3000 cm^2 (Table 1), the surface density of the material on the skin at ADI by the dose per unit body mass paradigm would be:

$$\text{Dermal dose} = \frac{0.35 \text{ mg}}{3000 \text{ cm}^2} = 120 \text{ ng/cm}^2$$

The calculated dose and a multiple or multiples of it (3- to 100-fold) might be a good range to test the percutaneous absorption of the compound on human skin. The range of dosing might also include the ranges of the most conservative and the greatest estimates of the ADI. Similarly, if human exposure data is available, the percutaneous absorption of the material can be determined in human or animal skin to determine the absorbed dose. The absorbed dose can then be converted to a dosage suitable for toxicological testing in an animal model.

B. EFFECTIVE DOSE AND CUTANEOUS METABOLISM

The *effective* dose is defined as the concentration of the active agent at the target organ. *In vitro* percutaneous absorption studies are used to measure the *absorbed* dose. Sometimes the actual active agent that acts on a target organ is a metabolite of the test material. If the percutaneously absorbed test material is metabolized in the skin to a toxic agent, this measurement, if significant, might be included in the determination of the effective dose of the risk agent. It has been observed that the specific activity of cutaneous enzymes and the rate of percutaneous absorption can affect the extent of metabolism during percutaneous absorption.[5] The effect of penetration rates on cutaneous metabolism of a toxic or therapeutic agent might significantly affect a physiologically based pharmacokinetic model incorporating dermal absorption and may need to be considered when extrapolating percutaneous absorption data to toxicological endpoints determined by a different dosing route.

III. PRESENTATION OF DATA AND DURATION OF EXPERIMENTS

A. WHAT SHOULD BE MEASURED?

The data presented from a percutaneous absorption study should include a description of the exact conditions of dosing and an accountability of the test material. At the end of the experiment, the surface should be washed to remove unabsorbed material. The skin is then removed and homogenized or digested. The mass balance of the experiment is obtained by totaling the amount of material from the surface wash, the skin homogenate, and the receptor fluid. Relatively hydrophilic materials will readily partition into an aqueous receptor fluid. For these types of compounds, most of the penetrating material will be in the receptor fluid. For lipophilic compounds, an appreciable amount of the absorbed test material remains in the skin at the end of the study. The value for the total absorption at the end of the experiment is obtained adding the quantity of material in the receptor fluid fractions and the amount remaining in the skin after washing. For a discussion of the rational for this procedure, consult Chapters 5 and 19 of this book.

B. STUDY DURATION — HOW SHOULD RESULTS BE REPORTED?

The duration of the study and the manner in which the results are presented are determined by the goals of the investigators and the audience to which they are reporting. For dermal risk assessment purposes, it is generally assumed that anything applied to the skin will be washed off in a period of 24 h, the average bathing frequency. For these reasons, it is convenient to run percutaneous absorption studies for a duration of 24 h. Moreover, toxicological units of dose are most commonly expressed as weight of a substance per unit of body weight per day (i.e., mg/kg/day). Frequently, however, percutaneous absorption data is not expressed this way in the literature. The percentage of the applied dose absorbed is more often encountered. The percentage of the applied dose which is absorbed, or percent dose absorbed, is a useful and convenient number for comparisons between different experimental conditions such as skin source, vehicle, duration, etc., where the dosage of the test material is essentially at the same concentration. Doses in the low $\mu g/cm^2$ range frequently can be increased with a proportionate increase in the percent dose absorbed. If dosage concentrations differ appreciably between a low dose and a high dose, the percent dose absorbed may greatly differ. A higher dose will still increase the amount absorbed but not necessarily in a linear fashion. For this reason, whenever percent dose absorbed is expressed, the dosing concentrations and conditions must be expressed so that the actual amount of test material penetrating can be determined. The area of application must be included in order for normalization of the data to a unit area basis so dosage concentrations in terms of surface density are necessary. Since dose concentrations affect the absorption of the test compound, the dosage conditions should be expressed even when mass or molar units are used to describe the quantity of penetrating material.

C. EXPRESSION OF KINETIC DATA

If the *in vitro* study is performed under sink conditions in the receptor fluid reservoir, rates of percutaneous absorption can be obtained. *In vitro* and *in vivo* studies which are relevant to exposure conditions frequently do not reach steady-state conditions. The rate of absorption may reach a pseudosteady state under certain circumstances such as with a transdermal drug delivery patch, but is more often a dynamic process. The resolution of rate of change of the absorption rate is limited only by the frequency of sampling and the sensitivity and precision of the analytical procedure. When flow-through diffusion cell systems are properly used, the concentration of the test material in the receptor fluid eluting from the cell is a true indication of absorption rate. Static cell measurements without periodic total replacement of the receptor fluid give a cumulative absorption value from which the rate of absorption is deduced. Cumulative absorption data is nicely augmented by an indication of the rates at which penetration occurred during the study. We typically determine the average rate of penetration in each of four 6-h fractions in a 24-h study. At a minimum, reporting the amount of the test material which penetrated in the first 6 h and the total absorption in 24 h would give the reader an indication of whether absorption was rapid or prolonged. If a steady state or pseudosteady state is reached during the study, that rate should be reported. The pharmacokinetic parameter, area under the curve (AUC), is, of course, the total absorption for the duration of the experiment. Absorption rate constants are dependent on dosing conditions. These conditions must be explicitly explained for the data to be of general use.

IV. STATISTICAL ANALYSIS OF DATA

It is difficult to draw conclusions in a study without a statistical analysis of the data. The exception to this may be the infrequent "all or none" type of response. Inferential statistics give an indication of the properties of a large group by sampling the properties of a small group. The mean and the standard deviation (SD) of the sampled group can describe the larger population if it is normally distributed and randomly sampled. Since percutaneous absorption studies are sometimes lengthy and tedious, requiring a lot of investigator attention to derive data values, and the source of skin for the studies is often from a small population, it is an idealization for most data to meet these assumptions. The investigator should measure as large a population as feasible and not bias results through their knowledge of the experimental treatments.

A. THE *N* IN PERCUTANEOUS ABSORPTION STUDIES

The number of experimental observations is referred to as the *N* of the experiment. An *N* of at least three observations is extremely desirable for conclusions to be made from the data. As the experimental *N* becomes larger, the sample mean becomes a more precise estimate of the population mean. An *N* in an experiment refers to the number of experimental subjects in a study, not the number of times the measurement was made. *In vitro* experimentalists frequently have adequate tissue for several measurements to be made with one experimental subject. If this is done, the results are to be averaged and the average treated as a single *N*. The *N* of the experiment should be reported along with the mean and a measure of dispersion of the data such as the SD or the standard error of the mean (SEM). It is more efficient to use the skin from the same subject under different treatments to minimize the effects of subject to subject variability on the differences between treatments. Graphical data are presented as the mean values with an indication of the error, such as error bars representing the SEM. Error bars on graphical data are meant to express the cumulative variations of the mean value and convey a certainty of the conclusions when significant differences between values exist. The variations associated with the mean value include analytical uncertainties, biological variation, and experimental error. Tabular data also requires an error assignment with the same number of significant digits as the mean to which it is associated. Both graphical and tabular data require the *N* of the experiment to be reported.

B. TESTING DIFFERENCES BETWEEN MEANS

In reporting experimental results, conclusions on the effects of different treatments must be supported by statistical evidence. The *t* test, first described by Gosset,[8] writing under the pseudonym of "Student", is the most widely used statistical test for determining if 2 mean values are different. Its use is only valid for the comparison between 2 means. The practice of using multiple *t* tests for comparisons of several means is unfortunately still common. The experiment-wise error rat, α, is the probability in finding a significant difference between means by chance. When only 2 means are compared, it is possible to fix α at a certain probability such as 5% ($\alpha = 0.05$). One is then certain that, in 95% of the time, the test for significant differences between the 2 mean values will be valid. When *t*-tests are used to compare more than two means, α does not remain fixed and the probability in finding a significant result by chance is greater than reported. The comparison of multiple means is accomplished only after an analysis of variance detects a significant difference among means. When an analysis of variance determines that all of the means in the study are not equal, an appropriate test is conducted to determine which means are different. The most popular test has been Fisher's least significant difference (LSD) procedure.[9] The Scheffé[10], Tukey's Honestly Significant Difference (HSD),[11] Duncan[12] (applied to unequal sample sizes by Kramer[13]), Newman,[14] and Keuls[15] tests are also suitable for multiple comparisons of means. When the object of the test is only to measure the difference between the treatment and a control value, the Dunnett procedure is usually employed.[16] The procedures of Scheffé, Tukey, and Duncan are explained in detail with numerical examples in Federer.[16] As mentioned earlier, *in vitro* studies allow different determinations to be made on skin sections obtained from the same source. This presents a unique opportunity for *in vitro* studies to increase their ability to discriminate differences between treatments by eliminating extraneous sources of variation. The paired comparison test can be employed when the differences between individual pairs of measurements are tested for significance. There are some site differences in skin permeability in humans and animals and so the skin sections used for such comparisons should be adjacent to each other.

The comparison of multiple treatments on absorption can also benefit from testing all of the treatments on skin from the same source. If an analysis of variance is conducted and the data are blocked by skin source, the variation between animals or human donors is accounted for by the procedure and smaller differences between treatment effects can be detected. If determinations are made on the same skin section before and after some treatment, for example the measurement of absorption on a skin section when freshly obtained and again after freezing and thawing, a repeated-measures design analysis of variance can be used. The repeated-measures design is simply a special case of factorial and nested-factorial design experiments but again is used to maximize the sensitivity of the analysis of variance.

V. CONCLUSIONS

The design of a percutaneous absorption study begins with consideration of the purpose of the data. A dose or range of doses which are relevant to the goals of the study are selected. The description of the experimental protocol includes all details of the amount, area, and manner of dosing. The reporting of the results includes a material balance of the test compound showing its distribution in the receptor fluid, skin, and skin wash at the end of the study or simply a total recovery with the amount in the receptor fluid and skin reported. The percutaneous absorption of a compound is best estimated by summing the amount of the material in the receptor fluid and the amount remaining in the skin after washing. Kinetic data describing the rate at which absorption took place can be a full graphical kinetic profile of the rates of penetration or minimally an indication of initial and overall average rates during the study. Metabolic data are helpful, if available, in determining the effective dose of the biologically active form of the test compound. Interpretations and conclusions of the study must be supported by a statistical analysis of the data. The data are reported with their mean value, the number of observations used to determine that value, and a measure of the dispersion of the data such as the SD or the SEM. For comparison between means, the appropriate statistical test is to be used. *In vitro* percutaneous absorption studies are unique in that the amount of skin obtained from a single source can be abundant and exploiting this in the experimental design can add sensitivity to the statistical test.

REFERENCES

1. **Kokoski, C. J., Henry, S. H., Lin, C. S., and Ekelman, K. B.,** Methods used in safety evaluation, in *Food Additives,* Branen, A. L., Davidson, P. M., and Salminen, S., Eds., Marcel Dekker, New York, 1990, 579.
2. *Federal Register,* 51, (33992), 1, 1986.
3. **Kliber, M.,** Metabolic turnover rate: a physiological meaning of the metabolic rate per unit body weight, *J. Theoret. Biol.,* 53, 199, 1975.
4. **O'Flaherty, E. J.,** Interspecies conversion of kinetically equivalent doses, *Risk Anal.,* 9, 587, 1989.
5. International Commission on Radiological Protection, *Report of the Task Force on Reference Man.,* ICRP Publ. No. 23, Pergamon, Elmsford, New York, 1975.
6. **Dourson, M. L. and Stara, J. F.,** Regulatory history and experimental support of uncertainty (safety) factors, *Reg. Toxicol. Pharmacol.,* 3, 224, 1983.
7. **Storm, J. E., Collier, S. W., Stewart, R. F., and Bronaugh, R. L.,** Metabolism of xenobiotics during percutaneous penetration: role of absorption rate and cutaneous enzyme activity, *Fund. Appl. Toxicol.,* 15, 132, 1990.
8. **Gosset, W. S.,** The probable error of a mean, *Biometrika,* 6, 1, 1908.
9. **Fisher, R. A., Ed.,** *The Design of Experiments,* 8th Ed., Oliver and Boyd, Edinburgh, 1966.
10. **Scheffé H.,** A method for judging all contrasts in the analysis of variance, *Biometrika,* 40, 87, 1953.
11. **Tukey, J. W.,** The problem of multiple comparisons, in *Experimental Design: Procedures for the Behavioral Sciences,* Brooks/Cole, Belmont, California, 1968.
12. **Duncan, D. B.,** Multiple range and multiple-F tests, *Biometrics,* 11, 1, 1955.
13. **Kramer, C. Y.,** Extension of multiple range tests to group means with unequal numbers of replications, *Biometrics,* 12, 307, 1956.
14. **Newman, D.,** The distribution of the range of samples from a normal population in terms of an independent estimate of standard deviation, *Biometrika,* 31, 20, 1939.
15. **Keuls, M.,** The use of the studentized range in connection with the analysis of variance, *Euphytica,* 1, 112, 1952.
16. **Dunnett, C. W.,** A multiple comparison procedure for comparing several treatments with a control, *J. Am. Stat. Assoc.,* 50, 1096, 1955.
17. **Federer, W. T., Ed.,** *Experimental Design,* Macmillan, New York, 1955.

Chapter 13

THE LPP THEORY OF SKIN PENETRATION ENHANCEMENT

B. W. Barry

TABLE OF CONTENTS

I. INTRODUCTION

When scientists, including clinicians, are faced with the problem of promoting the absorption of a drug through human skin, they usually have to consider the somewhat impermeable nature of the horny layer. The major difficulty which this skin barrier imposes on most attempts at transdermal delivery is compounded by the biological variability of human skin permeability, site to site and from patient to patient. Various tactics may be employed to surmount the skin barrier. One approach is to formulate a topical product so that the medicament exerts its maximum chemical potential in the base; i.e., the drug has a maximum tendency to leave the system and partition into the skin. Supersaturated systems may even be used. At least in principle, we can try to adjust the physicochemical environment of the penetrant molecule without modifying the nature of the stratum corneum.[1] However, it is very unusual to find a truly inert vehicle component and the horny layer usually changes on contact with a vehicle. A second method of approach, which is currently exciting interest, particularly for the delivery of peptide or protein drugs, although it has its problems, is iontophoresis, the electrically enhanced transport of bioactive materials.[2-7] A third technique which has had limited success uses ultrasound (phonophoresis) to drive drug molecules, contained in a coupling or contact agent, through the skin.[8]

However, at present the most popular solution for overcoming the intrinsic resistance of the horny layer and its biological variability is to incorporate penetration enhancers (accelerants or absorption promoters) into skin products or transdermal devices. As defined in this chapter, ideally, a penetration enhancer is a chemical which exhibits the *only* characteristic that it reversibly reduces the barrier nature of the stratum corneum without the accelerant damaging any viable cells.[9] Thus, we can list the desirable attributes of the ideal penetration enhancer as follows:

1. The material should be pharmacologically inert and should possess no action of itself at receptor sites in the skin or in the body generally.
2. The material should not be toxic, irritating, or allergenic.
3. On application, the onset of penetration-enhancing action should be immediate; the duration of the effect should be predictable and suitable.
4. When the material is removed from the skin, the tissue should immediately and fully recover its normal barrier property.
5. The barrier function of the skin should reduce in one direction only, so as to promote penetration into the skin. Body fluids, electrolytes, or other endogenous materials should not be lost to the atmosphere.
6. The enhancer should be chemically and physically compatible with a wide range of drugs and pharmaceutical adjuvants.
7. The substance should be an excellent solvent for drugs.
8. The material should spread well on the skin and it should possess a suitable skin "feel".
9. The chemical should formulate into lotions, suspensions, ointments, creams, gels, aerosols, transdermal devices, and skin adhesives.
10. It should be inexpensive, odorless, tasteless, and colorless so as to be cosmetically acceptable.

Promoters may, of course, be used in conjunction with a thermodynamic control approach, iontophoresis, or ultrasound.

There is extensive literature on penetration enhancement and the number of publications appears to increase exponentially. The literature to 1981 has been considered by Barry[9] and reviews have now been brought more up-to-date.[10,11] From these sources and recent papers the reader can readily appreciate the diversity of chemicals which may act as penetration enhancers — solvents (such as water, alcohols, dimethylsulfoxide, formamide and acetamide, pyrrolidones, propylene glycol), Azone and its derivatives, surfactants (anionic, cationic and nonionic), fatty acids and alcohols, terpenes and their derivatives, alkyl sulfoxides, phosphine oxides and sugar esters and miscellaneous materials such as urea and its long chain analogs, N,N-diethyl-m-toluamide, calcium thioglycolate, and anticholinergic agents. Many of these papers do not explain *how* a particular enhancer acts in promoting the permeation of a specific drug. When arguments are so adduced, different authors often provide conflicting hypotheses, occasionally within the same paper. The variations in the techniques used, the different enhancers employed, and the alternative penetrants selected for investigation all often tend to confuse the issue rather than to clarify the subject. Therefore the approach developed in the present chapter is to propose an overall concept which explains how penetration enhancers work, known as the Lipid-Protein-Partitioning Theory, and to select just a few examples to illustrate the principles involved.

II. THE LIPID-PROTEIN-PARTITIONING (LPP) THEORY OF PENETRATION ENHANCER ACTIVITY

The LPP theory proposes that enhancers usually work by one or more of three main mechanisms.[12-14] Accelerants can alter the *L*ipid domain of the stratum corneum, they may interact with the *P*rotein elements of the tissue, or they may also increase the *P*artitioning into the horny layer of a drug, a coenhancer or water or any combination of these three. To understand the relevance of these elements of the theory it is beneficial to review briefly the penetration pathways through human skin, the structure of the intercellular domain and the molecular locations for accelerant action between the corneocytes.

A. PENETRATION PATHWAYS THROUGH HUMAN SKIN

The possible macro pathways by which a drug may penetrate human skin are usually considered to be the transepidermal route (via the stratum corneum either inter- or intracellularly), the sweat glands, and hair follicles (Figure 1A). A range of factors controls the relative importance of each route, including the physicochemical properties of the penetrant, diffusional time scale, follicle and gland densities, properties of the stratum corneum, vehicle effects, metabolism, and hydration. For most situations at steady state the stratum corneum route dominates the transport process; we will concentrate on this pathway from now on.

Penetrants traversing the intact horny layer can use two micro pathways, the transcellular or intercellular routes (Figure 1B). Views as to the nature of the horny layer change as research workers probe its structure but a helpful starting point is to model the tissue as a wall-like entity with protein filled bricks and a lipid mortar. In fact, the corneocytes are much flattened and interdigitated and they are more like distorted pizzas enmeshed in semisolid lipid bilayers. Both the cellular keratin and the acellular lipid can affect drug partitioning and diffusion, and enhancers may act at both these sites. The chemical potential of a penetrant, its partition coefficient for neighboring phases, and its diffusion coefficients within these regions are factors which control the relative importance of these two pathways.[15,16] A crucial requirement for understanding the mechanism of drug permeation through the stratum corneum, and how penetration enhancers affect it, is to develop an in-depth knowledge of the physicochemistry of the intercellular cement,[17] the molecular architecture of which is beautifully designed for its barrier role. Further detail of this lipid bilayer structure is provided below.

B. ARCHITECTURE OF THE INTERCELLULAR REGION

Despite extensive investigations undertaken over recent years, the molecular composition and organization of the regions between the cornecytes is still not completely defined. What

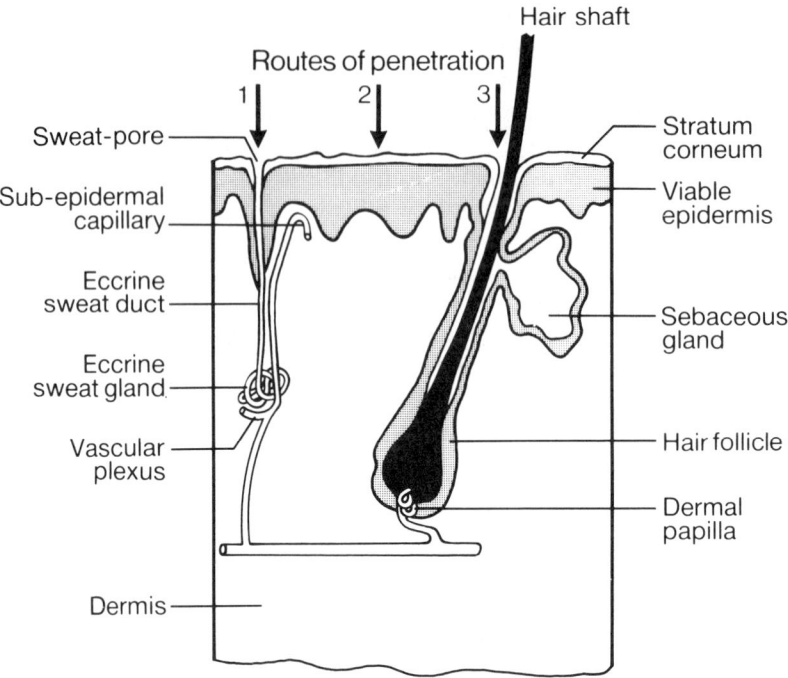

FIGURE 1A. Cross section of human skin showing possible macro routes for drug penetration across human skin — via the intact horny layer or through the hair follicles or eccrine sweat glands.

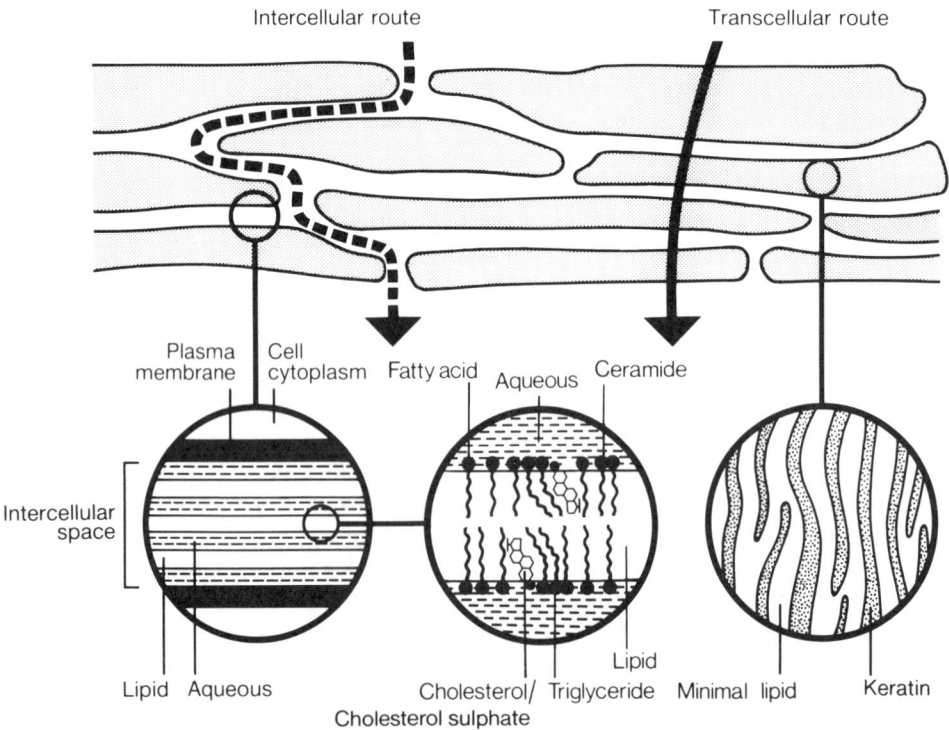

FIGURE 1B. Possible micro routes of drug permeation through human stratum corneum — intercellular or transcellular pathways; after Elias.[20]

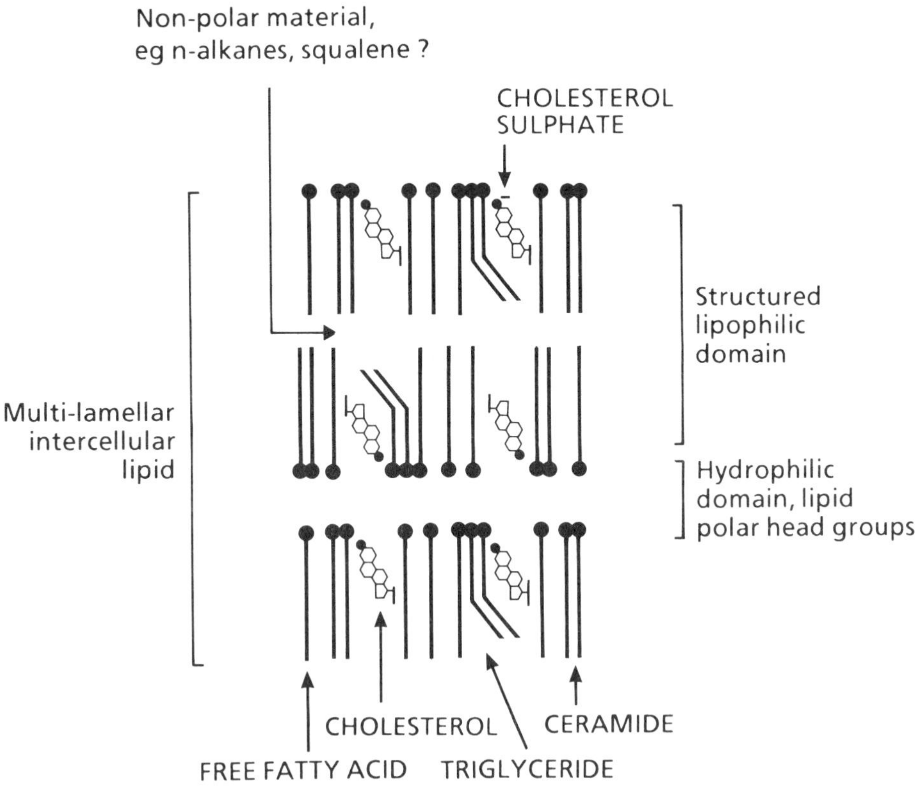

FIGURE 2. One model proposed for the structure of the lipid bilayers formed throughout the intercellular domains of human stratum corneum.

is known is that a complex mixture of essentially neutral lipids[18] forms bilayers whose structure may be idealized as in Figure 2.[19-23] Although this figure emphasizes the solid-like packing of the lipid chains, there is in fact a variety of packing geometries ranging from all trans to some gauche to disordered liquid tails. Thus, regions of the intercellular domain may contain crystalline alkyl chains (arranged as an orthorhombic perpendicular subcell), gel structures, liquid crystals, and liquid hydrocarbon chains. The various proportions of these microphases will depend on variables such as hydration level, temperature, and the presence of salts and foreign chemicals — especially in the present context, the nature and concentration of accelerants. Because the lipid bilayer tails are long compared with those in viable cell membranes and may be saturated or unsaturated, we have postulated that they interact in two main formats. Straight chains tend to produce 'solid' packing in the depth of the bilayer; chains with *cis*-double bonds or side groups inhibit tight packing. The plate-like rings of cholesterol will orientate near the polar head groups partially immobilizing this region but allowing more movement deep in the bilayer because of decreasd packing there. Ionic cholesterol sulfate may be important in stabilizing the bilayer,[19] together with carboxylic head groups. The presence of different regions in the lipid with distinct energies of interaction may account for the two major lipid peaks seen in differential scanning calorimetry and thermogavimetric analysis.[24]

The above plan of intercellular architecture can be developed further by including recent findings relating to lipid constitution. Small amounts of lipid are immobilized via covalent bonding to corneocyte surface protein[25-28] and omega-hydroxyceramides may act as inter-cellular rivets where horny cell envelopes directly contact each other.[29,30] Figure 3 illustrates these ideas, but still further complexity is possible. Proteins have been detected in the

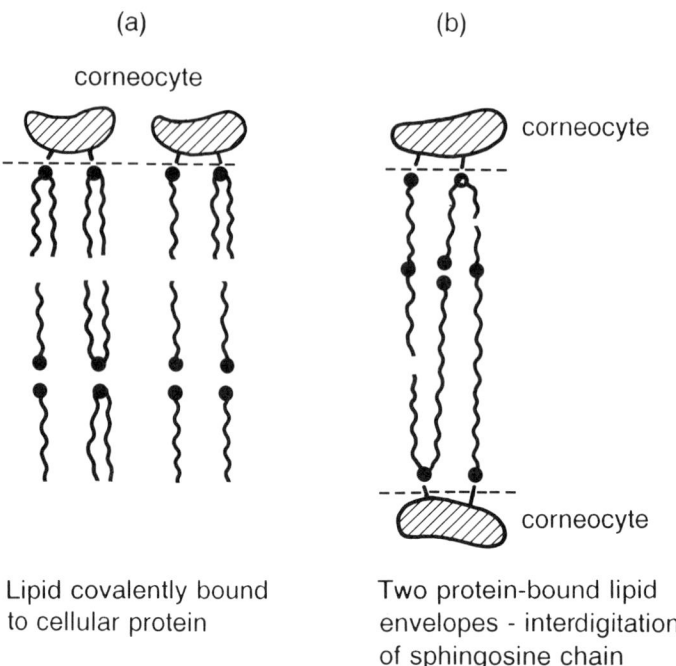

(a)   (b)

FIGURE 3.   Models of intercellular structures which include protein-bound lipid; (a) lipid covalently bound to cellular protein, (b) two protein-bound lipid envelopes with interdigitation of sphingosine chains.

intercellular domain and Figure 4 shows suggested structures, with protein associated with lipid. White and his co-workers[31] have suggested that the unit cell is unlikely to be a simple bilayer; they proposed various possibilities including two apposed asymmetric bilayers, symmetric bilayers with asymmetrically distributed protein, or a combination of the two. Our picture incorporating these modifications is also shown in Figure 4.

C. POSSIBLE SITES FOR ACCELERANT ACTION WITHIN THE INTERCELLULAR REGION

Considering the picture developed above of the structure of the intercellular domain, we can suggest three main sites at which an enhancer may modify the bilayer structure so as to promote permeation[32,33] (Figure 5).

Many accelerants have the molecular structure and charge density suitable for interacting with polar head groups of the lipid. They could then disturb hydration spheres and head group interactions and fluidize this region, thus aiding in particular the diffusion of polar (hydrophilic) penetrants (Site A). More aqueous fluid may enter the tissue, increasing interlamellar volume and thus the volume available for polar diffusion (Site B). Disruption of interfacial structure may reflect into increased disorder of lipid chains; a nonpolar penetrant will then more readily permeate through the route (Site C).

An enhancer may change the overall chemical constitution of the aqueous region and thus have a direct action on Site B. Thus, if the stratum corneum absorbs high concentrations of solvents such as ethanol, propylene glycol, or DMSO and its analogs, the resulting medium may increase the operating partition coefficient for oil soluble molecules such as the steroids (e.g., estradiol and hydrocortisone) thus enhancing their permeation.

Because of their structures, many accelerants should have a direct action at Site C, the purely lipid region. They can insert between the lipid hydrophobic tails, disorganizing the packing, and easing the diffusion of a nonpolar penetrant. This increased fluidity may produce

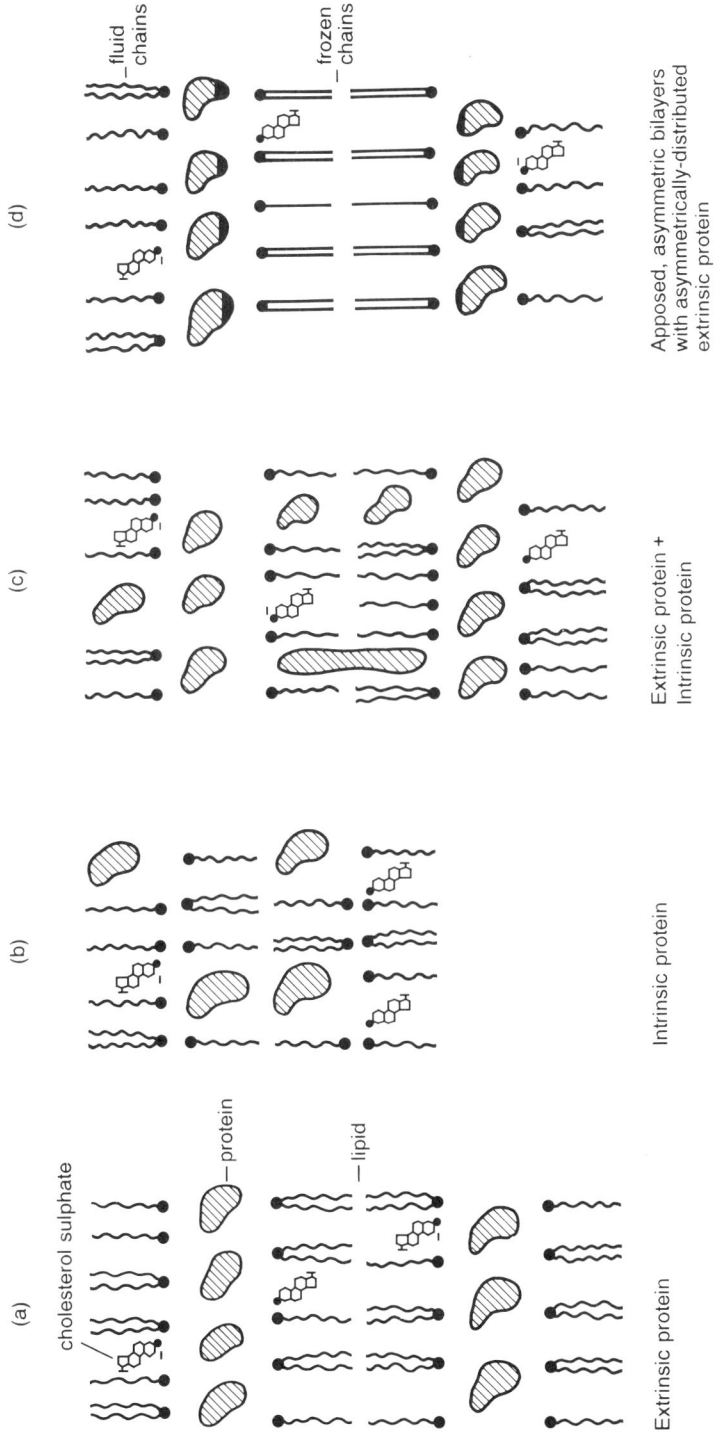

FIGURE 4. Models of intercellular structures which include protein: (a) with extrinsic protein, (b) with intrinsic protein, (c) with extrinsic and intrinsic protein, (d) apposed, asymmetric bilayers with asymmetrically distributed protein.

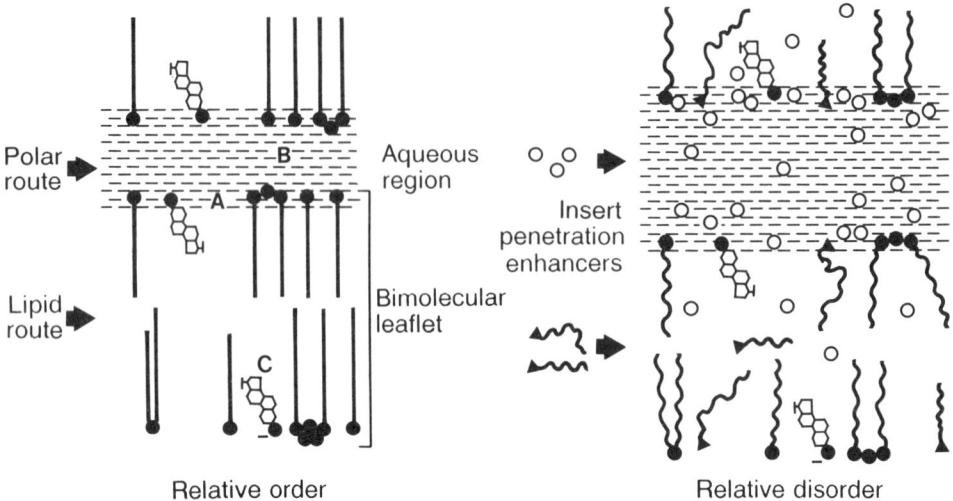

FIGURE 5.   Sites for enhancers to act in the intercellular space of the horny layer. For the enhancers, linear chains represent Azone, bent chains correspond to *cis*-unsaturated promoters, and the circles stand for small polar solvents such as DMSO and its analogs: ethanol, the pyrrolidones, and propylene glycol.

further disorder in the polar head group region and so enhance polar route penetration. Accelerants with bulky polar heads may also disrupt site A directly.

Co-solvents, such as propylene glycol and ethanol, may act synergistically to increase the horny layer concentration of enhancers such as the terpenes, Azone, and oleic acid by modifying the properties of sites A and B.[14]

D. THE INTRACELLULAR ROUTE

Although it has become customary to ignore the intracellular route nowadays, the volume fraction of the corneocytes in relation to the entire stratum corneum is large. If a drug does partition into the cell, then probably we can disregard the intracellular lipid because of its low content. The intracellular polar route would be represented by semicrystalline keratin and the most relevant enhancers acting here would include water, aprotic solvents, pyrrolidines, and surfactants. These accelerants would interact with polar head groups on the keratin (the alkyl chains of surfactants could additionally interact with hydrophobic residues), relaxing the binding forces and altering the conformations of the helices and possibly, in extreme cases, forming pores through the tissue.

III. EXPERIMENTAL TECHNIQUES FOR PROBING ENHANCER ACTIVITY

There are several major ways for investigating how some chemicals promote the permeation of drugs through the skin, including both *in vivo* and *in vitro* techniques. With suitable adjustments, essentially all the methods available for monitoring percutaneous absorption may be modified so as to gather suitable data on enhancer activity. For brevity, this chapter will review only some of these protocols which have been used in our own laboratories and which have provided the examples of enhancer activity quoted in Section IV.

A. PERMEATION THROUGH HUMAN CADAVER SKIN

Depending on the polarity of a penetrant (and thus the ease with which it is cleared by a hydrophillic receptor fluid) either dermatomed human skin samples or heat-separated epidermal sheets provide good membranes for assessing enhancer activity. Animal tissues should be avoided. We use two main protocols — the infinite dose technique (pseudosteady state diffusion) and the finite dose method (solvent-deposited film). In the first approach, fully hydrated membranes are treated with a solution of the test penetrant (often saturated aqueous radioactive 5-fluorouracil), the permeability coefficient (K_p) is determined, the 5-FU is washed off, the membrane is treated with enhancer (typically for 12 h), the enhancer is removed and the K_p is redetermined. An Enhancement Ratio (E.R.) may be calculated from:[34]

$$E.R. = \frac{K_p \text{ after application of penetration enhancer}}{K_p \text{ before application of penetration enhancer}} \tag{1}$$

The finite dose design compares the permeation of a drug delivered from a solvent-deposited film through skin both untreated and pretreated with promoters. With this approach, steady-state diffusion rates cannot easily be measured, but the amount of drug penetrating at set times can be determined. The advantages of the technique are that physiochemical conditions are allowed to change, as they would for a patient undergoing treatment, and the tissue does not hydrate excessively. The main disadvantage of the protocol is the complexity in interpreting the data at a fundamental level. However, once again we can define an enhancement ratio by:[35]

$$E.R. = \frac{\text{Amount of drug penetrating accelerant treated skin in time t}}{\text{Amount of drug penetrating untreated skin in time t}} \tag{2}$$

Over the years we have selected penetrant molecules of different polarities to assess the effects of many enhancers altering the polar and lipid routes through the horny layer, e.g., 5-FU, mannitol, estradiol, hydrocortisone, and progesterone. We used accelerants such as Azone, dimethylsulfoxide, dimethylformamide, 2-pyrrolidone, N-methyl-2-pyrrolidone, oleic acid, decylmethylsulfoxide, sodium lauryl sulfate, propylene glycol, urea analogs, several essential oils, and a wide range of terpenes.[34-40]

Both these techniques separate the enhancer from the applied formulation and thus the interpretation of the experiment is not compromised by the enhancer modifying the chemical potential of the drug in the vehicle. Only the effect of the accelerant on the membrane is monitored. Further experiments are required to check the effects of co-administration of drug and enhancer from a topical product or transdermal device in the clinical situation.

B. VASOCONSTRICTOR BIOASSAYS WITH TOPICAL STEROIDS

This bioassay, originally introduced to correlate with clinical anti-inflammatory activity, can be modified to assess the bioavailability of a topical steroid as affected by penetration enhancers.[41,42] We simply score the intensity of blanching induced on the forearm of volunteers and construct blanching curves. With a steroid at approximately constant chemical potential in a vehicle (e.g., 10% saturation) we can define a bioavailability which is similar to an enhancement ratio from:

$$\text{Bioavailability} = \frac{\text{AUC of steroid in enhancer solution}}{\text{AUC of steroid in DMI}} \tag{3}$$

AUC represents the area under the curve and DMI is dimethylisosorbide — a control solvent which is not an enhancer.

C. DIFFERENTIAL SCANNING CALORIMETRY (DSC) OF HUMAN STRATUM CORNEUM

DSC experiments, together with thermogravimetric and thermomechanical techniques, infrared spectroscopy, and X-ray diffraction, provide evidence on the structure of the stratum corneum and how enhancers modify phase transitions within the tissue.[24-27,43-46] So far, DSC has proved to be the most fruitful of these techniques in regard to elucidating enhancer mechanisms. With a typical sweep time of 10°/per minute, a hydrated sample of human horny layer in hermetically sealed pans usually produces four main peaks which we can classify as:

Endotherm T1	Lipid melting; from sebaceous lipid/fat contamination?
Endotherm T2 + T3	Lipid melting
Endotherm T4	Protein denaturation of intracellular keratin

Endotherm T1 is of no importance as regards assessing the mode of action of penetration enhancers. There is some controversy with respect to the underlying nature of T2 and T3.[24,27] These peaks could arise from the intercellular lipid melting in two stages,[24] or the bilayer melting followed by disruption of protein-lipid complexes at cell membranes,[27] or a third possibility is that the two transitions follow from heat changes to asymmetric bilayers, or a combination of all three mechanisms.[14] The important practical point is that if an enhancer modifies T2 and T3 then the process suggests a lipid interaction and we can use this fact as evidence that the accelerant is working at least in part by changing packing in the intercellular lipid domain.

IV. SOME EXAMPLES OF ENHANCER ACTIVITY AS INTERPRETED VIA THE LPP THEORY

Using the concepts outlined previously, we can attempt to rationalize how various enhancers modify the stratum corneum by referring to the LPP theory for our guidelines. The theory concentrates on three main themes — lipid modification, protein interaction, and partitioning alterations in the horny layer. The LPP concept is a useful tool in helping to focus our minds on possible modes of action of accelerants. The scheme indicates the most useful avenue of approach for further work to follow; such additional investigations are best aimed at deducing fundamental molecular mechanisms rather than authors simply concluding experimental reports with empirical observations.

Figure 6 illustrates the formulas of some common enhancers which represent the major classes of accelerants investigated up until fairly recently. Figure 7 provides the structures for a range of terpenes which we are at present studying.[40] During our discussion as to how various enhancers operate, it may be useful to refer to Figure 8 which is a composite diagram showing how typical enhancers modify the status of components in the horny layer.

A. EXAMPLES OF LIPID INTERACTION

In our experience, very many penetration enhancers exert some, if not most, of their activities by modifying the organization of the intercellular lipid. This interaction is most directly monitored by comparing DSC traces of treated stratum corneum with those from control tissue. One of the most dramatic effects is revealed when we compare an untreated, 20% hydrated stratum corneum control sample with the same specimen modified by Azone (Figure 9). Depending on the treatment conditions, the lipid endotherms T1, T2, and T3 reduce in size or disappear entirely whereas the protein peak remains essentially unaltered.[24,43,44] These traces are strong evidence that Azone operates by lipid interaction only; it is too lipophilic to enter the corneocyte and interact with the keratin. Figure 8 illustrates Azone inserting between the lipid chains and thus inhibiting their crystallization.

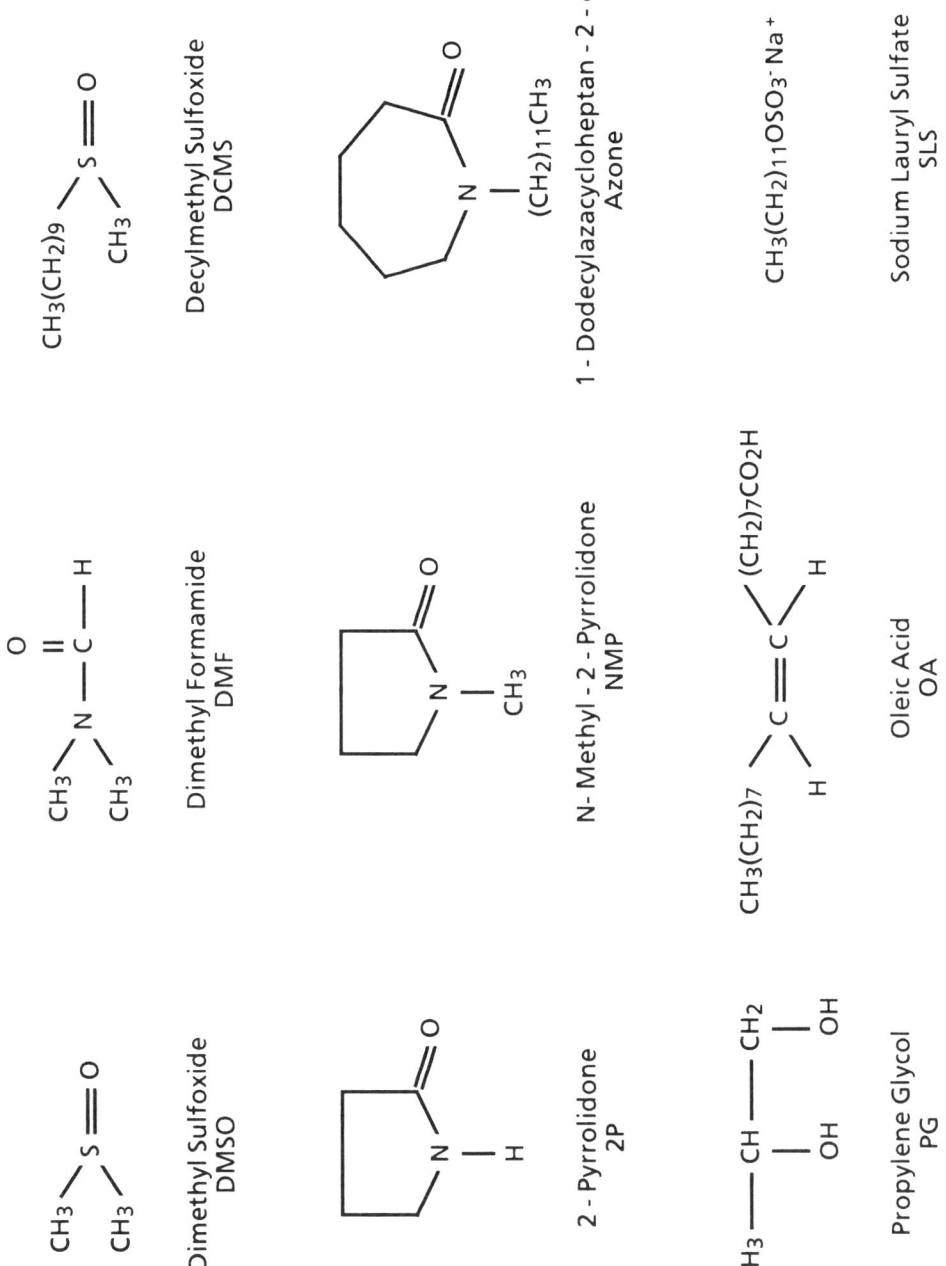

FIGURE 6.   Formulas of some common penetration enhancers.

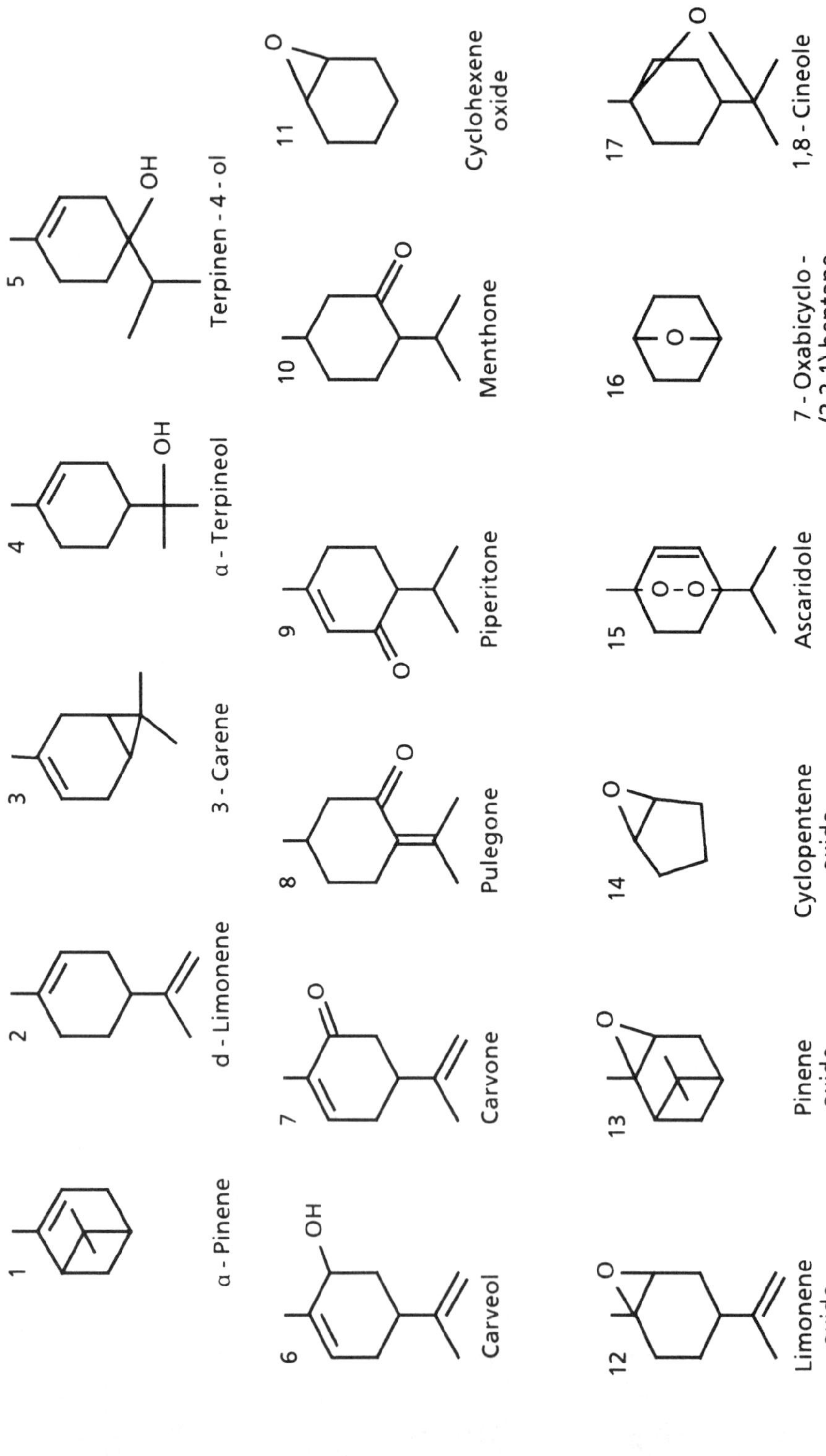

FIGURE 7. Formulas of cyclic terpene enhancers.

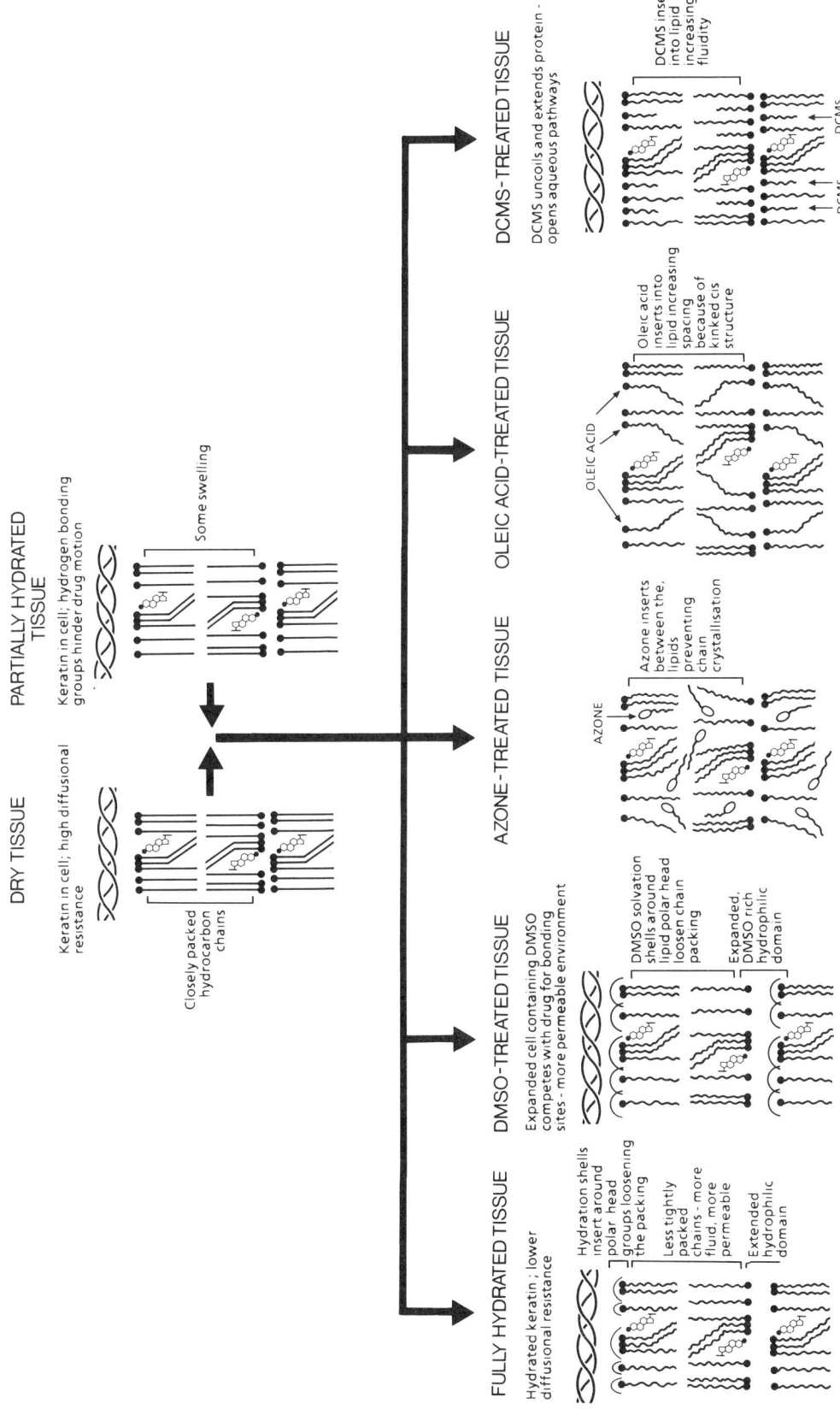

FIGURE 8. Proposed action mechanisms of skin accelerants water, dimethylsulfoxide, Azone, oleic acid, and decylmethylsulfoxide according to the LPP theory.

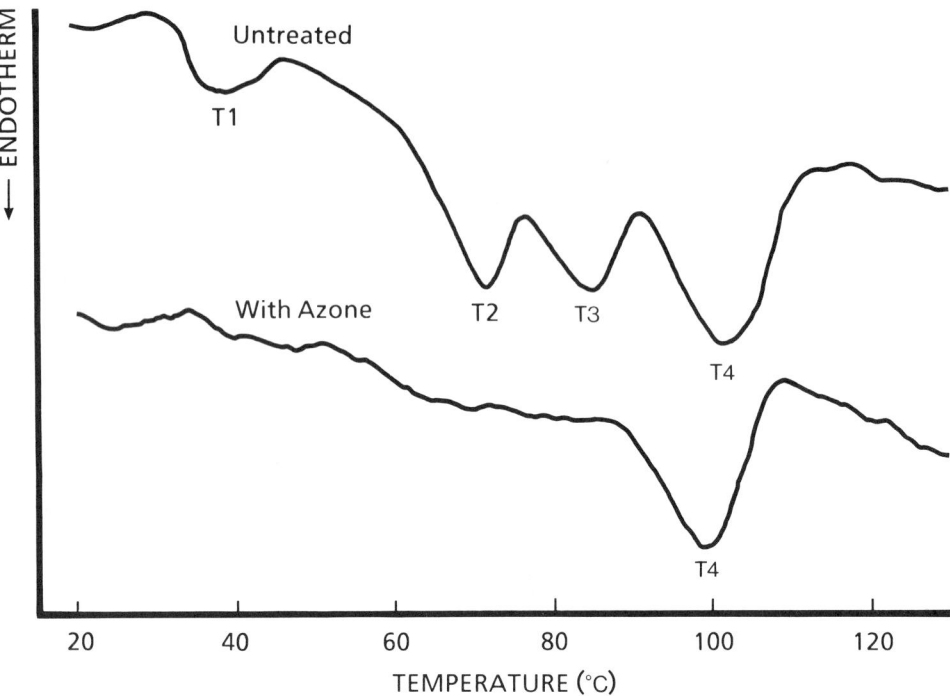

FIGURE 9.   Differential scanning calorimetry of human horny layer and the effect of Azone.

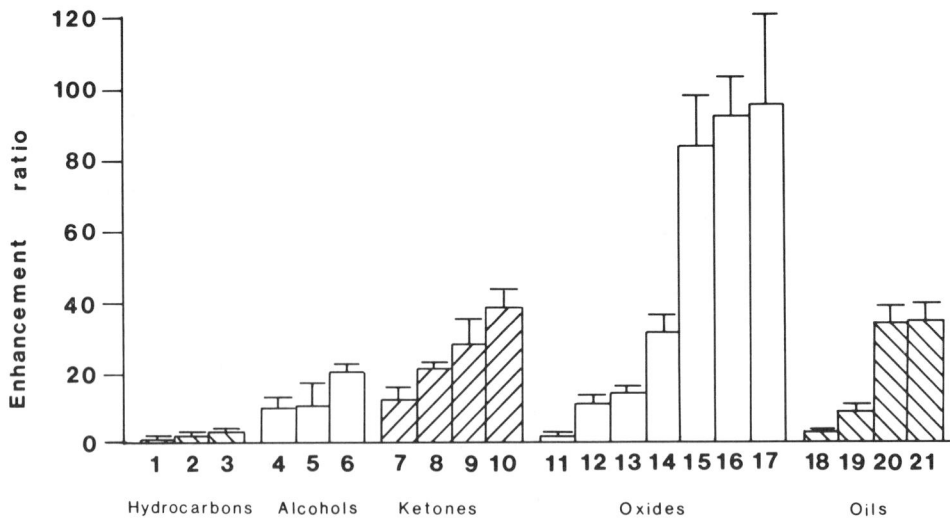

FIGURE 10.   Enhancement ratios for terpenes. Numbers correspond to those in Figure 7 with additions for oils: 18, anise; 19, ylang-ylang; 20, chenopodium; 21, eucalyptus.

Cyclic terpenes provide a second example of a class of accelerants which reduces the barrier resistance of the stratum corneum by interacting with the lipid.[39,40] Using 5-FU as a test penetrant, we obtained enhancement ratios via the infinite dose technique (Equation 1) and Figure 10 illustrates the data. The terpenes were classified into hydrocarbons, alcohols, ketones, oxides or commercial oils. The hydrocarbons were poor enhancers whereas the oxides covered a wide range of activity. Using permeation, partitioning, and DSC studies

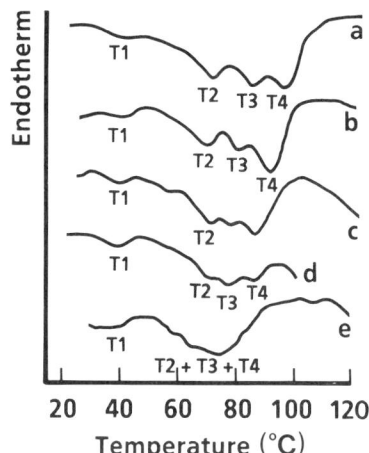

FIGURE 11. DSC thermograms of stratum corneum: (a) control; (b) 0.1%; (c) 0.2%; (d) 0.5% and (e) 1% aqueous sodium lauryl sulfate.

it was concluded that lipid modification was the main mode of action. The high log P values (octanol/water) implied that there would be little terpene-keratin interaction within the corneocyte so that possible protein modification available as an option according to the LPP theory could be discarded.

B. EXAMPLES OF PROTEIN MODIFICATION

Surface active molecules, particularly ionic species, tend to interact well with the keratin in the corneocyte. In one of our DSC studies, stratum corneum samples were treated with 0.1, 0.2, 0.5, and 1.0% aqueous solutions of sodium lauryl sulfate and representative thermograms are illustrated in Figure 11.[24] As the concentration of the sulfate increased, T3 and T4 temperatures decreased and eventually the peaks merged with T2 to produce one extended broad endotherm. The surfactant swells and disrupts the entire membrane—we conclude that both lipid and protein structures are affected. The sulfate may both uncoil and extend keratin (hence reducing T4 sizes) and also insert in the lipid structure.

Similarly, decylmethylsulfoxide provided DSC traces which indicated some interaction with keratin. The sulfoxide lowered T2, T3, and T4, this action supporting the idea that the compound enhances permeation by altering protein conformation and forming water channels within the cell.[47] However, once again the major effect appears to be on the lipid structure. The sulfoxide may act by inserting its rather short chain into the bilayer structure, as depicted in Figure 8, increasing spacing in the lipid packing and consequently reducing the rigidity of the environment.[45]

Small polar enhancers such as dimethylsulfoxide, dimethylformamide, and the pyrrolidones also affect the protein helices but the changes in the DSC traces are complicated by lipid interactions. The solvents lower the temperature transition of T4 and broaden the peak; the effects are best seen at low concentrations when lipid interactions are not so prominent. Alternatively, treatment of delipidized stratum corneum illustrates the effect well.[24] Figure 8 illustrates relevant molecular interactions.

C. EXAMPLES OF PARTITIONING PROMOTION

The third component of the LPP concept relates to partitioning effects when the theory proposes that one action of a penetration enhancer may be to increase the concentration of a penetrant, or a co-enhancer, in the stratum corneum. The increased level of the drug in the membrane would then promote permeation by raising its concentration gradient and thus increasing the flux. A raised level of co-enhancer may increase further the concentration of the drug in the membrane, or reduce the barrier resistance of the tissue by lipid or protein interaction, or achieve both modifications.

Small polar promoters such as DMSO and its analogs, the pyrrolidones, ethanol, and propylene glycol, may accumulate in the horny layer to such an extent that they change its solubility characteristics so increasing the partitioning of a co-enhancer and/or a drug. We suggest that the synergy often shown by propylene glycol combined with many accelerants such as Azone, long chain alcohols and acids, and terpenes may depend in large part on a mechanism by which the glycol increases partitioning of the main promoter into the stratum corneum.

An instructive example of propylene glycol acting as a penetration enhancer by increasing the partitioning into the skin of co-enhancers comes from work with urea analogs.[38] Urea and three of its analogs (Figure 12) were assessed as skin penetration enhancers for the model penetrant 5-FU using the enhancement ratio technique (Equation 1). So as to keep the chemical potentials of the analogs constant, each was applied to the skin as a saturated solution in dimethylisosorbide, light liquid paraffin, or propylene glycol. It was found that urea and the vehicles alone, including propylene glycol, were ineffective as enhancers. The urea analogs behaved similarly at saturation in any one vehicle, suggesting that their intrinsic activities as accelerants were comparable. However, the most obvious effect was that the analogs only markedly enhanced drug penetration (sixfold) when delivered from propylene glycol (Figure 13). Presumably, the glycol promotes partitioning of the lipophilic ureas into the skin where they disrupt lipid packing and increase the diffusivity of the 5-FU.

Cyclic terpenes as examples of enhancers which interact strongly with bilayer lipids is considered in Section IV.A. Terpenes also provide good examples of synergy when delivered to the skin from propylene glycol vehicles.[37] Figure 14 compares the enhancement ratios (Equation 1) of 5-FU derived by treating human skin either with neat terpenes, 0.5 *M* terpenes in propylene glycol or saturated solutions (or 90% solutions for miscible terpenes) in the glycol. *If* the propylene glycol was having no effect, and under ideal conditions, we would expect each neat terpene to produce a maximum effect as it is presented to the skin at its greatest chemical potential. On dilution to 0.5 *M* in propylene glycol we would predict a fall in E.R., yet for carvone and pulegone this parameter increased. Saturated solutions (or approximations to saturation for miscible terpenes), being at the same thermodynamic activity as the liquid terpene, should provide E.R.s equal to pure terpenes. However, these propylene glycol/terpene mixtures increased the E.R.s up to fourfold for carveol, carvone, pulegone, and 1,8-cineole even though propylene glycol alone had a negligible promoter effect. We explain these results mainly on the basis that propylene glycol is promoting the passage of the terpenes into the stratum corneum where they exert their accelerant activity.

V. SUMMARY OF THE MAIN ASPECTS OF THE LPP THEORY

The theory assumes the following:

1. That at steady state most molecules permeate human skin across the intact stratum corneum and shunt route penetration is negligible. (The concepts would also apply to that component of penetrant mass using the follicular route and passing through the horny layer of the follicle).
2. The rate limiting step in the percutaneous absorption process lies in permeation across the stratum corneum. Thus, effects within the vehicle (dissolution of crystals, diffusion, evaporation, dilution by transepidermal water, etc.) are not rate-determining, the horny layer is essentially intact and clearance into the viable tissues and blood is fast and unaffected by the enhancer.

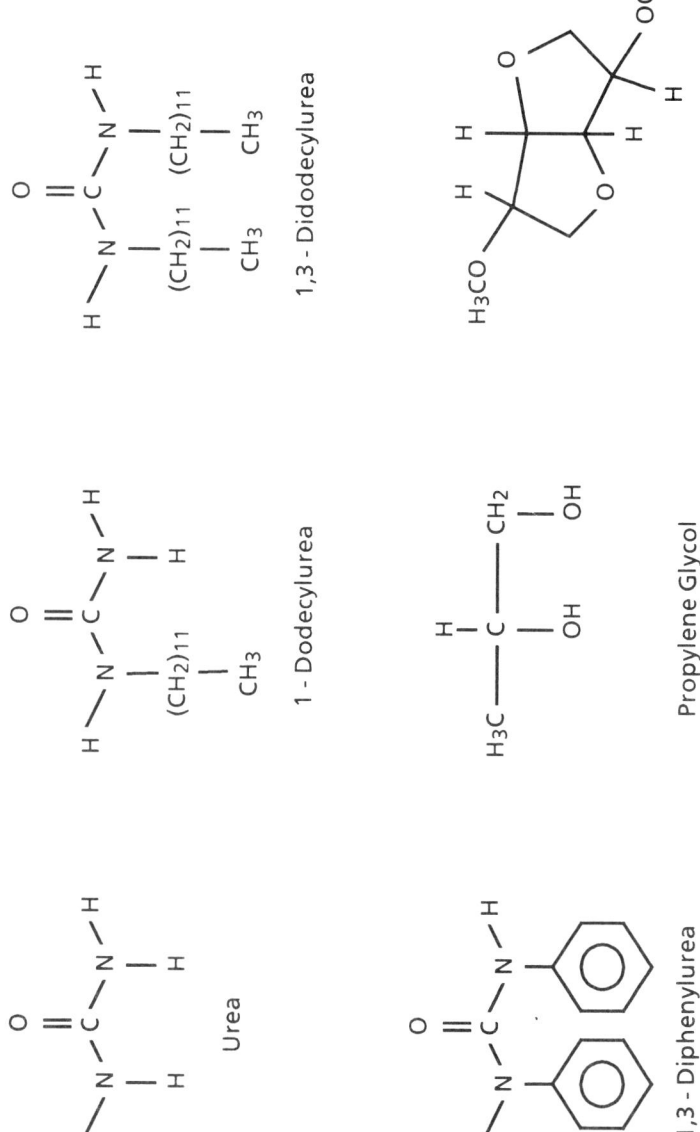

FIGURE 12. Structural formulas of urea analogs and vehicles.

UREA ANALOGS - PENETRATION ENHANCER

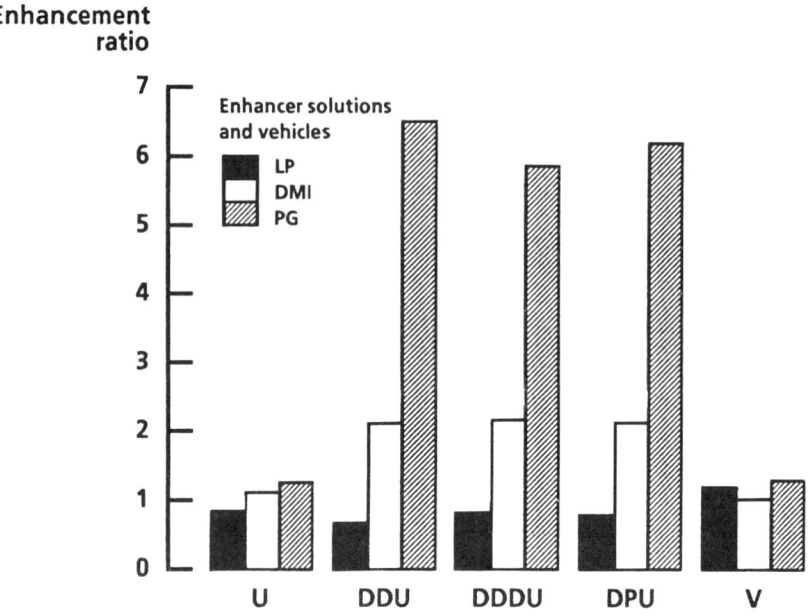

FIGURE 13.   Enhancement ratios for 5-fluorouracil after stratum corneal treatment with saturated solutions of urea analogs in different vehicles. Key: U, urea; DDU, 1-Dodecylurea; DDDU, 1,3-Didodecylurea; DPU, 1,3-Diphenylurea; V, vehicle alone.

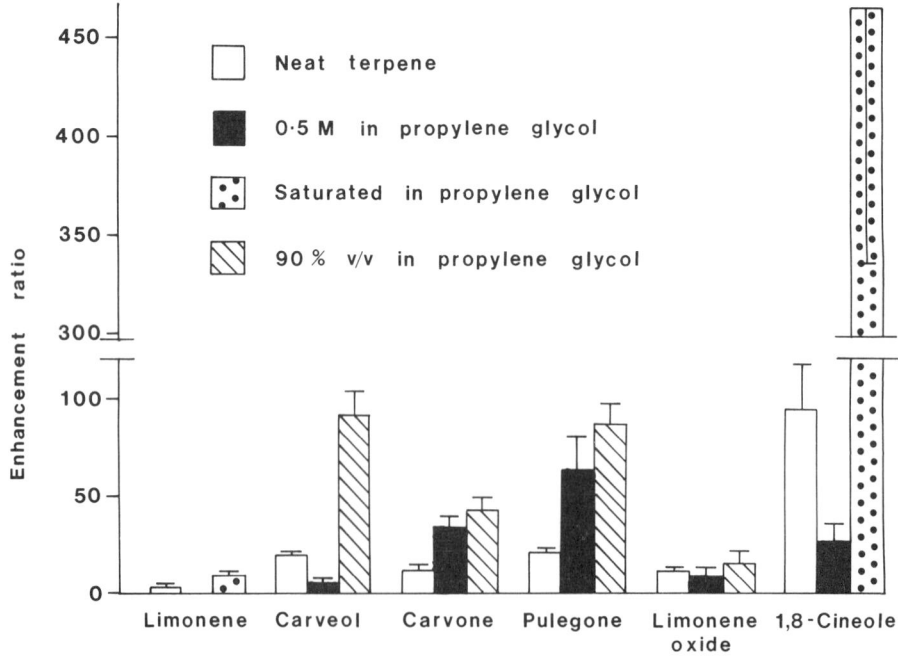

FIGURE 14.   Enhancement ratios of 5-fluorouracil after horny layer treatment with neat terpenes, 0.5 *M* terpenes in propylene glycol and saturated or 90% solutions in propylene glycol.

3. Enhancer effects are essentially reversible. Thus, the theory does not consider, for example, the corrosive effects of strong acids and alkalies, osmotic shock produced by asymmetric concentrations of materials like DMSO, and lipid extraction by solvents such as chloroform/methanol.

4. Increases in permeation by maximizing drug chemical potential in the vehicle, by producing supersaturation in the vehicle, and by iontophoresis, ultrasound or heat are all excluded.

The main conclusions of the theory regarding the mechanisms whereby enhancers which we have studied reduce the barrier function of the skin are

1. Most enhancers so far investigated appear to disturb intercellular lipid packing, increasing fluidity.
2. Many accelerants interact with intracellular protein.
3. A combination of processes 1 and 2 leads to effective enhancers.
4. Many small polar accelerants have the correct physicochemical properties to accumulate in the horny layer, altering its solubility properties, and thus promoting the partitioning of a drug, a co-enhancer, water or a combination of two or three of these.
5. The permeability of the horny layer increases with raised water content and any treatment which promotes hydration usually facilitates drug absorption.

ACKNOWLEDGMENTS

The author appreciates the collaboration of Dr. M. Goodman and A. C. Williams who generated much of the experimental data considered in this chapter.

REFERENCES

1. **Barry, B. W.,** Optimizing percutaneous absorption, in *Percutaneous Absorption, 2nd ed.,* Bronaugh, R. L. and Maibach, H. I., Eds., Marcel Dekker, New York, 1989, 531.
2. **Sloan, J. B. and Soltani, K.,** Iontophoresis in dermatology, *J. Amer. Acad. Dermatol.,* 15, 671, 1986.
3. **Tyle, P.,** Iontophoretic devices for drug delivery, *Pharm. Res.,* 3, 318, 1986.
4. **Bellantone, N. H., Rim, S., Francoeur, M. L., and Rasadi, B.,** Enhanced percutaneous absorption via ionotophoresis. I. Evaluation of an *in vitro* system and transport of model compounds, *Int. J. Pharm.,* 30, 63, 1986.
5. **Chien, Y. W., Siddiqui, O., Shi, W.-M., Lelawongs, P., and Liu, J.-C.,** Direct current iontophoretic transdermal delivery of peptide and protein drugs, *J. Pharm. Sci.,* 78, 376, 1989.
6. **Banga, A. K. and Chien, Y. W.,** Iontophoretic delivery of drugs: fundamentals, developments and biomedical applications, *J. Controlled Release,* 7, 1, 1988.
7. **Burnett, R.,** Iontophoresis, in *Transdermal Drug Delivery,* Hadgraft, J. and Guy, R. H., Eds., Marcel Dekker, New York, 1989, 247.
8. **Tyle, P. and Agrawala, P.,** Drug delivery via phonophoresis, *Pharm. Res.,* 6, 355, 1989.
9. **Barry, B. W.,** *Dermatological Formulations. Percutaneous Absorption,* Marcel Dekker, New York, 1983, 160.
10. **Woodford, R. and Barry, B. W.,** Penetration enhancers and the percutaneous absorption of drugs: an update, *J. Toxicol. Cut. & Ocular Toxicol.,* 5, 165, 1986.
11. **Walters, K. A.,** Penetration enhancers and their use in transdermal therapeutic systems, in *Transdermal Drug Delivery,* Hadgraft, J. and Guy, R. H., Eds., Marcel Dekker, New York, 1989, 197.
12. **Barry, B. W.,** Action of skin penetration enhancers—the lipid-protein-partitioning theory, *Int. J. Cosmet. Sci.,* 10, 281, 1988.
13. **Goodman, M. and Barry, B. W.,** Lipid-protein-partitioning (LPP) theory of skin enhancer activity. Finite dose technique, *Int. J. Pharm.,* 57, 29, 1989.

14. **Barry, B. W.,** Lipid-protein-partitioning theory of skin penetration enhancement, *J. Controlled Release,* in press, 1991.
15. **Barry, B. W.,** Mode of action of penetration enhancers in human skin, *J. Controlled Release,* 6, 85, 1987.
16. **Barry, B. W.,** Penetration enhancers: mode of action in human skin, *Pharmacol. Skin,* 1, 121, 1987.
17. **Williams, M. L. and Elias, P. M.,** The extracellular matrix of stratum corneum: role of lipids in normal and pathological function, *CRC. Crit. Rev. Therp. Drug Carrier Syst.,* 3, 95, 1987.
18. **Lampe, M. A., Williams, M. L., and Elias, P. M.,** Human epidermal lipids: characterization and modulations during differentiation, *J. Lipid Res.,* 24, 131, 1983.
19. **Williams, M. L. and Elias, P. M.,** The extracellular matrix of stratum corneum: role of lipids in normal and pathological function, *CRC Crit. Rev. Ther. Drug Carrier Syst.,* 3, 95, 1987.
20. **Elias, P. M.,** Epidermal lipids, membranes, and keratinization, *Int. J. Dermatol.,* 20, 1, 1981.
21. **Lampe, M. A., Burlingame, A. L., Whitney, J. A., Williams, M. L., Brown, B. E., Roitman, E., and Elias, P. M.,** Human stratum corneum lipids: characterization and regional variations, *J. Lipid Res.,* 24, 120, 1983.
22. **Imokawa, G., Akasaki, S., Hattori, M., and Yoshizuka, N.,** Selective recovery of deranged water-holding properties by stratum corneum lipids, *J. Invest. Dermatol.,* 87, 758, 1986.
23. **Cox, P. and Squier, C. A.,** Variations in lipids in different layers of porcine epidermis, *J. Invest. Dermatol.,* 87, 741, 1986.
24. **Goodman, M. and Barry, B. W.,** Action of penetration enhancers on human stratum corneum as assessed by differential scanning calorimetry, in *Percutaneous Absorption, 2nd ed.,* Bronaugh, R. L. and Maibach, H. I., Eds., Marcel Dekker, New York, 1989, 567.
25. **Golden, G. M., Guzek, D. B., Harris, R. R., McKie, J. E., and Potts, R. O.,** Lipid thermotropic transitions in human statum corneum, *J. Invest. Dermatol.,* 86, 255, 1986.
26. **Knutson, K., Potts, R. O., Guzek, D. B., Golden, G. M., McKie, J. E., Lambert, W. J., and Higuchi, W. I.,** Macro- and molecular physical-chemical considerations in understanding drug transport in the stratum corneum, in *Advances in Drug Delivery Systems,* Anderson, J. M. and Kim, S. W., Eds., Elsevier, Amsterdam, 1986, 67.
27. **Potts, R. O.,** Physical characterization of the stratum corneum: the relationship of mechanical and barrier properties to lipid and protein structure, in *Transdermal Drug Delivery,* Hadgraft, J. and Guy, R. H., Eds., Marcel Dekker, New York, 1989, 23.
28. **Wertz, P. W. and Downing, D. T.,** Stratum corneum: biological and biochemical considerations, in *Transdermal Drug Delivery,* Hadgraft, J. and Guy, R. H., Eds., Marcel Dekker, New York, 1989, 1.
29. **Wertz, P. W., Madison, K. C., and Downing, D. T.,** Covalently bound lipids of human stratum corneum, *J. Invest. Dermatol.,* 92, 109, 1989.
30. **Wertz, P. W., Swarzendruber, D. C., Kitko, D. C., Madison, K. C., and Downing, D. T.,** The role of the corneocyte lipid envelopes in cohesion of the stratum corneum, *J. Invest. Dermatol.,* 93, 169, 1989.
31. **White, S. H., Mirejovsky, D., and King, G. I.,** Structure of lamellar lipid domains and corneocyte envelopes of murine stratum corneum. An X-ray diffraction study, *Biochemistry,* 27, 3725, 1988.
32. **Barry, B. W.,** Penetration enhancers; mode of action in human skin, *Pharmacol. Skin,* 1, 121, 1987.
33. **Barry, B. W.,** Action of skin penetration enhancers—the lipid-protein-partitioning theory, *Int. J. Cosmet. Sci.,* 10, 281, 1988.
34. **Goodman, M. and Barry, B. W.,** Action of penetration enhancers on human skin as assessed by the permeation of model drugs 5-fluorouracil and estradiol: infinite dose technique, *J. Invest. Dermatol.,* 91, 323, 1988.
35. **Goodman, M. and Barry, B. W.,** Lipid-protein-partitioning (LPP) theory of skin enhancer activity: finite dose technique, *Int. J. Pharm.,* 57, 29, 1989.
36. **Barry, B. W. and Bennett, S. L.,** Effect of penetration enhancers on the permeation of mannitol, hydrocortisone and progesterone through human skin, *J. Pharm. Pharmacol.,* 39, 535, 1987.
37. **Barry, B. W. and Williams, A. C.,** Human skin penetration enhancement: the synergy of propylene glycol with terpenes, *16th Int. Symp. Controlled Release Bioactive Mater.,* Pearlman, R. and Miller, J. A., Eds., Controlled Release Society, Chicago, U.S.A., 1989, 33.
38. **Williams, A. C. and Barry, B. W.,** Urea analogues in propylene glycol as penetration enhancers in human skin, *Int. J. Pharm.,* 56, 43, 1989.
39. **Williams, A. C. and Barry, B. W.,** Essential oils as novel skin penetration enhancers, *Int. J. Pharm.,* 57, R7, 1989.
40. **Williams, A. C. and Barry, B. W.,** Terpenes and the lipid-protein-partitioning theory of skin penetration enhancement, *Pharm. Res.,* 8, 17, 1991.
41. **Barry, B. W., Southwell, D., and Woodford, R.,** Optimization of bioavailability of topical steroids: penetration enhancers under occlusion, *J. Invest. Dermatol.,* 82, 49, 1984.
42. **Bennett, S. L., Barry, B. W., and Woodford, R.,** Optimization of bioavailability of topical steroids: non-occluded penetration enhancers under thermodynamic control, *J. Pharm. Pharmacol.,* 37, 298, 1985.

43. **Goodman, M. and Barry, B. W.,** Differential scanning calorimetry (DSC) of human stratum corneum: effect of Azone, *J. Pharm. Pharmaol.,* 37, Suppl. 80P, 1985.

44. **Goodman, M. and Barry, B. W.,** Differential scanning calorimetry of human stratum corneum: effect of penetration enhancers Azone and DMSO, *Anal. Proc.,* 26, 397, 1986.

45. **Goodman, M. and Barry, B. W.,** Action of skin penetration enhancers Azone, oleic acid, and decyl-methylsulphoxide; permeation and differential scanning calorimetry (DSC) studies, *J. Pharm. Pharmacol.,* 38(Suppl.), 71P, 1986.

46. **Van Duzee, B. F.,** Thermal analysis of human stratum corneum, *J. Invest. Dermatol.,* 65, 404, 1975.

47. **Cooper, E. R.,** Effect of decylmethyl sulfoxide on skin penetration, in *Solution Behavior of Surfactants: Theoretical and Applied Aspects,* Mittal, K. L. and Fendler, E. J., Eds., Plenum Press, New York, 1982, 1505.

Chapter 14

IN VITRO PERCUTANEOUS ABSORPTION OF ALACHLOR AND ATRAZINE: EFFECT OF FORMULATION DILUTION WITH WATER AND ABSORPTION INTERACTIONS

Ronald C. Wester, Daniel A. W. Bucks, and Howard I. Maibach

TABLE OF CONTENTS

I. INTRODUCTION

Skin absorption of alachlor is of interest because of its wide use. Label precautions advise the use of protective equipment and prompt washing if on skin; however, accidents and mishandling result in skin exposure. Alachlor, a substituted acetanilide [$C_{14}H_2ONClO_2$ (2-chloro-2'-6'-diethyl-N-(methoxymethyl) acetanilide)] is a pre-emergent herbicide. The structure is shown in Figure 1. The active ingredient is 45% of the formulation, the major inactive ingredient being a hydrocarbon solvent. The formulation is used on corn, soybeans, peanuts, dry beans, grain sorghum, lima beans, and other listed crops. Alachlor also is used commercially as a mixture with atrazine. This chapter addresses the properties of alachlor regarding penetration, binding, and removal from skin using the commercial formulation. The skin absorption of atrazine and interactions with alachlor are also determined.

II. METHODS

Percutaneous absorption of ^{14}C-labeled alachlor or ^{14}C-labeled atrazine through human skin was determined using flow-through design glass penetration cells (LG-1084-LPC, Laboratory Glass Apparatus Inc., Berkeley, CA) and radiotracer methodology. In the first study, three dilutions of [14C] alachlor (Lasso® formulation, Monsanto Co., St. Louis, MO) with distilled water were utilized: 1:20 (v/v), 1:40 (v/v), and 1:80 (v/v). In the second study, the skin absorption of [14C]-alachlor or [14C]-atrazine as Lariat® formulation (Monsanto Co., St. Louis, MO) diluted 1:9 or 1:64 with water was determined with and without the presence of each other.

Human thigh skin was obtained at autopsy from various donors and dermatomed to a thickness of 0.38 mM. The dermatomed skin, stored in Eagles minimum essential cell culture medium at 4°C, was used within 5 d. Circular pieces of epidermis were clamped between the two ground glass portions of the penetration cells. Top sections of the cells were open to the environment and allowed application of each alachlor dilution directly to the exposed skin surface area of 5.7 cm². The receptor volume of each cell was 3 ml. The water-jacketed cells were maintained at 37°C using a recirculating constant temperature water bath. Since alachlor has a large log P octanol/water partition coefficient of 3.52[1] and low solubility in saline, human plasma (Irwin Memorial Blood Bank, San Francisco, CA) was used as the receptor fluid.[2] Plasma, pumped through the receptor chambers of the cells at 5 ml/h, was collected in 5 ml aliquots directly into scintillation vials. Plasma in the receptor chambers was continually mixed by magnetic stir bars.

For each dilution, 4 cells were dosed with 0.5 ml or 1.0 ml. As such, there was no "insoluble" alachlor deposited on the skin surface. After 8 h, the skin surface was gently washed three times with distilled water to remove material merely adsorbed on the surface. Other exposure times were 30 min. and 4 h. for 0.5 ml doses of each dilution and 1 h and 4 h for 1.0 ml doses of each dilution. The plasma receptor fluid, washes, and skin were analyzed for amount of rabiolabel by liquid scintillation counting.

The binding behavior of ^{14}C alachlor from the above dilutions to powdered stratum corneum was determined.[3] Powdered stratum corneum was prepared as follows: callus was cut into fine pieces with scissors and then pulverized in a mortar and pestle containing dry ice. Particles of stratum corneum that would pass through a 48 mesh sieve but would be retained by an 80 mesh sieve were used (180 to 300 μm). In a plastic microcentrifuge tube, 1.2 mg of powdered stratum corneum was mixed with 1.2 ml of solution by vortexing. The durations of contact were as follows: 1:20 dilution (30 min, 2 and 4 h); 1:40 dilution (30 min, 2, 4, and 8 h); and 1:80 dilution (1, 4, and 8 h). After a given contact time, the mixture was separated by centrifugation and the supernate removed. The stratum corneum pellet was resuspended 3 times in distilled water to remove material adsorbed on the surface. Four

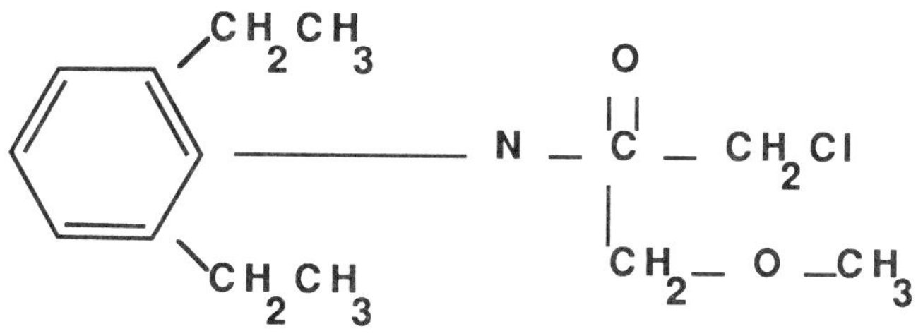

FIGURE 1.   Structure of alachlor. (From Bucks, D. A. W., Wester, R. C., Mobayen, M. M., Yang, D., Maibach, H. I., and Coleman, D. E., *Toxicol. Appl. Pharmacol.*, 100, 417, 1989. With permission.)

TABLE 1A
Newman-Keuls Multiple Range Test of the % of Applied Alachlor in the Receptor Solution

	0.5 ml 1:20	1 ml 1:20	0.5 ml 1:40	1 ml 1:40	0.5 ml 1:80	1 ml 1:80
0.5 ml 1:20		N.S.	<0.01	<0.01	<0.01	<0.01
1 ml 1:20			<0.01	<0.01	<0.01	<0.01
0.5 ml 1:40				N.S.	<0.01	<0.01
1 ml 1:40					<0.01	<0.01
0.5 ml 1:80						N.S.
1 ml 1:80						

Note: *In vitro* percutaneous absorption of alachlor (in the Lasso formulation) from three dilutions. Two volumes (0.5 and 1.0 cc) were employed. This table summarizes the statistical analysis of the cumulative % of the applied dose collected in the receptor solution (human plasma) after 8 h. Note the significant enhancement in % dose absorbed afforded by increasing dilution and no effect on increased volume of application. N.S. represents an alpha value > 0.05, i.e., no significant difference between the means.

tubes were prepared for each test. Using the methodology above and the 1:20 dilution, the ability to decontaminate powdered stratum corneum was determined by substitution of the second and third distilled water rinses with either a 10% or 50% soap and water solution (Ivory® liquid soap, Proctor and Gamble Co., Cincinnati, OH) following a 0.5-h exposure period.

III. RESULTS

The *in vitro* percutaneous absorption of diluted ^{14}C-alachlor was assessed by the rate and extent of appearance of radiotracer in the receptor fluid (human plasma).[4] In Figure 2, the percent of applied dose per hour collected in the receptor fluid vs. time is plotted for the three dilutions. After the peak rate of absorption was observed at 3 to 5 h, the skin penetration of alachlor decreased rapidly. At the peak time period, the maximal rate of absorption (percent of applied dose per hour) for each alachlor dilution was with the 0.5 ml dose: (1) for 1:20, 0.19%/h; (2) for 1:40, 0.61 $\mu g/cm^2/h$; and (3) for 1:80, 0.96%/h. Whereas the maximal flux for each alachlor dilution was with the 1.0 ml dose; (1) for 1:20, 4.2 $\mu g/cm^2/h$; (2) for 1:40, 7.6 $\mu g/cm^2/h$; (3) for 1:80, 9.6 $\mu g/cm^2/h$. The results of the Newman-Keuls multiple range test results indicate significant increases ($p < 0.01$) in mass absorbed between: (1) 1.0 ml of the 1:40 dilution and 0.5 ml of the 1:20 dilution and (2) 1.0 ml of the 1:80 and 0.5 ml of the 1:40 dilution (Table 1). Note that the same mass of alachlor was applied to the skin in the two later comparisons above.

TABLE 1B
Newman-Keuls Multiple Range Test of the Mass of
Alachlor in the Receptor Solution

	0.5 ml 1:20	1 ml 1:20	0.5 ml 1:40	1 ml 1:40	0.5 ml 1:80	1 ml 1:80
0.5 ml 1:20		N.S.	N.S.	<0.01	<0.01	<0.01
1 ml 1:20			N.S.	<0.01	N.S.	<0.01
0.5 ml 1:40				N.S.	N.S.	<0.01
1 ml 1:40					<0.01	N.S.
0.5 ml 1:80						<0.01
1 ml 1:80						

Note: *In vitro* percutaneous absorption of alachlor (in the Lasso formulation) from three dilutions. Two volumes (0.5 and 1.0 cc) were employed. This table summarizes the statistical analysis of the comparison of the cumulative mass of the applied dose collected in the receptor solution (human plasma) after 8 h. Note the significant enhancement in mass absorption afforded by increasing dilution. N.S. represents an alpha value > 0.05, i.e., no significant difference between the means.

From Bucks, D. A. W., Wester, R. C., Mobayen, M. M., Yang, D., Maibach, H. I., and Coleman, D. E., *Toxicol. Appl. Pharmacol.,* 100, 417, 1989. With permission.

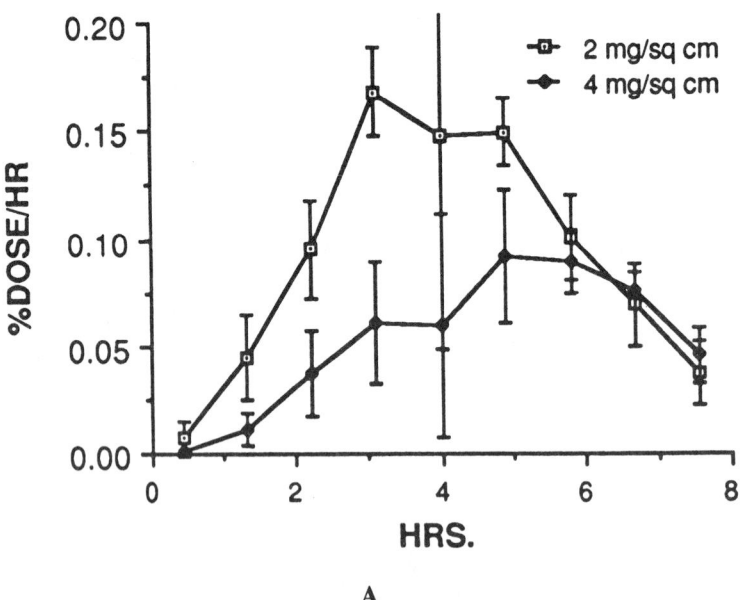

FIGURE 2A—C. Average (± SD) percent of the applied dose collected in the plasma receptor solution per hour (N = 4). Values are plotted at the midpoint of collection interval. (From Bucks, D. A. W., Wester, R. C., Mobayen, M. M., Yang, D., Maibach, H. I., and Coleman, D. E., *Toxicol. Appl. Pharmacol.,* 100, 417, 1989. With permission.)

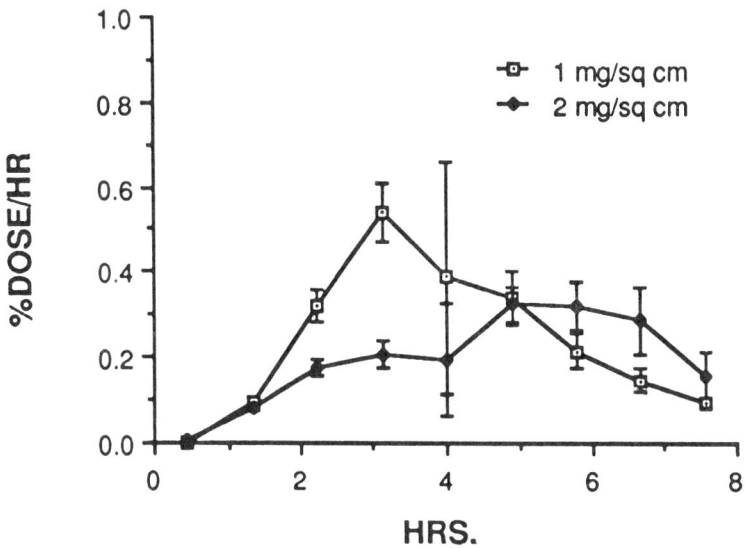

FIGURE 2B.

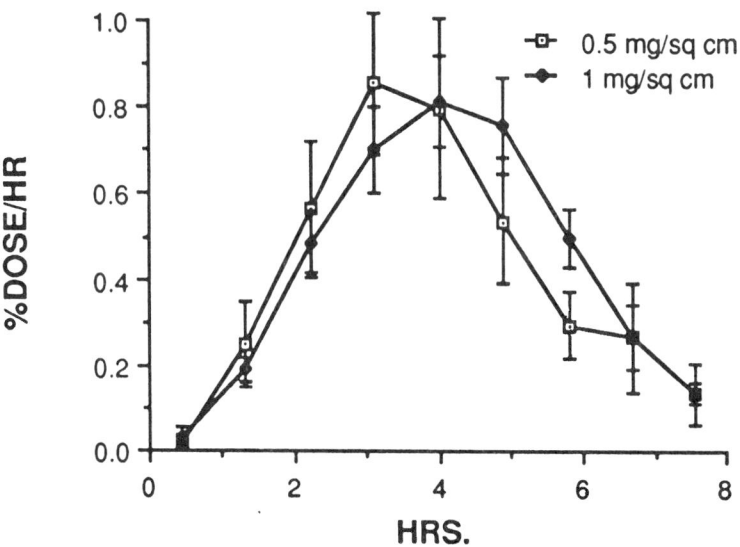

FIGURE 2C.

   The deposition of alachlor (both percent applied dose and mass) from the above dilutions and volumes of application after 8-h exposure into the skin, wash, and receptor solution are presented in Table 2. Dose accountability was excellent (88% or greater). The data show that only a small percentage of the ^{14}C-labeled alachlor penetrated the skin. A much greater amount was present in the skin digests and distilled water washes than in the receptor solution. In general, an increased volume of application led to an increased amount of alachlor recovered in the washes. A one-way analysis of variance of the amount of alachlor present in the skin, after distilled water washing 8-h postapplication, indicated no significant

TABLE 2A
Dose and Mass Accountability of Alachlor

Accountability of Applied Dose
[% Applied Dose, Mean (SD)]

	Dose mg/cm^2	PA	Washes	Skin	Total
0.5 cc of	2	0.82	52	43	96
1:20		(0.13)	(25)	(18)	(9)
1.0 cc of	4	0.47	77	23	100
1:20		(0.11)	(12)	(12)	(3)
0.5 cc of	1	2.1	44	49	95
1:40		(0.4)	(9)	(4)	(8)
1.0 cc of	2	1.8	52	49	104
1:40		(0.1)	(20)	(18)	(3)
0.5 cc of	0.5	3.7	24	60	88
1:80		(0.7)	(10)	(7)	(5)
1.0 cc of	1	3.9	51	46	101
1:80		(0.3)	(20)	(21)	(3)

TABLE 2B
Dose and Mass Accountability of Alachlor

Accountability of Applied Dose
[% Applied Dose, Mean (SD)]

	Dose total mg	PA	Washes	Skin	Total
0.5 cc of	11.4	0.09	6.0	5.0	11
1:20		(0.02)	(2.8)	(2.1)	(1)
1.0 cc of	22.8	0.11	18	5.4	23
1:20		(0.03)	(3)	(2.7)	(1)
0.5 cc of	5.7	0.13	2.6	2.9	5.7
1:40		(0.02)	(0.5)	(0.2)	(0.4)
1.0 cc of	11.4	0.21	6.2	5.9	12
1:40		(0.02)	(2.3)	(2.2)	(0)
0.5 cc of	2.85	0.11	0.73	1.8	2.5
1:80		(0.02)	(0.3)	(0.2)	(0.2)
1.0 cc of	5.7	0.23	3.0	2.7	6.0
1:80		(0.02)	(1.2)	(1.2)	(0.2)

Note: The effect of concentration and volume of application on the *in vitro* percutaneous absorption (PA) of alachlor (in the Lasso EC formulation) were assessed using human skin. Three dilutions (1:20, 1:40, and 1:80) and two volumes of application (0.5 and 1.0 cc) were employed. This table (A and B) summarizes dose and corresponding mass accountability; accountability of applied mass (mg alachlor, Mean, SD).

From Bucks, D. A. W., Wester, R. C., Mobayen, M. M., Yang, D., Maibach, H. I., and Coleman, D. L., *Toxicol. Appl. Pharmacol.*, 100, 417, 1989. With permission.

TABLE 3
Alachlor Binding to Powdered Stratum
Corneum: Effect of Duration of Contact
[% Applied Dose, Mean (SD)]

Treatment	Alachlor 1:20 dilution (N = 4) (hours)		
	0.50	2.00	4.00
Vehicle	91.80	91.35	91.80
(SD)	0.20	0.20	0.10
1st Wash	90.10	89.52	90.10
(SD)	0.05	0.05	0.08
2nd Wash	88.80	8.23	88.68
(SD)	0.09	0.05	0.05
3rd Wash	87.70	87.19	87.70
(SD)	0.04	0.05	0.02

Treatment	Alachlor 1:40 dilution (N = 4) (hours)			
	0.50	2.00	4.00	8.00
Vehicle	94.53	94.55	94.40	93.32
(SD)	0.22	0.12	0.08	0.50
1st Wash	92.65	92.67	92.43	91.10
(SD)	0.05	0.05	0.07	0.06
2nd Wash	91.07	91.12	90.71	89.15
(SD)	0.05	0.05	0.05	0.05
3rd Wash	89.59	89.62	89.06	87.35
(SD)	0.05	0.00	0.05	0.02

Treatment	Alachlor 1:80 dilution (N = 4) (hours)		
	2.00	4.00	8.00
Vehicle	94.74	94.36	94.07
(SD)	0.18	0.05	0.08
1st Wash	91.31	90.93	90.61
(SD)	0.08	0.05	0.08
2nd Wash	87.95	87.78	87.33
(SD)	0.20	0.07	0.05
3rd Wash	84.67	84.60	84.06
(SD)	0.07	0.06	0.05

Note: Carbon 14-labeled alachlor, in three dilutions of the Lasso EC formulation with water (1:20, 1:40, and 1:80; v/v), was mixed with powdered human stratum corneum, let set for 30 min, and then centrifuged. The supernate (vehicle) was removed and the stratum corneum was washed three times with distilled water.

From Bucks, D. A. W., Wester, R. C., Mobayen, M. M., Yang, D., Maibach, H. I., and Coleman, D. E., *Toxicol. Appl. Pharmacol.*, 100, 417, 1989. With permission.

difference ($p > 0.05$) with the dilutions and volumes employed. Skin levels were observed to be stable within 1 h of exposure (much earlier than maximal flux levels, data not shown).

The effect of duration of contact on alachlor binding to powdered stratum corneum is summarized in Table 3. Binding to powdered stratum corneum was not observed to be time-dependent for the 0.5- to 8-h exposure periods tested. Powdered stratum corneum demonstrated a very high capacity for the alachlor (>90%). Each distilled water washing removed 2% of the applied dose. No effect of concentration on binding was observed; therefore,

TABLE 4
Partitioning: Alachlor in Lasso with Powdered Human Stratum Corneum

[^{14}C] Alachlor percent dose

Stratum cornum	90.3 ± 1.2
Lasso supernatant	5.1 ± 1.2
Water only wash of S.C.	4.6 ± 1.3
10% Soap and water wash	77.2 ± 5.7
50% Soap and water wash	90.0 ± 0.5

Note: [^{14}C] Alachlor in Lasso® EC formulation (1:20 dilution) mixed with powdered human stratum corneum, let set for 30 min, then centrifuged. Stratum corneum wash with (1) water only; (2) 10% soap and water; (3) 50% soap and water.

From Bucks, D. A. W., Wester, R. C., Mobayen, M. M., Yang, D., Maibach, H. I., and Coleman, D. E., *Toxicol. Appl. Pharmacol.*, 100, 417, 1989. With permission.

TABLE 5
In Vitro Percutaneous Absorption of [^{14}C]-Alachlor and [^{14}C]-Atrazine in Human Skin

	Percent dose (Mean ± SD)			
Formulation	Skin	Surface wash	Receptor fluid	Total
1	2.65 ± 1.28	81.49 ± 11.81	3.07 ± 3.95	87.21 ± 13.15
2	5.71 ± 6.35	79.35 ± 14.02	4.19 ± 3.36	89.25 ± 14.88
3	2.60 ± 1.75	77.92 ± 28.47	1.50 ± 0.63	82.02 ± 28.64
4	0.35 ± 0.1	84.23 ± 10.45	6.15 ± 5.88	90.73 ± 7.46
5	0.69 ± 0.34	97.76 ± 1.74	8.09 ± 9.81	106.54 ± 8.67
6	0.35 ± 0.23	96.83 ± 12.77	6.31 ± 7.58	103.49 ± 6.23

Note: Formulations are as follows
1. [^{14}C]-alachlor with atrazine, 1:9 v/v Lariat dilution
2. [^{14}C]-alachlor with atrazine, 1:64 v/v Lariat dilution
3. [^{14}C]-alachlor, no atrazine, 1:9 v/v Lariat dilution
4. [^{14}C]-atrazine with alachlor, 1:9 v/v Lariat dilution
5. [^{14}C]-atrazine with alachlor, 1:64 v/v Lariat dilution
6. [^{14}C]-atrazine, no alachlor, 1:9 v/v Lariat dilution

these experiments reflect: (1) alachlor partitioning between powdered stratum corneum and vehicle; (2) alachlor partitioning between powdered stratum corneum and vehicle; and (3) decontamination of powdered stratum corneum with water. Soap and water (10 and 50% solutions) removed 77 and 90% of alachlor from powdered stratum corneum, respectively, (Table 4).

Table 5 gives the percent applied dose of [^{14}C]-alachlor or [^{14}C]-atrazine in human skin, surface wash, human plasma receptor fluid, and total accountability for the variety of formulation combinations listed. Figure 3 shows the effect of atrazine and dilution with water on alachlor skin absorption. The addition of atrazine had no effect (p > 0.05) on the skin absorption of alachlor. However, as noted previously, there is a tendency of alachlor skin absorption to increase when the formulation is further diluted with water.

Figure 4 shows the effect of alachlor and water dilution on atrazine skin absorption.

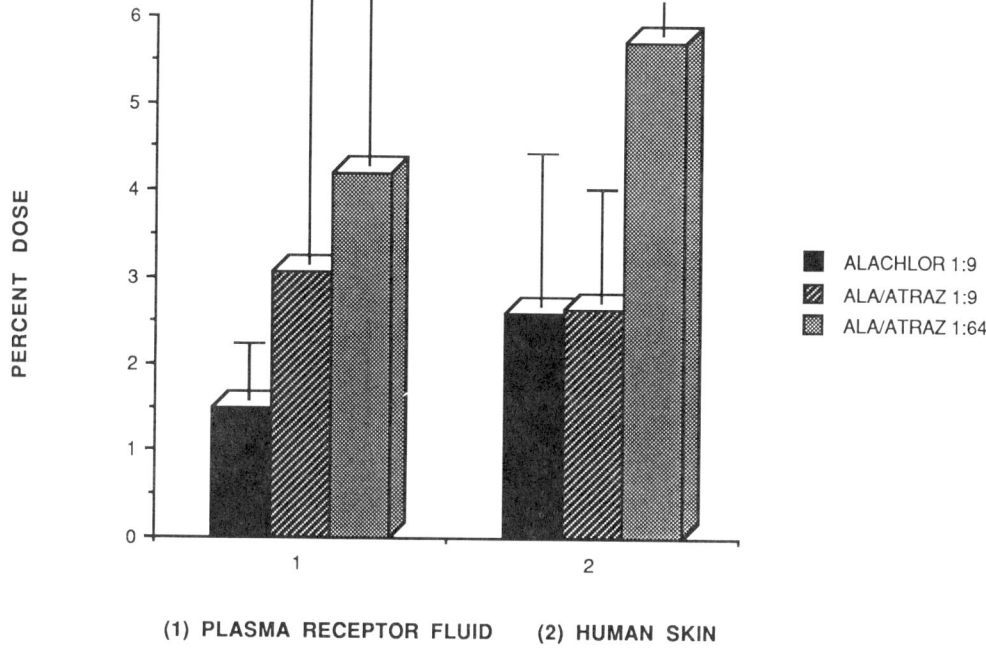

FIGURE 3. Effect of atrazine and dilution with water on alachlor *in vitro* human skin absorption (Mean ± SD).

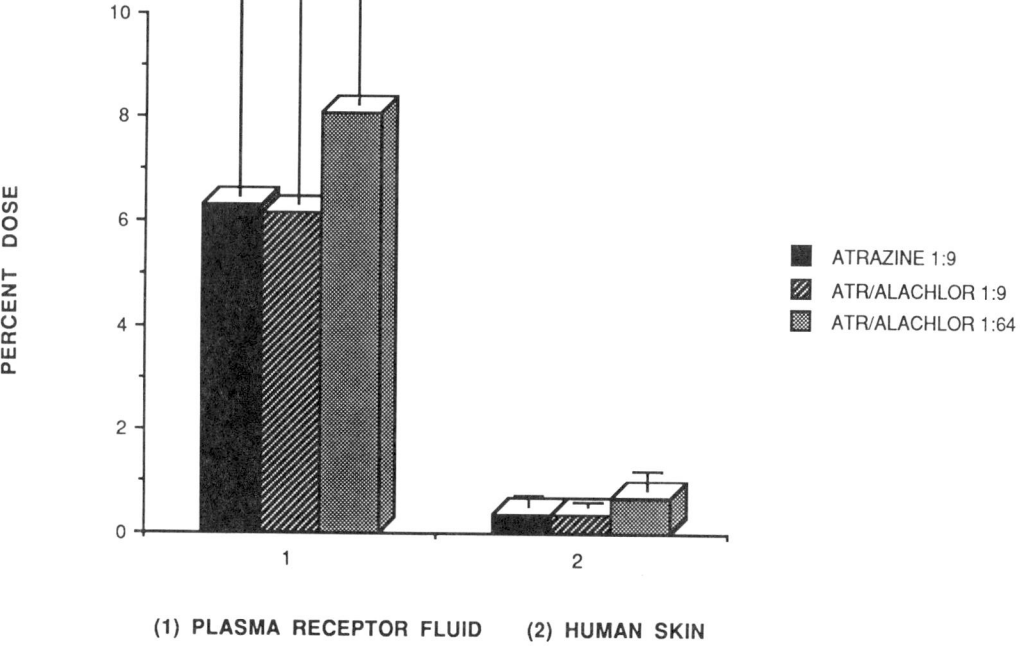

FIGURE 4. Effect of alachlor and dilution with water on atrazine *in vitro* human skin absorption (Mean ± SD).

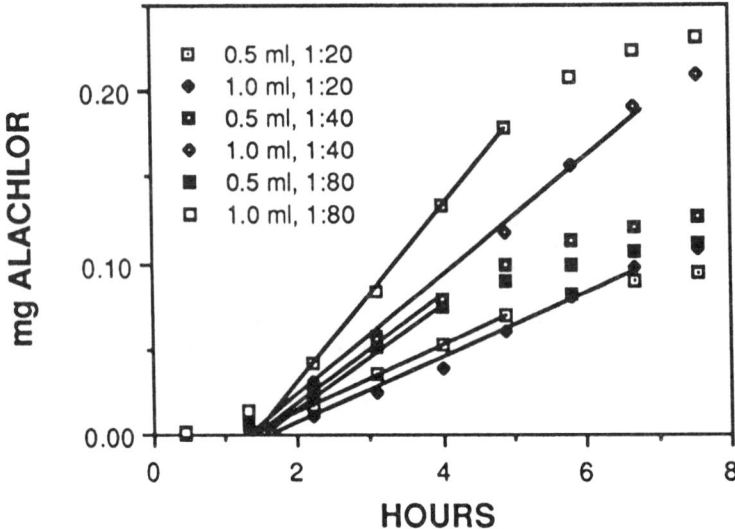

FIGURE 5. Cumulative mass of alachlor collected in the plasma receptor solution over the 8-h exposure period. The point of intersection on the X-axis is the lag time. (From Bucks, D. A. W., Wester, R. C., Mobayen, M. M., Yang, D., Maibach, H. I., and Coleman, D. E., *Toxicol. Appl. Pharmacol.*, 100, 417, 1989. With permission.)

Neither the addition of alachlor nor dilution with water ($p > 0.05$) changed the *in vitro* absorption of atrazine into and through human skin.

IV. DISCUSSION

In vitro percutaneous absorption of alachlor in the commercial formulations was studied when diluted with water and when mixed with atrazine. The study was done using flow-through *in vitro* cells and radiotracer methodology. Application volumes of 0.5 and 1.0 ml were used to address the effect of formulation dilution by comparison of the data resulting from the same mass of chemical applied to the skin from different dilutions, the effect of volume of application of each vehicle on the percutaneous absorption, and the interactions of atrazine in combination with alachlor. The percentage of the applied alachlor dose absorbed was low (0.5 to 4%), corresponding to levels of 0.09 to 0.23 mg alachlor for 8-h skin application time. This corresponds very well with the results of Kronenberg et al.[5] for the *in vivo* percutaneous absorption of alachlor in Rhesus monkeys. They reported 0.11 mg of alachlor absorbed following topical application of a 1:29 dilution of alachlor (in Lasso EC) with water.

A lag time of 1.2 to 1.8 h was observed (Figure 5). The peak rate of absorption was 3- to 5-h postapplication and flux rates declined thereafter. Vehicle depletion of alachlor might be responsible for the decrease in rate. The excised skin from the *in vitro* penetration studies and the powdered stratum corneum from the binding experiments both demonstrated high capacity of alachlor from all dilutions employed. The binding of alachlor to stratum corneum is not unique. Other chemicals containing aromatic rings and/or chlorine behave in a similar manner.[3]

When the same mass of alachlor was applied to the skin, both total percent of applied dose and total mass of alachlor absorbed (as measured in the receptor fluid) significantly increased ($p < 0.01$, see Tables 1 and 2) with increasing dilution. Comparison of absorption from 0.5 ml of the 1:20 dilution with 1.0 ml of the 1:40 dilution and comparison of absorption from 0.5 ml of the 1:40 dilution with 1.0 ml of the 1:80 dilution demonstrated the more dilute formulation, i.e., lower alachlor concentration, resulted in the greater level of absorption (both in percent of applied dose and total mass of alachlor in receptor solution). A greater increase in $K_{SC/V}$, the partition coefficient of the compound between the stratum corneum and vehicle, with dilution the solubilizing agent in the formulation with water compared to the expected decrease in flux by halving alachlor concentration would explain the data. Poulsen[6] first hypothesized that changes in solubilizer concentration in vehicle would change the partition coefficient of the penetrant between skin and vehicle and accordingly result in corresponding changes in observed drug absorption. As stated above, skin stratum corneum demonstrated a high capacity for alachlor. This would decrease dramatically the concentration of alachlor in each dilution. As expected, alachlor skin levels were established before steady state flux rates of penetration. No significant differences in final alachlor skin levels were observed between the different dilutions.

The addition of atrazine demonstrated no influence on the absorption of alachlor; nor was there any suggestion of synergism with respect to atrazine and alachlor absorption. Dilution with water had no effect on atrazine absorption.

Data from this study should be interpreted with caution. Cadaver skin from a single anatomic site was used in these *in vitro* absorption studies. The physical-chemical interactions between alachlor and cadaver skin may be different when compared to interactions with living human skin, and the *in vitro* apparatus and receptor fluid may influence results. The volume of alachlor per skin area was in excess of volumes normally encountered in agricultural usage. The data suggest that alachlor may bind primarily to the stratum corneum of the skin. The chemical could be effectively removed from the stratum corneum by washing with soap and water. In conclusion, alachlor penetrates the skin at low levels. Increasing the alachlor dilution with water increased significantly ($p < 0.01$) the percentage (3.9%) and the total mass (0.23 mg) penetrating skin. These data cannot be extrapolated to infinite dilution. Atrazine had no effect on alachlor skin absorption. Future studies may be needed to determine whether these data can be extrapolated to normal *in vivo* human skin.

REFERENCES

1. **Hansch, C. and Leo, A., Eds.,** *Substituent Constants for Correlations in Chemistry and Biology,* Wiley-Interscience, New York, 1979.
2. **Wester, R. C., Maibach, H. I., Surinchak, J., and Bucks, D. A. W.,** Predictability of *in vitro* diffusion systems: effects of skin types and ages on percutaneous absorption of triclocarban, in *Percutaneous Absorption,* Bronaugh, R. and Maibach, H., Eds., Marcel Dekker, New York, 1985, 223.
3. **Wester, R. C., Mobayen, M., and Maibach, H. I.,** *In vivo* and *in vitro* binding to powdered stratum corneum as methods to evaluate skin absorption of environmental chemical contaminants from ground and surface water, *J. Toxicol Environ. Health,* 21, 367, 1987.
4. **Bucks, D. A. W., Wester, R. C., Mobayen, M. M., Yang, D., Maibach, H. I., and Coleman, D. L.,** *In vitro* percutaneous absorption and stratum corneum binding of alachlor: effect of formulation dilution with wtaer, *Toxicol. Appl. Pharmacol.,* 100, 417, 1989.
5. **Kronenberg, J. M., Fuhremann, T. W., and Johnson, D. E.,** Percutaneous absorption and excretion of alachlor in rhesus monkeys, *Fundam. Appl. Toxicol.,* 10, 664, 1988.
6. **Poulsen, B. J.,** Diffusion of drugs from topical vehicles: an analysis of vehicle effects, in *Advances in Biology of Skin.* Vol. XII, Mantagna, W., van Scott, E. J., and Stoughton, R. B., Eds., Publication No., 511 from The Oregon Regional Primate Research Center, 1972, 495.

Chapter 15

IN VIVO — IN VITRO CORRELATIONS FOR HYDROPHOBIC COMPOUNDS

Robert L. Bronaugh and Steven W. Collier

TABLE OF CONTENTS

I. INTRODUCTION

Probably the greatest discrepancy between *in vitro* and *in vivo* absorption values is found when comparisons are made with hydrophobic compounds. As discussed in an earlier chapter on receptor fluid (Chapter 5), hydrophobic compounds are particularly difficult to measure by *in vitro* methods requiring either the use of more lipophilic receptor fluids or the inclusion of absorbed material retained in skin as part of the absorbed dose. This chapter will again address these problems but will focus on studies that have compared *in vitro* results to data obtained in corresponding *in vivo* experiments.

The absorption of hydrophobic compounds is also difficult to measure by *in vivo* techniques which rely on measurement of absorbed dose in the urine and feces. These compounds are often distributed into body fat and therefore are slowly excreted from the body. Urine and feces samples must sometimes be collected for months before excretion of test compound is completed. The amount of test compound excreted in a one week experiment can be a very small percentage of the total amount of compound absorbed. Correction for total absorption with a parenteral correction factor can produce large errors in absorption estimation since the error obtained in the topical measurement can be magnified many fold when the data is multiplied by a large correction factor.

When there is a lack of agreement between *in vivo* and *in vitro* absorption results, it is frequently assumed that the *in vitro* results are in error and that an accurate measurement cannot be obtained by *in vitro* techniques. It should be kept in mind that errors can occur with both *in vivo* and *in vitro* measurements. An attempt at understanding the reasons for the lack of agreement of test results can lead to a more accurate interpretation of the data. If the *in vitro* and the *in vivo* studies are performed correctly, with all variables taken into account, then good agreement in absorption values should be obtained.

An early *in vivo* — *in vitro* comparison of hydrophobic compounds was made by Tsuruta. He measured the permeation of chlorinated organic solvents through mouse skin using *in vivo* methods[1] and through rat skin by standard *in vitro* diffusion cell techniques.[2] He observed that there was qualitative agreement in the absorption results obtained by the two methods; the compounds were ranked in the same order of absorption by the *in vivo* and the *in vitro* method. However there were quantitative differences; the most hydrophobic compounds in the series (the tri- and tetra-chloro derivatives) were much less readily absorbed in the *in vitro* studies. Tetrachloroethylene was absorbed through mouse skin *in vitro* at a rate of only 0.6 nmoles/min/cm^2 skin but permeated through rat skin *in vivo* at 24.4 nmoles/min/cm^2. It was suggested that these differences might be due to a lack of solubility of the test compounds in the normal saline receptor fluid.

II. USE OF A LIPOPHILIC RECEPTOR FLUID

A. EARLY STUDIES — HAIRED ANIMAL AND HUMAN SKIN

It was clear that some modification of current methods for measuring percutaneous absorption had to be made to permit accurate *in vitro* measurement of hydrophobic compounds. In the early 1980s we compared a number of lipophilic receptor fluids for possible use in facilitating the partitioning of hydrophobic compounds from skin in *in vitro* studies.[3] We determined that of the receptor fluids evaluated, a 6% solution of PEG 20 oleyl ether (Volpo 20®, Croda Inc., New York, NY) was most satisfactory. The importance of using a split-thickness preparation of skin was also observed. In these studies a dermatome (Padgett Dermatome®, Kansas City, MO) was used to prepare sections of skin approximately 350 μm in thickness to remove dermal tissue that serves as an artificial reservoir for hydrophobic compounds in *in vitro* experiments. This membrane still contains 250 to 300 μm of dermal tissue but was the thinnest section of haired rat skin that could be made without damaging

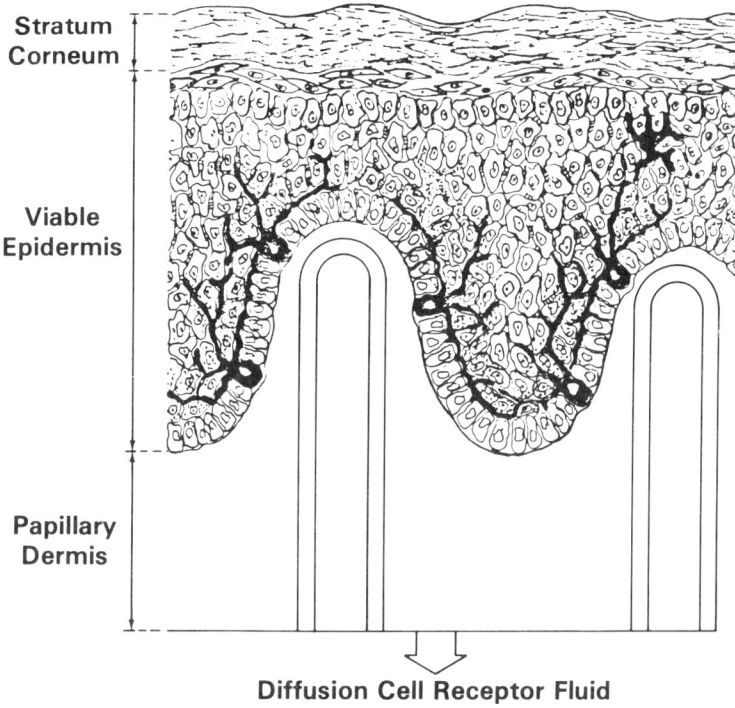

Stratum Corneum

Viable Epidermis

Papillary Dermis

Diffusion Cell Receptor Fluid

FIGURE 1.   Diagram of the barrier layer of skin. In an *in vitro* study, absorbed material can partition from skin into the receptor fluid. In an *in vivo* study, blood circulating through the capillary loops in the papillary dermis carries away much of the absorbed material.

its barrier properties. Chemicals that are absorbed through skin *in vivo* are taken up by capillaries directly beneath the epidermis and therefore do not diffuse through the dermal tissue (Figure 1).

The *in vivo* percutaneous absorption of AETT (acetyl ethyl tetramethyl tetralin) and cinnamyl anthranilate was determined in Osborne-Mendel rats by application of these compounds to back skin in a petrolatum vehicle. The amount of test compound excreted in the urine and feces was determined as a measure of percutaneous absorption.[3] As shown in Figure 2, a marked improvement in the *in vivo—in vitro* correlations with cinnamyl anthranilate was obtained when a PEG 20 oleyl ether solution was utilized as the receptor fluid. *In vitro* absorption was increased to 61% of the absorption values obtained in the *in vivo* experiments. Similar results were obtained for AETT.

The percutaneous absorption of the fragrance ingredients safrole and cinnamyl anthranilate were measured *in vivo* (monkey) and *in vitro* (human).[4] *In vivo* absorption was measured by adjusting urinary excretion by a parenteral correction factor. *In vitro* experiments were performed in flow-through cells with 6% PEG 20 oleyl ether as the receptor fluid. Occlusion of the site of application was found to enhance the absorption of both of these compounds primarily by the prevention of evaporation of these volatile compounds. Safrole is extremely volatile and its *in vitro* absorption increased from 15% to 38% following occlusion. *In vitro—in vivo* correlations of absorption (occluded skin) were good for cinnamyl anthranilate: 39% *in vivo;* 53% *in vitro.* Comparisons of safrole absorption were poor (13% *in vivo;* 38% *in vitro*) due likely to the difficulty in accurately measuring absorption of volatile compounds.

Yang, Roy, and co-workers[5,6,7] have examined the *in vivo-in vitro* comparability of data obtained with hydrophobic polynuclear aromatic compounds such as anthracene[5] and benzo(a)pyrene.[6,7] They have utilized techniques previously developed in our laboratories,

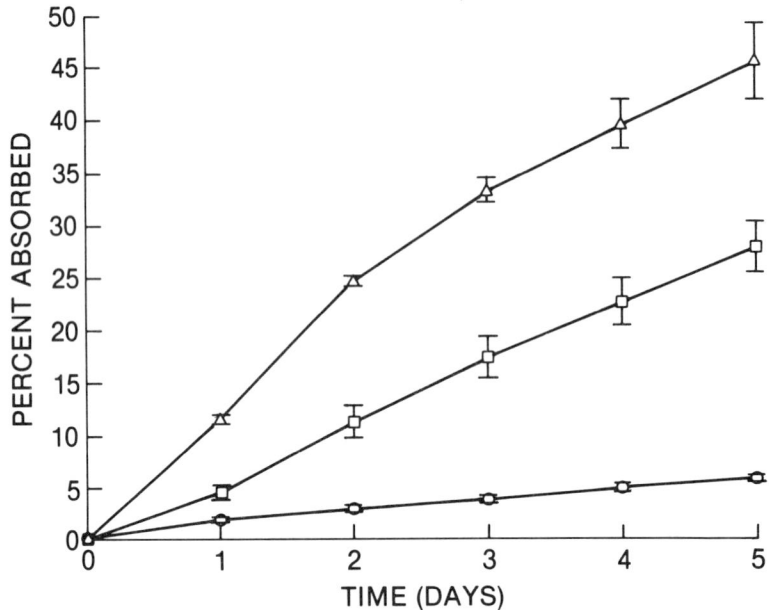

FIGURE 2.   Absorption of cinnamyl anthranilate from a petrolatum vehicle. Key: ($\triangle$) *in vivo* results; ($\bigcirc$) *in vitro* (saline in receptor; ($\square$) *in vitro* (6% PEG 20 oleyl ether in receptor). At 5 d, 13% of the *in vivo* absorption was obtained by *in vitro* techniques, using saline in the diffusion cell receptor. Absorption was increased to 61% of the *in vivo* results by substituting 6% PEG 20 oleyl ether as the receptor fluid.

namely 350 μm dermatome section of rat skin and a receptor fluid consisting of a 6% PEG 20 oleyl ether solution.

The skin permeation of C^{14}-labeled anthracene was measured in Sprague-Dawley rats over a 6 d period. The compound was applied in a hexane:acetone (1:7) vehicle at a concentration of 9.3 μg/cm². *In vitro* absorption was measured in Franz-type diffusion cells with dermatomed skin and surfactant receptor fluid as described above. The *in vivo* values were determined by combining the urinary and fecal excretion each day. At the end of the 6 d experiment only 1.3% of the applied dose remained in the tissues. Total percutaneous absorption was 52% of the applied dose by the *in vivo* method and 56% with *in vitro* techniques.

Using the same methodology, Yang, Roy, and Mackerer[6] found almost identical absorption of benzo(a)pyrene through rat skin by *in vivo* (46.2%) and *in vitro* (49.9%) methods. When normal saline was utilized as the receptor fluid, only 2.1% of the applied dose was absorbed as indicated from receptor fluid levels. Good agreement of *in vivo* and *in vitro* benzo(a)pyrene data was also obtained when the test compound was applied to skin in a mixture of soil.[7]

Different receptor fluids were evaluated for *in vitro* studies of the insecticide cypermethrin in the human and rat.[8] Epidermal membranes and also full thickness skin was assembled in static diffusion cells. The lipophilic receptor fluids used were: 50% aqueous ethanol, 6% PEG 20 oleyl ether, 20% fetal bovine serum, and 6% PEG 400. When full thickness rat or human skin was used, no absorption of cypermethrin could be detected in the receptor fluid. With rat epidermis, the test compound could not be detected in a saline or a PEG 400 receptor fluid but was readily measurable in the other more lipophilic fluids. Values most similar to *in vivo* results were obtained with 50% aqueous ethanol but no evaluation was made of barrier integrity following exposure of skin to this receptor fluid. Approximately 60% of the in vivo absorption was obtained with PEG 20 oleyl ether solution. Absorption

TABLE 1
Absorption of Hydrophobic Fragrance Ingredients Through the Skin
of the Haired Rat (Cortisone Control)

Compound and receptor fluid	% Dose absorbed (5 d)	Cortisone permeability constant (cm/h $\times$ 10^5)
Cinnamyl anthranilate		
Normal saline (3)	4.9 $\pm$ 0.7	5.7 $\pm$ 0.5
6% PEG-20 oleyl ether (3)	46.4 $\pm$ 3.9	6.6 $\pm$ 0.3
AETT		
Normal saline (4)	0.28 $\pm$ 0.01	4.4 $\pm$ 0.4
6% PEG-20 oleyl ether (4)	8.5 $\pm$ 0.8	5.4 $\pm$ 0.5

Note: Values are the Mean $\pm$ SE of the number of static cell determinations in parentheses. Compounds were applied to haired rat skin (300 μm thick) in a petrolatum vehicle (25 mg/cm^2). *In vivo* absorption values (see Reference 3) were cinnamyl anthranilate = 45.6%; AETT = 18.9%.

of cypermethrin through human skin was only detected when 50% aqueous ethanol was the receptor fluid. Human skin was more than 20 times less permeable to the insecticide than rat skin.

B. THIN DERMATOME SECTIONS—HAIRLESS ANIMAL SKIN

An important factor in the accurate measurement of absorption of hydrophobic compounds is the preparation of the skin membrane. In initial studies we observed that one could not use full-thickness skin in the diffusion cell and expect to observe enhanced penetration of these compounds due to partitioning into lipophilic receptor fluids. A dermatome was used to remove approximately two thirds of the dermis of rat skin by preparing a 350 μm dermatome section. However, use of mild lipophilic receptor fluids that did not damage barrier properties of skin did not result in absorption values as high as those in *in vivo* experiments.

In order to correct this problem, we reasoned that a thinner barrier layer with less dermal tissue must be prepared.[9] With haired Osborne-Mendel rat skin, 300 μm sections of dermatomed skin were prepared and assembled in diffusion cells. The barrier integrity of membranes was assessed with a quick check of ^{3}H-water permeation. Approximately half of the thin membranes prepared at this thickness were damaged and had to be rejected from further study. Intact membranes were used to assess the effectiveness of 6% PEG 20 oleyl ether as a receptor fluid for hydrophobic compounds.

The effect of the thinner 300 μm dermatome section on cinnamyl anthranilate and AETT absorption is shown in Table 1. The PEG 20 oleyl ether surfactant solution did not effect the barrier properties of the membranes as verified by the lack of change in cortisone permeation. No significant difference was seen in the *in vivo* and *in vitro* absorption of cinnamyl anthranilate indicating the importance of removing as much dermis as possible from the *in vitro* membrane used in permeation studies. *In vitro* absorption of the more hydrophobic AETT molecule through the 300 μm membranes still did not agree exactly with in vivo absorption measurements.

The absorption of benzo(a) pyrene was measured through Osborne-Mendel rat skin (Table 2). The compound was applied to skin at a concentration of 5 μg/cm^2 in an acetone vehicle. Good agreement was achieved between *in vitro* values using 300 μm dermatomed skin and surfactant (56.0%) compared with the *in vivo* results (48.3%). The maintenance of viability of skin in the diffusion cell was not required for comparability of permeation of radioactivity through skin, however no information is obtained on the biotransformation of benzo(a)pyrene when absorption studies are conducted in this way.

TABLE 2
Benzo[a]pyrene Permeation Through Rat Skin
(Acetylsalicylic Acid Control)

Receptor fluid	Percent applied; dose absorbed	
	Benzo[a]pyrene	Acetylsalicylic Acid
Normal saline	3.7 ± 0.1	19.4 ± 2.0
6% PEG 20 oleyl ether	56.0 ± 0.9	22.6 ± 2.7

Note: Values are the Mean ± SE of three determinations. Compound
was applied in an acetone vehicle to haired rat skin (300 μm
thick). Permeation of the two compounds was measured simultaneously in dual-label experiments. *In vivo* benzo[a]pyrene
absorption was 48.3 ± 2.1% (5 rats).

TABLE 3
AETT Absorption Through Fuzzy Rat Skin (Cortisone Control)

Receptor fluid	Skin thickness (μm)	AETT (% absorbed)	Cortisone ($K_p \times 10^5$)
Saline (5)	300	2.1 ± 0.1	5.6 ± 0.6
6% PEG-20 oleyl ether (12)	300	30.5 ± 1.9	5.9 ± 0.7
3% Bovine serum albumin (5)	300	5.4 ± 0.7	5.3 ± 0.8
Saline (5)	200	1.7 ± 0.1	6.5 ± 0.4
PEG-20 oleyl ether			
0.25% (4)	200	10.8 ± 0.5	5.4 ± 1.0
0.5% (8)	200	22.2 ± 1.2	6.5 ± 0.6
1.0% (8)	200	34.5 ± 1.2	8.8 ± 1.1
6.0% (11)	200	39.5 ± 2.9	9.4 ± 1.2
3% Bovine serum albumin (4)	200	10.3 ± 0.7	7.1 ± 0.5
Minimum essential media (MEM) (4)	200	1.3 ± 0.2	7.4 ± 1.3
MEM + 3% bovine albumin (4)	200	3.5 ± 0.3	7.1 ± 1.0

Note: Values are the Mean ± SE of the number of determinations in parentheses. AETT
and cortisone were applied to the skin surface in petrolatum (5 mg/cm^2). *In vivo*
absorption of AETT was 23.8%.

Thin, undamaged skin sections of hairless animal skin (fuzzy rat) could be prepared at
a thickness of approximately 200 μm. The barrier integrity of these thin membranes was
damaged by the use of 6% PEG 20 oleyl ether (Table 3). When a concentration of 0.5%
surfactant was used, barrier integrity was maintained as evidenced by the lack of significant
change in cortisone permeation. The permeation of AETT through dermatomed fuzzy rat
skin (22.2%) agreed closely with the corresponding *in vivo* value (Table 3, 23.8%). DDT
skin absorption in the fuzzy rat was determined using the 200 μm sections of dermatomed
skin and 0.5% PEG 20 oleyl ether (Table 4). Good agreement was obtained between the *in
vitro* (60.6%) and the *in vivo* (69.5%) absorption results. This is in contrast to previous
studies with human skin and a normal saline receptor fluid that resulted in underestimation
of absorption by the *in vitro* method.[10]

III. VIABLE SKIN

Viable epidermal membranes have recently been prepared from hairless guinea pig skin.[11]
These membranes allow for accurate measurement of absorption of hydrophobic compounds
by sampling receptor fluid, since the artificial dermal reservoir has been removed. The

TABLE 4
Absorption of DDT Through Fuzzy Rat Skin

Receptor fluid	Applied dose absorbed (%)
Normal saline (4)	1.8 ± 0.1
0.5% PEG-20 oleyl ether (4)	59.4 ± 5.1

Note: Values are the Mean ± SE of the number of determinations in parentheses. DDT was applied in an acetone vehicle to a skin section 200 μm in thickness. *In vivo* absorption of the compound was 69.5 ± 1.7% (5 rats).

amount remaining in skin at the end of the experiment must also be included as absorbed material. The addition of 4% bovine serum albumin to the Hanks' balanced salt solution receptor fluid promotes partitioning of test compounds from skin into the receptor fluid. This partitioning is further aided by increased mixing of the receptor contents with stirring bars. The permeation of the water insoluble compound AETT into the receptor fluid was enhanced 60-fold compared to its absorption through full-thickness hairless guinea pig skin. Good agreement was obtained between *in vivo* and *in vitro* absorption results.

IV. CONCLUSIONS

In vitro percutaneous absorption studies must be performed with special care when hydrophobic compounds are studied. A thin section of skin with most or all of the dermis removed facilitates measurements of rates of permeation during the course of the experiment since the dermal tissue can no longer serve as an "artificial" reservoir. A lipophilic receptor fluid is essential to enable water insoluble test compounds to partition from skin into the diffusion cell receptor solution. Lack of agreement between *in vivo* and *in vitro* results may indicate a problem in methodology with one of the two types of studies. It should not be assumed that the *in vitro* study is in error.

REFERENCES

1. **Tsuruta, H.,** Percutaneous absorption of organic solvents: I. Comparative study of the *in vivo* percutaneous absorption of chlorinated solvents in mice, *Ind. Health,* 13, 227, 1975.
2. **Tsuruta, H.,** Percutaneous absorption of organic solvents: II. A method for measuring the penetration rate of chlorinated solvents through excised rat skin, *Ind. Health,* 15, 131, 1977.
3. **Bronaugh, R. L. and Stewart, R. F.,** Methods for *in vitro* percutaneous absorption studies. III. Hydrophobic compounds, *J. Pharm. Sci.,* 73, 1255, 1984.
4. **Bronaugh, R. L., Stewart, R. F., Wester, R. C., Bucks, D., Maibach, H. I., and Anderson,** Comparison of percutaneous absorption of fragrances by humans and monkeys, *J. Fd. Chem. Toxic.,* 23, 111, 1985.
5. **Yang, J. J., Roy, T. A., and Mackerer, C. R.,** Percutaneous absorption of anthracene in the rat: comparison of *in vivo* and *in vitro* results, *Toxicol. Ind. Health,* 2, 79, 1986.
6. **Yang, J. J., Roy, T. A., and Mackerer, C. R.,** Percutaneous absorption of benzo(a)pyrene in the rat: comparison of *in vivo* and *in vitro* results, *Toxicol. Ind. Health,* 2, 409, 1986.
7. **Yang, J. J., Roy, T. A., Krueger, A. J., Neil, W., and Mackerer, C. R.,** *In vitro* and *in vivo* percutaneous absorption of benzo(a)pyrene from petrolatum crude-fortified soil in the rat, *Bull. Environ. Contam. Toxicol.,* 43, 207, 1989.
8. **Scott, R. C. and Ramsey, J. D.,** Comparison of the *in vivo* and *in vitro* percutaneous absorption of a lipophilic molecule (cypermethrin, a pyrethroid insecticide), *J. Invest. Dermatol.,* 89, 142, 1987.

9. **Bronaugh, R. L. and Stewart, R. F.,** Methods for *in vitro* percutaneous absorption studies. VI. Preparation of the barrier layer, *J. Pharm. Sci.,* 75, 487, 1986.

10. **Bronaugh, R. L.,** *In vitro* methods for percutaneous absorption of pesticides, in *Dermal Exposure to Pesticide Use,* Honeycutt, R., Zweig, G., and Ragsdale, N. N., Eds., American Chemical Society, Washington, D.C., 1985, 33.

11. **Bronaugh, R. L., Collier, S. W., and Stewart, R. F.,** *In vitro* percutaneous absorption of a hydrophobic compound through viable hairless guinea pig skin, *Toxicologist,* 9, 61, 1989.

Chapter 16

IN VITRO ABSORPTION: SKIN FLAP MODEL

J. E. Riviere

TABLE OF CONTENTS

I. INTRODUCTION

Numerous *in vitro* and *in vivo* animal models have been developed to assess the rate, extent, and mechanisms of drug and xenobiotic percutaneous absorption. Most of these have been covered in the earlier chapters of this text. All approaches, including *in vivo* human studies, have advantages as well as significant limitations. For example, one is precluded from doing human studies with compounds having significant systemic toxicity (or unknown toxicity). Invasive procedures to study drug metabolism or the distribution of drug within human skin using biopsy procedures is not possible. For certain mechanistic studies, extensive tissue and blood sampling may not be warranted in human trials. Thus, *in vitro* human and animal and *in vivo* animal studies are often used. However, limitations inherent to these techniques are present. It is this author's belief that a thorough understanding of the mechanisms of percutaneous absorption cannot be achieved in a single model system for all chemicals. Individual systems have specific advantages and limitations which offer unique perspectives on various aspects of dermal penetration. Some are designed to rapidly screen compounds for their ability to penetrate stratum corneum; others to assess cutaneous metabolism.

It is not the purpose of this chapter to compare the relative merits of all existing experimental models for studying percutaneous absorption. Rather, the development and utilization of a novel perfused skin preparation, the isolated perfused porcine skin flap (IPPSF) will be presented. A thorough understanding of this model, which is considered by some to be a hybrid *in vitro/in vivo* model, should illustrate the fundamental differences between *in vitro* and *in vivo* experimental paradigms since in some ways it shares the strengths and weaknesses of both approaches.

II. WHY AN ISOLATED PERFUSED SKIN PREPARATION

Skin is a complex organ in mammals whose roles include:

1. Providing a barrier to the outside environment (prevents compound entry into as well as fluid loss from the body)
2. Temperature regulation (sweating and altered blood flow.
3. Thermal insulation
4. Neurosensory functions
5. Immunological recognition
6. Endocrine factor release
7. Metabolism
8. External chemical communications
9. Protecting the body from harmful ultraviolet radiation[1]

When skin models are developed for studying percutaneous absorption, often only the barrier function is considered important. For many compounds, this has fortunately been acceptable as the stratum corneum is often the rate limiting barrier to percutaneous absorption. However, for other chemicals or new drug delivery strategies, this assumption may not be valid and other functions of the skin may play important roles.

The microanatomy of skin reflects these varied functions. In addition to the keratinocytes which primarily comprise the viable epidermis, other cells present include melanocytes, Langerhans cells, and Merkel cells. The dermis consists of two layers of connective tissue that merge into one another and consists of several cell types (e.g., lymphocytes, fibroblasts, mast cells). The dermis also contains appendages such as hair follicles, sweat and sebaceous glands which penetrate the epidermis and offer an alternate route for compound penetration.

The skin is very well vascularized and innervated. A review of the anatomy of skin[1] should be consulted for further details.

If an *in vitro* model is to accurately reflect all these aspects of *in vivo* skin structure and function, then most of these cells and appendages should be present and maintained in a functional state. Present diffusion cell models were never designed with these constraints in mind. Similarly, recently developed cell and organ culture approaches often contain differentiated epidermal cells on a fibroblast/collagen matrix. Importantly, appendages, other cell types, and the microvasculature are not present. These factors are the primary design constraints which support the development of a more biologically relevant *in vitro* skin preparation.

Isolated organ perfusions have been a useful tool for pharmacological and toxicological studies of the kidney, liver, lung, intestine, and heart. These organs have lent themselves to perfusion techniques because of both an easily isolatable vascular supply and a "closed" anatomical structure (encapsulated or surrounded by serosa) amenable to perfusion. The skin does not possess either of these attributes. Early attempts at perfusing skin did not attempt to create closed vascular systems and were never optimized for detailed absorption or disposition studies.[2] These limitations may be overcome if a two-stage reconstructive surgical procedure is utilized to create a "closed" skin preparation with an easily isolatable vasculature. If such a tubed, pedicle flap is employed, skin may be perfused under ambient environmental conditions without concern for dermal dehydration. Such preparations may then be harvested by cannulating the artery. Certain anatomic regions of the body are well suited to isolated perfusion protocols because of the presence of direct cutaneous vasculature. The caudal superficial epigastric artery and veins on the ventral abdomen of many animal species appears to be an excellent site for such tubed skin flap preparations. Specific details of the surgery will be presented later.

The next consideration is the selection of an appropriate animal species. For ethical reasons, human skin is not an available option. If the objective of dermal penetration studies is to quantitatively predict *in vivo* human percutaneous absorption, most authors agree that the skin of monkeys and pigs are the optimal animal species[3-9] with various hairless rodents and guinea pigs also being viable alternatives. Of these species, monkeys and pigs would be better candidates for isolated flap models primarily due to their larger size which facilitates surgery, and provides a large surface area of skin for testing prototype human transdermal patches. The large dosing area also facilitates detection of small fluxes of absorbed compounds and/or their metabolites. In our model, pig skin was finally selected because of this species wider availability and significantly lower acquisition and maintenance costs.

It should be stressed that many authors consider pig skin to be the best animal model for human skin because of similarity in structure and function. Both species have skin with a similar gross appearance and texture, sparse hair coat, a relatively thick epidermis, similar arrangements of dermal collagen and elastic fibers, and a comparable microcirculation. Differences primarily relate to the presence of only apocrine sweat glands in swine and eccrine sweat glands in humans.[1,10-17] Biochemically, the enzyme histochemistry of human and porcine epidermis is very similar.[18,19] *In vitro* tissue slice studies of glucose and fatty acid metabolism have demonstrated similar activities and epidermal/dermal distribution in both species.[20-22] Similarly, pig and human skin have very similar epidermal lipid compositions and distributions.[23-25] The rate of epidermal cell turnover in hairless mice, pigs, and man, assessed using initiated thymidine incorporation into nucleic acids or glycine incorporation into cellular proteins, is 30 d in humans, 26 to 27 d in pigs, compared to only 3 to 4 d in hairless mice.[26]

The final advantage of an isolated perfused skin preparation is that it is a humane alternative *in vitro* animal model. The pig serves as a skin donor and can be used for other purposes after harvest of the preparation. Importantly, systemically toxic or irritating chemicals may be studied after the skin has been harvested from the animal.

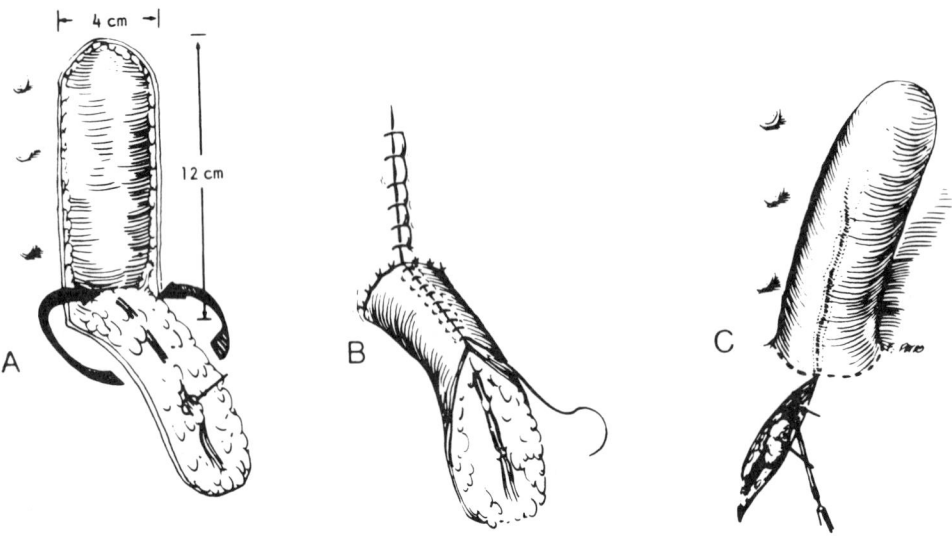

FIGURE 1.   Surgical procedure used to create IPPSF. In stage one (A & B), a skin flap is created and tubed and later harvested in stage two (C) by cannulating the superficial epigastric artery.

III. EXPERIMENTAL TECHNIQUE

The IPPSF is a single-pedicle, axial pattern tubed skin flap obtained from the ventral abdomen of female weanling swine.[27-30] Two flaps per animal, each lateral to the ventral midline, can be created in a single surgical procedure. As depicted in Figure 1, the procedure involves two steps: creation of the flap in Stage I and harvest in Stage II. Briefly, pigs weighing approximately 20 to 30 kg are premedicated with atropine sulfate and xylazine hydrochloride, induced with ketamine hydrochloride and inhalational anesthesia is maintained with halothane. Each pig is prepared for routine aseptic surgery in the caudal abdominal and inguinal regions and a 4×12 cm area of skin, known from previous dissection and *in vivo* angiography studies to be perfused primarily by the caudal superficial epigastric artery and its associated paired venae comitantes, is demarcated. Following incision and scalpel dissection of the subcutaneous tissue, the caudal incision is apposed and sutured and the tubed skin flap edges trimmed of fat and closed. The remaining deep and superficial subcutaneous tissues and skin incision are then closed as well. Two days later, a second surgical procedure is used to cannulate the caudal superficial epigastric artery and harvest each of these skin flaps. The two day period between flap creation and harvest was determined to be optimal from the standpoint of lack of overall flap leakiness, normal histologic appearance (minimal variation in epidermal thickness), normal vascularization, and animal housing economics.[28] The IPPSF is then transferred to the perfusion chamber described below. The remaining wound is flushed and allowed to heal and pigs are returned to the housing facility. A complete description of the surgical procedure is reported elsewhere.[30]

The isolated perfused organ apparatus, depicted in Figure 2, is a custom designed temperature and humidity regulated chamber made specifically for this purpose (Diamond Research, Raleigh, NC). A computer monitors perfusion pressure, flow, pH, and temperature ($\pm 0.1°C$). Flexibility is afforded in the experimental design by allowing both temperature and relative humidity to be maintained at specific set points (normally 37°C and 60 to 80% RH). Media is gassed with 95% oxygen and 5% carbon dioxide by passage through a silastic oxygenator. Normal perfusate flow through the skin flap is maintained at 1 ml/min/flap (3 to 7 ml/min/100 g) with a mean arterial pressure ranging from 30 to 70 mmHg. These values are very consistent with *in vivo* values reported in the literature. Both recirculating and

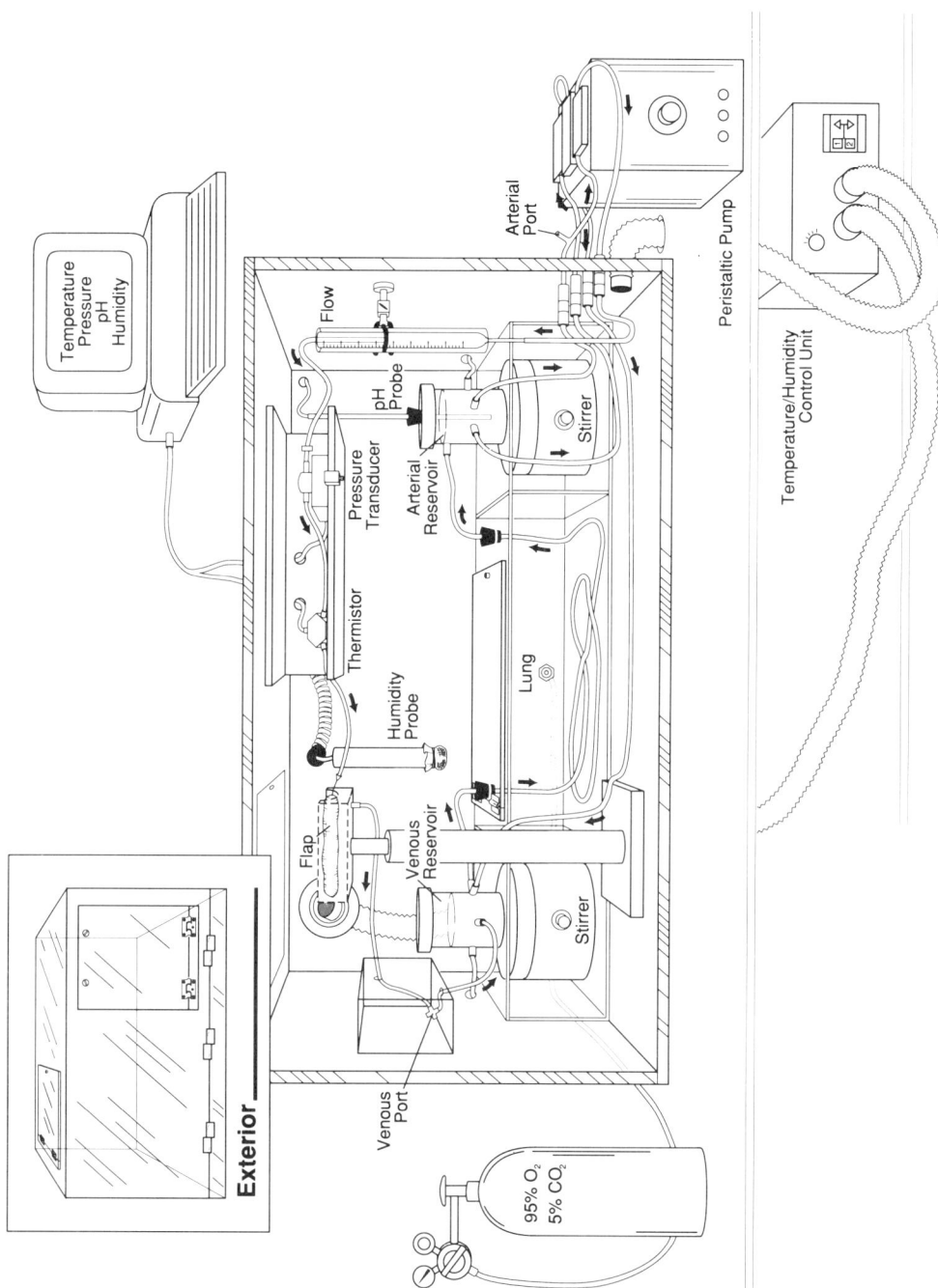

FIGURE 2. Specialized perfusion chamber used to maintain viability of IPPSFs.

nonrecirculating configurations are possible. In the recirculating mode (Figure 2), perfusate is constantly shunted between an "arterial" reservoir, the silastic oxygenator and the "venous" reservoir at a higher flow rate to maintain adequate mixing. In addition, each reservoir rests on a magnetic stirrer. The pH is measured in the arterial reservoir and maintained by infusion of appropriate buffers. A separate circuit, with ports for both arterial and venous sampling delivers oxygenated arterial perfusate at the regulated flow rate into the arterial cannula implanted in the IPPSF. The skin flap sits on a cradle which collects venous drainage from the preparation and returns it to the venous reservoir.[27,31]

The perfusion media is a Krebs-Ringer bicarbonate buffer (pH 7.4, 350 mosmole/kg), containing albumin (45 g/l) and supplied with glucose (80 to 120 mg/dl) as the primary energy source. Since the IPPSF is not a sterile organ preparation, antimicrobials (penicillin G and amikacin) are included to prevent bacterial overgrowth from the microflora normally present on the skin surface. Heparin is included to prevent coagulation in the skin flap's vasculature from residual formed blood elements[27,31] In the nonrecirculating mode, perfusate is pumped through the oxygenator to the arterial reservoir and then to the IPPSF. This is the primary configuration for percutaneous absorption studies. A fraction collector may be used to automatically collect venous perfusate over defined time intervals. This single-pass system has advantages for assessing topical bioavailability and metabolite profiles when recirculation would confound the analysis (e.g., cutaneous reuptake of drug).

IV. TISSUE VIABILITY AND BIOCHEMISTRY

To date, our laboratory has perfused over 1100 IPPSFs. Biochemical function of skin has been assessed in the IPPSF by monitoring glucose utilization (arterial and venous glucose extraction), lactate production, lactate dehydrogenase (LDH) leakage, perfusate flow, pressure, vascular resistance and pH.[27,31-33] Because of the high concentration of LDH normally found in skin, leakage into the perfusate should be an excellent marker of epidermal integrity. Terminal LDH concentrations are usually less than 10 IU/L, increasing an order of magnitude in nonviable preparations. In addition, samples are collected for light and transmission electron microscopy after the completion of an experiment. Light microscopy on samples taken at the end of a perfusion experiment demonstrate essentially normal appearing porcine skin. Similarly, detailed electron microscopic studies have confirmed this finding.[28] Glucose utilization, flow, pressure, vascular resistance, and pH have been determined to be sufficient for routine viability assessment during an experiment. Average glucose utilization is approximately 20 to 40 mg/h for a typical 30 g skin flap, dropping to less than 10 mg/h in nonviable preparations. Figure 3 depicts glucose utilization and Figure 4 mean perfusate flow over time in a large subset of perfused preparations. These figures illustrate the relative stability of perfusion parameters throughout the cause of a typical absorption experiment. Although these parameters are primarily used to assess viability in percutaneous absorption studies, they are also sensitive indicators for detecting direct cutaneous toxicity, another major use of the IPPSF which is not addressed in this chapter focused on percutaneous absorption.[34,35] Lactate production is linearly related to glucose consumption at a molar ratio of approximately 1.7. This finding is in agreement with numerous studies on cutaneous glucose utilization which suggests that 70 to 80% of cutaneous glucose is metabolized via glycolysis with lactate as the primary endproduct. A final sensitive real time indicator of IPPSF viability and function is the vascular resistance profile over the course of an experiment. Vascular resistance is calculated as the pressure/flow and is relatively stable throughout the course of an experiment unless toxicity occurs or vasoactive drugs are administered.[33,35] When vasodilators such as tolazoline are administered, vascular resistance decreases. Table 1 lists the mean flap characteristics and viability parameters for a large subset of normal IPPSFs.

MEAN GLUCOSE UTILIZATION (N=221)

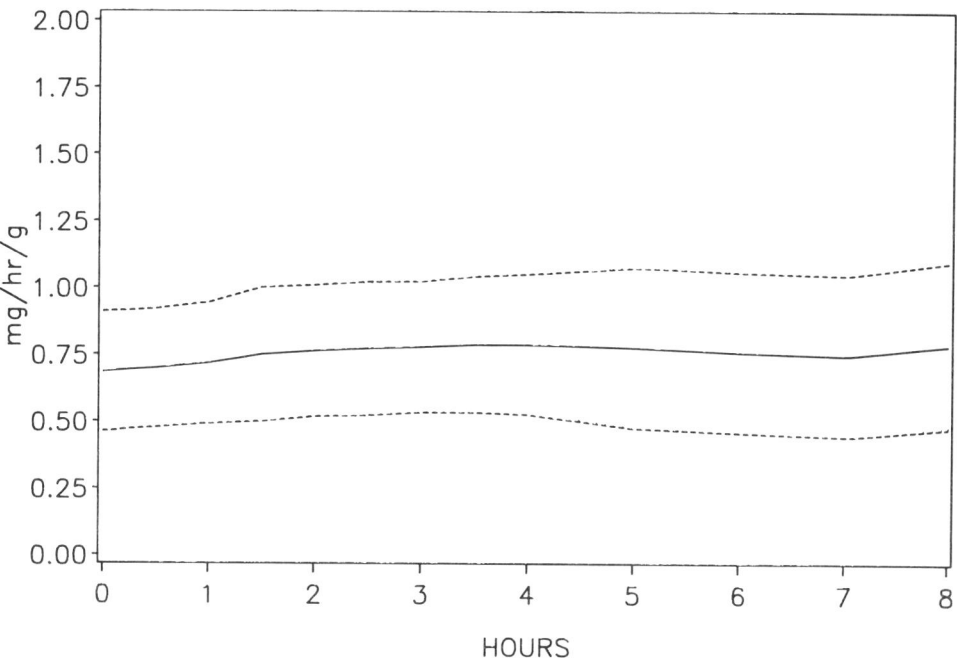

FIGURE 3.   Mean (± SD) glucose utilization over an 8-h period in a series of 221 normal IPPSFs.

MEAN FLOW (N=221)

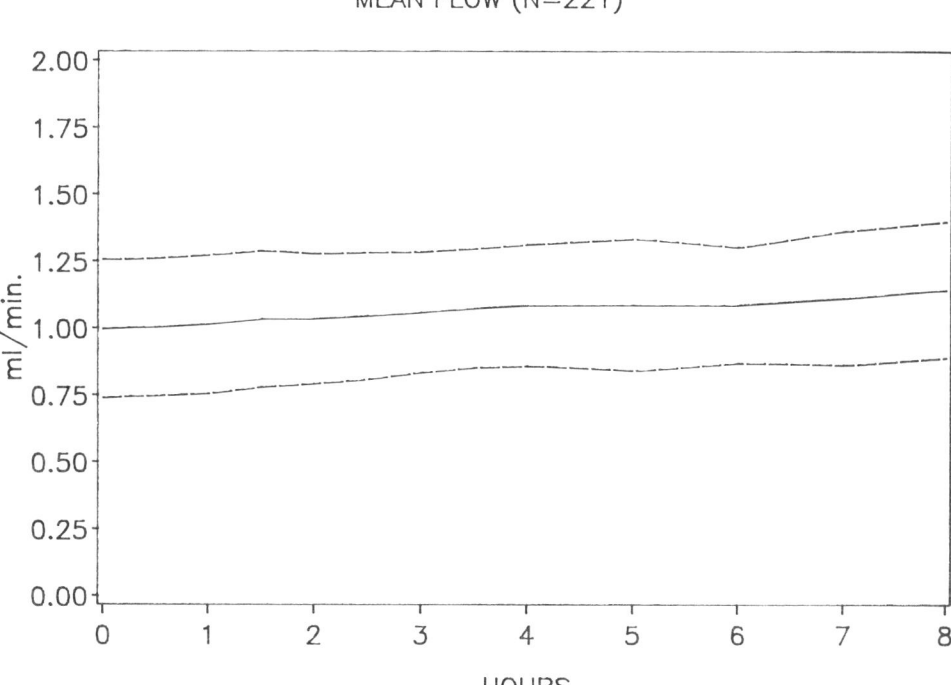

FIGURE 4.   Mean (± SD) perfusate flow over an 8-h period in a series of 221 normal IPPSFs.

TABLE 1
Characteristics of Normally Perfused IPPSFs
Used in Percutaneous Absorption Studies

Parameter	Mean	SD	N
Flap weight (g)	27.8	6.2	301
Perfusate flow (ml/min)	1.18	0.35	299
Perfusate pressure (mm Hg)	38	22	269
Glucose utilization (mg/h/g flap)	0.77	0.23	255
Perfusate osmolality (mosmole/kg)	298	9	41
Perfusate pH	7.45	0.12	255
Chamber relative humidity (%)	65	12	255
Chamber temperature (°C)	36.6	1.06	256

V. PERCUTANEOUS ABSORPTION STUDIES

Percutaneous absorption studies are conducted by placing the study chemical on the surface of the IPPSF and assaying venous effluent over time for absorbed compound. Compound may be applied neat or diluted in vehicle under ambient or occluded conditions. Various types of patches or transdermal delivery systems may also be employed. A dosing area is prepared (5 cm²) with a flexible, inert plastic border and compound applied within this area. IPPSFs are usually allowed to equilibrate for 1 to 2 h prior to compound application and most absorption studies are conducted in a nonrecirculating (single-pass) system. When experiments are designed to model uptake of compound from perfusate into the skin (systemic distribution, outward transdermal migration, inverse penetration, etc.), drug is added to the arterial reservoir and infused into the IPPSF. Analysis of the venous efflux profile allows one to study the kinetics of cutaneous uptake.[29,36-39]

Figures 5, 6, and 7 depict the venous efflux profiles of ^{14}C-labeled lindane, parathion, and carbaryl after topical application (n = 4 IPPSFs/compound) of 40 μg/cm² to a 5 cm² area of skin. These data, whose variability is representative of most IPPSF experiments, demonstrate the reproducibility of the model and its ability to differentiate between compounds by comparing the shapes of the venous efflux profiles.

In order to derive the most information from the IPPSF, pharmacokinetic models have been developed to analyze these venous efflux profiles. The first approach employed utilized a compartmental approach[29] which adequately described the profiles but was not optimal in predicting *in vivo* absorption. A physiological-based approach was then adopted which took advantage of the unique sampling sites available in an isolated perfused organ preparation. A mathematically identifiable model was initially developed to model cutaneous uptake experiments,[36] which was then expanded to study percutaneous absorption.[37] This model, formulated for a recirculating experimental design, is depicted in Figure 8. The model is formulated with four basic "compartments": (1) the cutaneous capillary bed which receives drug from the arteries with a flux of J_{01} and looses drug in the veins with an overall flux of J_{10}; (2) the nonvascular tissue of skin which receives topically penetrated drug at a rate of K_{42}; (3) a skin "depot" which tends to sequester chemical; and (4) the skin surface. Except for flow through the capillary bed, all other movements of compounds are described by linear first-order fractional rate constants (K_{12}, K_{21}, K_{42}, K_{23}, K_{32}, K_{42}, K_{40}). Note that if K_{32} is zero, compartment three functions as a sink. The rate of surface loss (K_{40}) may also be estimated. Similarly, in a nonrecirculating experiment where reperfusion does not occur, J_{01} is absent.

The principal reason for developing such a model is to extract the most information possible from the IPPSF venous efflux-time profile. Because of the ability to repeatedly obtain samples of both arterial and venous perfusate as well as directly measure perfusate

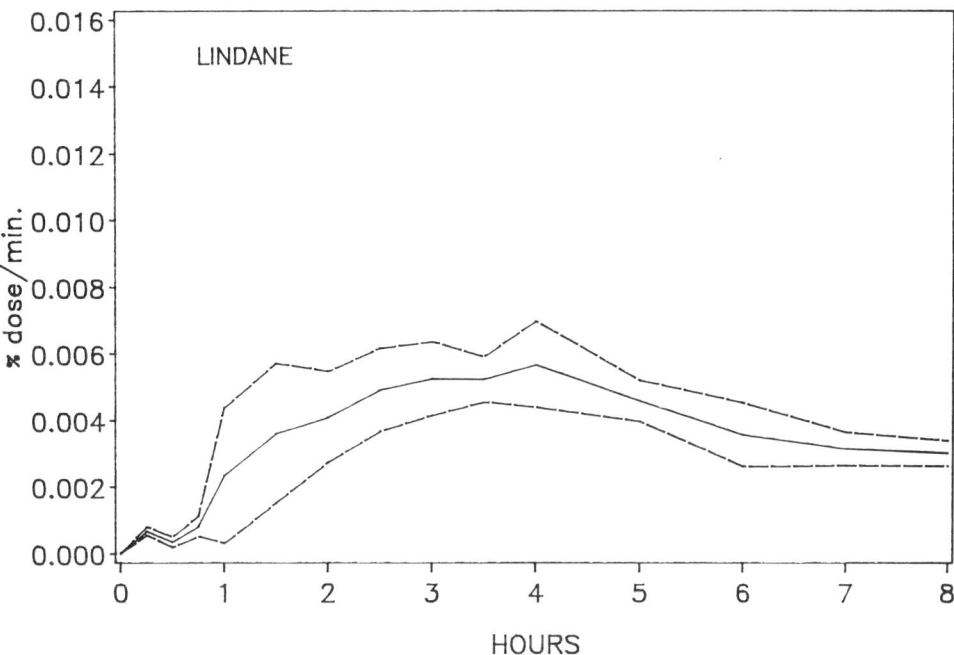

FIGURE 5.   Percutaneous absorption profile of lindane in four IPPSFs (Mean ± SD) after application of 40 μg/cm².

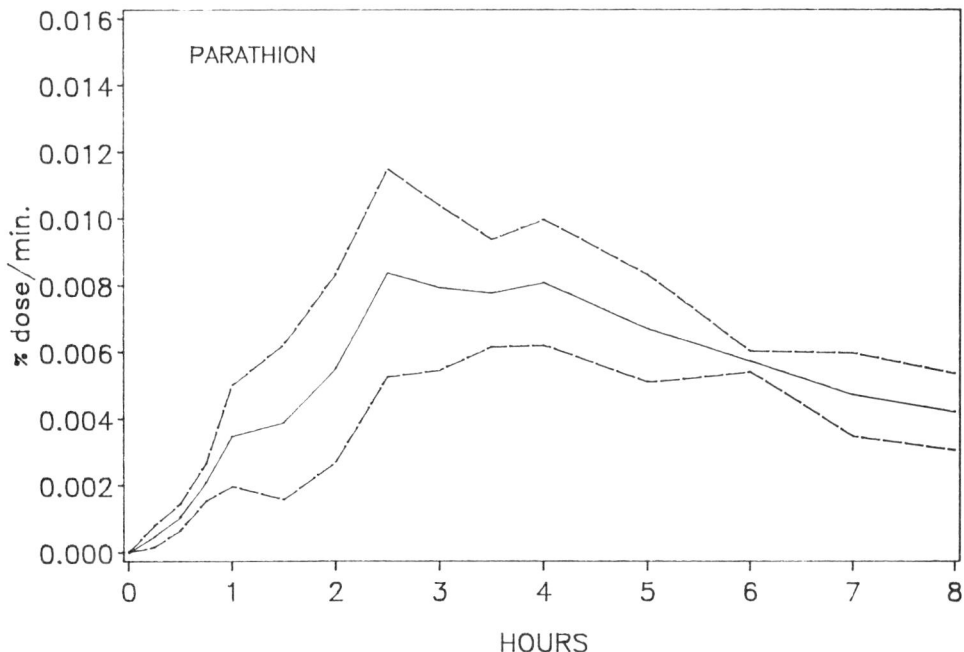

FIGURE 6.   Percutaneous absorption profile of parathion in four IPPSFs (Mean ± SD) after application of 40 μg/cm².

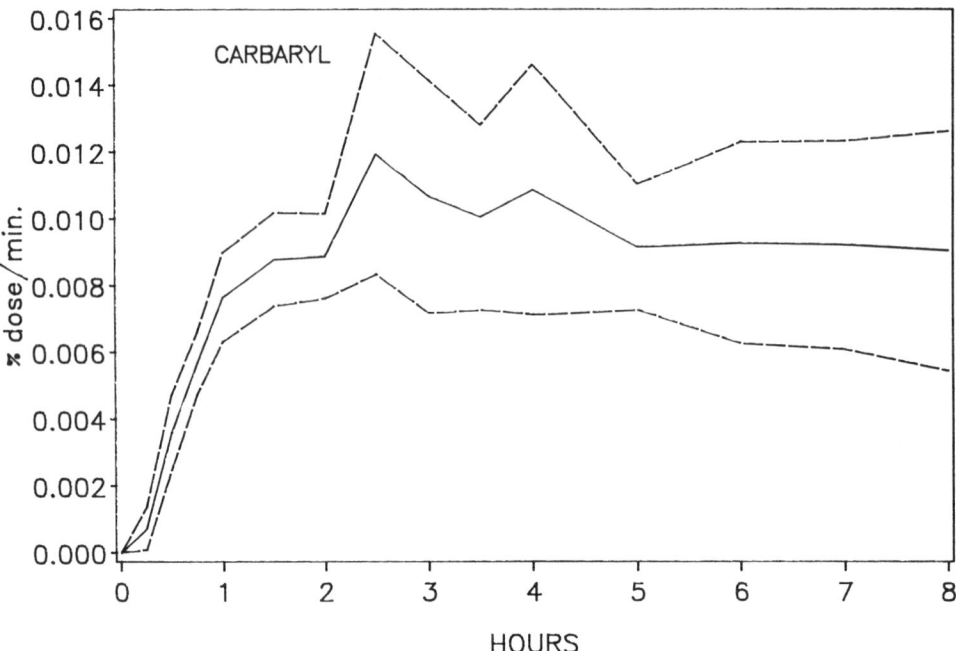

FIGURE 7. Percutaneous absorption profile of carbaryl in four IPPSFs (Mean ± SD) after application of 40 μg/ cm².

flow, relatively sophisticated yet conservative pharmacokinetic models may be formulated and then fit to the observed data. The parameters from these models (e.g., volumes, inter-compartmental rate constants) may be correlated to specific physiologic processes. The full models may then be used to extrapolate the percutaneous absorption flux profile to any point in time. Predicted tissue concentrations may be checked against actual assayed values. It must be stressed that the use of pharmacokinetic models in IPPSF preparations is different than their use in other *in vitro* and *in vivo* situations because of the large number of experimental sampling points available in an isolated perfused organ model. The use of such models also facilitates *in vivo* extrapolations, improves the quantifiability of mechanistic studies, and allows the multifactorial processes in dermal penetration to be independently evaluated (e.g., penetration vs. metabolism vs. depot formation).

The pharmacokinetic model in Figure 8 was used to analyze the IPPSF venous efflux profiles obtained after applying 40 μg/cm² (¹⁴C-radiolabeled) of benzoic acid, caffeine, carbaryl, diisopropylflurophosphate, malathion, parathion, progesterone, and testosterone to IPPSFs (n = 4/compound) in 8-h experiments. The cumulative amount absorbed after 6 d is then estimated using the computed model parameters for each compound. The same compounds under identical conditions were also administered *in vivo* to pigs and the dose fraction absorbed was determined.[40] The *in vivo* studies required parallel intravenous studies to determine fractions eliminated in excreta and total collection of urine and feces for 6 d until no further compound excretion was detected. The correlation of IPPSF predicted to actual *in vivo* absorption was excellent (R = .973)[37] and better than when IPPSF fluxes were analyzed using a classic compartmental model.[29]

When the IPPSF data is compared to previously published *in vivo* human absorption data for these same chemicals,[40-44] the linear correlation seen in Figure 9 is obtained. These data support the previous evidence that for certain compounds, the pig (and by extension the IPPSF) is an excellent model for predicting human absorption.

Additional studies have been conducted evaluating the cutaneous metabolism of parathion

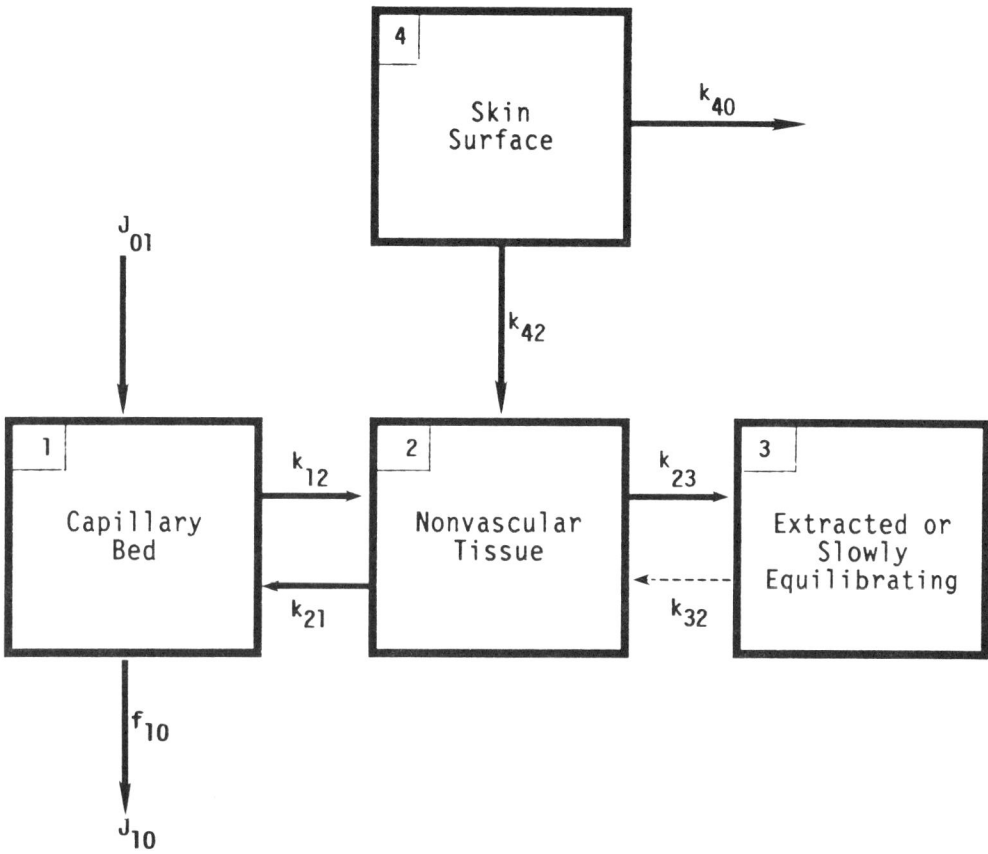

FIGURE 8. Physiological-based pharmacokinetic model for describing IPPSF percutaneous absorption data (Ks are rate constants, Js are fluxes).

seen during percutaneous absorption.[39] The organophosphate insecticide parathion is known to undergo in the liver a cytochrome P-450 mediated desulfuration reaction to form the biologically active metabolite paraoxon. This can be further metabolized to p-nitrophenol. When parathion was applied to IPPSFs in a recirculating experiment (Figure 10), these same metabolic transformations were seen to occur in the skin with approximately 80% of parathion being metabolized to paraoxon. Recent studies in nonrecirculating IPPSFs have indicated that only 25% of parathion undergoes this first-pass cutaneous metabolism, suggesting that a significant amount of metabolism occurs upon reperfusion in a recirculating experiment. When parathion is administered in an occluded patch, the same fraction of parathion penetrates but less paraoxon and more p-nitrophenol is produced. Finally, when flaps are pretreated with the cytochrome P-450 inhibitor 1-aminobenzotriazole, metabolism is drastically reduced. These preliminary data suggest that the IPPSF may prove to be a useful model to probe the mechanisms of pesticide absorption and cutaneous metabolism under different chemical and environmental influences.

Another application of the IPPSF involves studies in transdermal iontophoresis.[38] The most interesting result obtained involves the modulation of transdermal lidocaine flux seen when vasoactive drugs are coiontophoresed. As can be seen from the mean flux profiles depicted in Figure 11, lidocaine flux is greatly enhanced when a vasodilator such as tolazoline is coiontophoresed. A reduced flux is seen when a vasoconstrictor such as norepinephrine is coadministered. A similar enhancement with tolazoline is seen *in vivo* in pigs but is *not* seen when *in vitro* experiments are conducted using dermatomed skin. This direct involve-

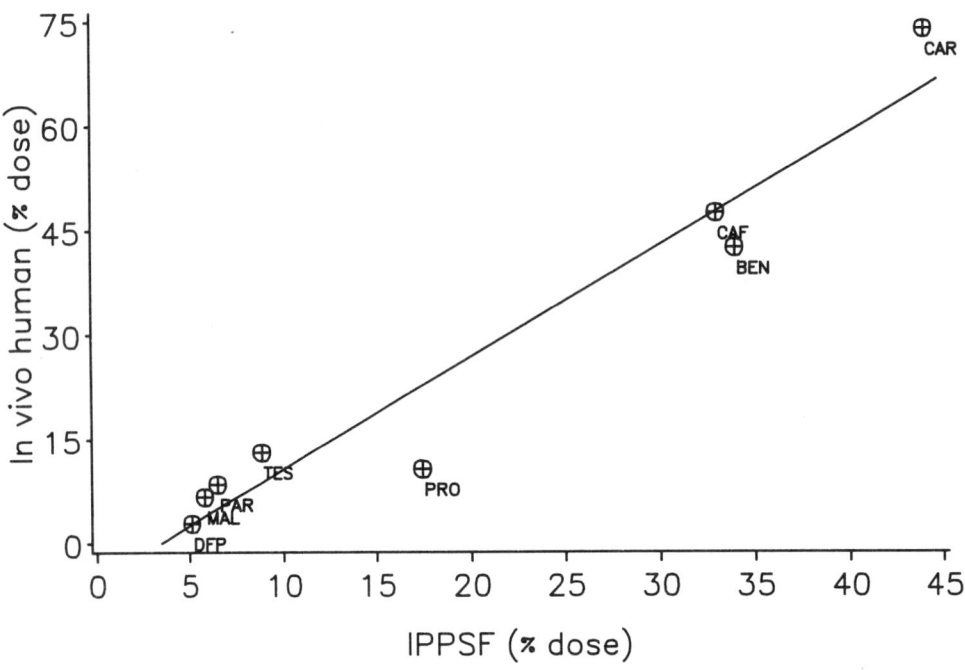

FIGURE 9.   IPPSF predicted versus reported *in vivo* human percutaneous absorption data for eight compounds (BEN = benzoic acid; CAF = caffeine; CAR = carbaryl; DFP = Diisopropylfluorophosphate; MAL = malathion; PAR = parathion; PRO = progesterone; TES = testosterone).

ment of the cutaneous microcirculation in percutaneous absorption is one of the advantages of utilizing a preparation such as the IPPSF. Another advantage seen in these studies was that a unique morphological alteration seen with high dose lidocaine iontophoresis *in vivo* was also seen in the IPPSF[34] allowing the pharmacological data to be integrated with these morphological findings.

VI. CONCLUSIONS

This brief overview illustrates the creation and utilization of the IPPSF for percutaneous absorption studies. Although by definition it is an *in vitro* model, in many respects it more closely resembles the *in vivo* situation. For compounds in which porcine and human absorption profiles are similar, the IPPSF should be a useful model to study factors which modulate absorption. Mechanistic studies may also be conducted in an *in vitro* model which more closely resembles the *in vivo* setting. This would be especially true for compounds which are vasoactive. The pharmacokinetics of percutaneous absorption and cutaneous metabolism may also be easily investigated. An especially promising area of research is the effects of altered environmental conditions, such as temperature and humidity, on the absorptive flux and cutaneous metabolic profile of topically applied compounds. Because these conditions may affect multiple points in the absorptive process (stratum corneum, epidermal cell activity, capillary flow), a model system which possesses all components is required to accurately model the overall process and to study the interactions between each component. Potential advantages are listed in Table 2.

It must be stressed that the IPPSF is *not* designed to be a replacement for other *in vitro* models. Diffusion cells studies using either animal, human, or human organ culture membranes will always have a critical role in assessing percutaneous absorption. Similarly, *in vivo* animal and human studies, when feasible, will be conducted. Economics will be a

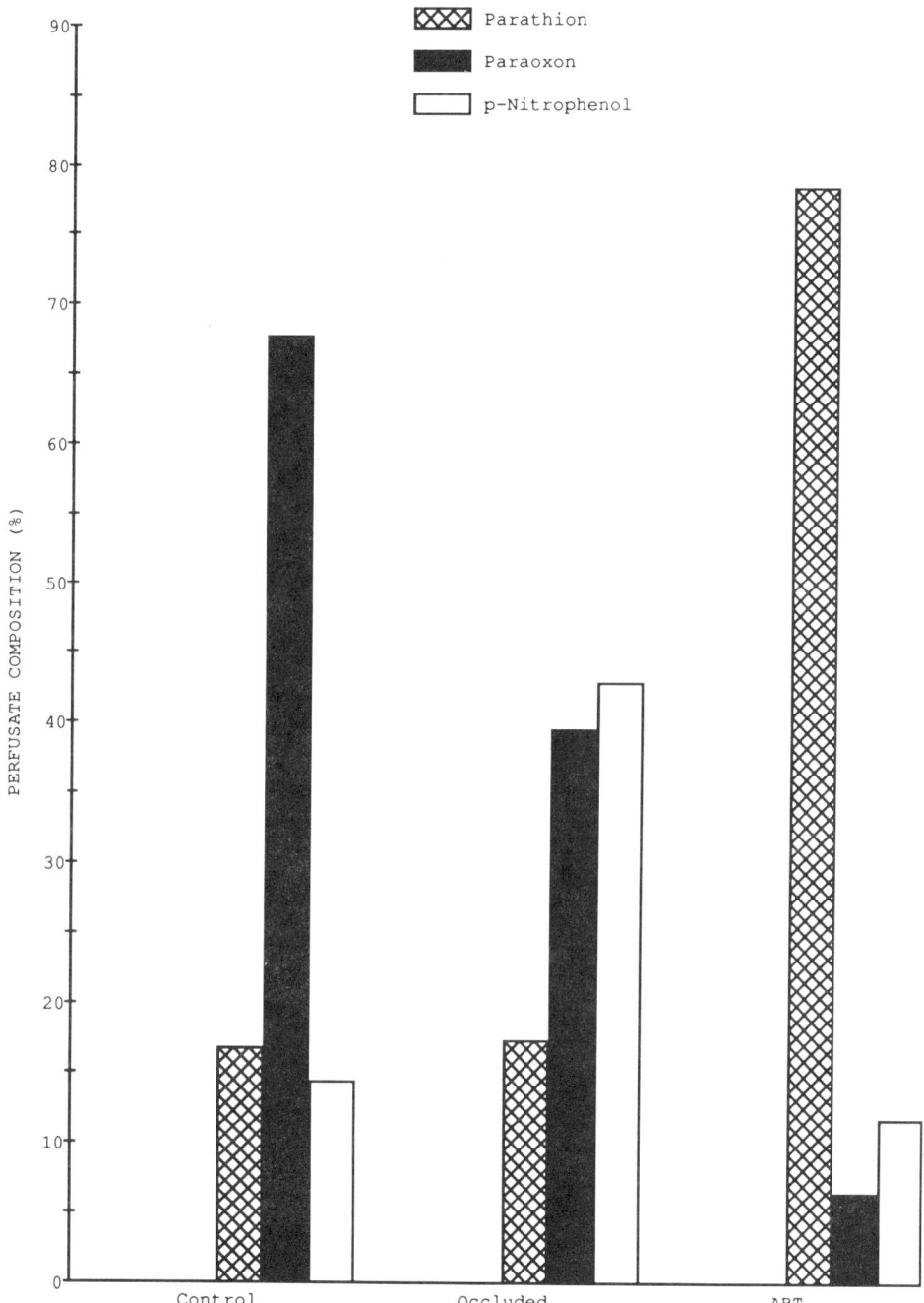

FIGURE 10. Metabolic profile of absorbed parathion and metabolites in IPPSFs after normal or occluded dosing, and after pretreatment with ABT (1-aminobenzotriazole), a cytochrome P-450 inhibitor.

major factor since IPPSF studies are more comparable to *in vivo* costs. The only situation this author can see which might dictate more extensive use of the IPPSF in screening studies is when the compound or drug delivery process modulates the cutaneous microcirculation. In these cases, classic *in vitro* techniques may be misleading and *in vivo* strategies do not allow for precise monitoring of low level drug or xenobiotic metabolite fluxes. Prior validation of the IPPSF on a compound by compound basis with human data would be prudent

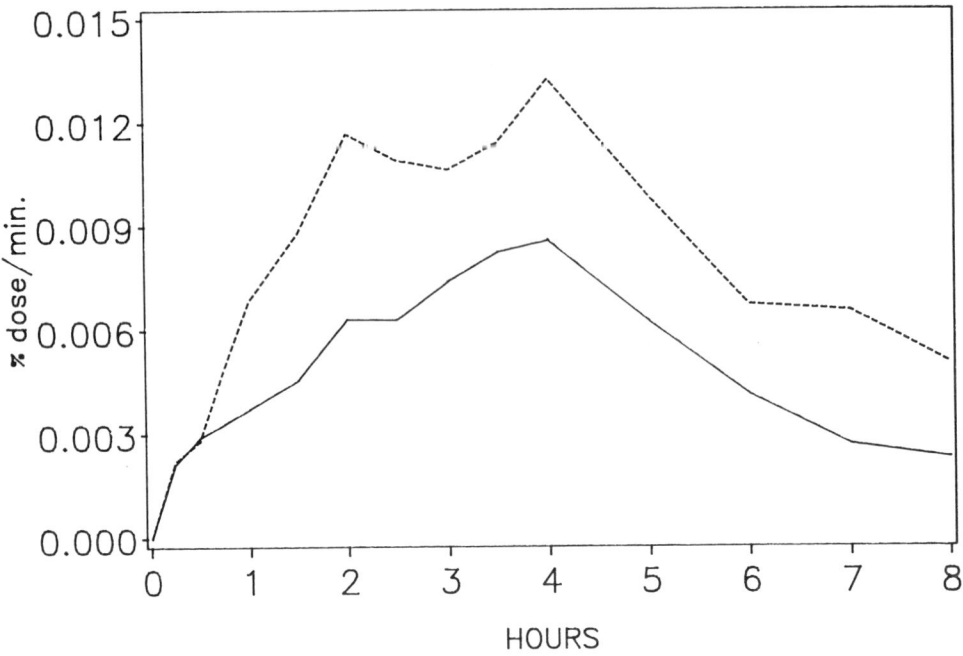

FIGURE 11.   Percutaneous absorption profile of lidocaine hydrochloride administered by transdermal iontophoresis (—, lidocaine alone; -----, lidocaine and tolazoline coiontophoresed).

TABLE 2
Potential Advantages of the Isolated Perfused Porcine Skin Flap for Percutaneous Absorption Studies

Advantages
   Anatomical integrity
   Viability
   Functional cutaneous microcirculation
   Ease of sampling arterial and venous perfusates
   Controlled environment allows for manipulation of perfusate and
    environmental (temperature, humidity) conditions
   Allows cutaneous metabolism to be studied without systemic
    metabolic interference
   Relatively large skin surface area
   Local toxicity may be detected

if it is to be employed to design transdermal delivery systems. A paradox is that for compounds which because of toxicity (potential or real) cannot be given to humans, the IPPSF may be an ideal model to estimate the degree of human absorption. As indicated in the beginning of this chapter, a single model should not solely be relied on for risk assessment purposes, a hierarchy or battery of different systems should be employed and quantitative strategies developed to provide an integrated risk assessment profile.

REFERENCES

1. **Monteiro-Riviere, N. A.,** Comparative anatomy, physiology, and biochemistry of mammalian skin, in *Dermal and Ocular Toxicology: Fundamentals and Methods,* Hobson, D. W., Ed., Telford Press, chap. 1, in press.

2. **Kjaersgaard, A. R.,** Perfusion of isolated dog skin, *J. Invest. Dermatol.,* 22, 135, 1954.

3. **Bartek, M. J., LaBudde, J. A., and Maibach, H. I.,** Skin permeability *in vivo*: comparison in rat, rabbit, pig, and man, *J. Invest. Dermatol.,* 58, 119, 1972.

4. **Bartek, M. J. and LaBudde, J. A.,** Percutaneous absorption, *in vitro,* in *Animal Models in Dermatology,* Maibach, H. I., Ed., Churchill Livingstone, New York, 1975, 103.

5. **Wester, R. C. and Maibach, H. I.,** Animal models for percutaneous absorption, in *Models in Dermatology,* Vol. 2, Maibach, H. I. and Lowe, N. J., Eds., Karger, Basel, 1985, 159.

6. **Bronaugh, R. L., Stewart, R. F., and Congdon, E. R.,** Methods for *in vitro* percutaneous absorption studies. II. Animal models for human skin, *Toxicol. Appl. Pharm.,* 62, 481, 1982.

7. **Marzulli, F. N., Brown, D. W. C., and Maibach, H. I.,** Techniques for studying skin penetration, *Toxicol. Appl. Pharmacol. Suppl.,* 3, 79, 1969.

8. **Reifenrath, W. G., Chellquist, E. M., Shipwash, E. A., Jederberg, W. W., and Kreuger, G. G.,** Percutaneous penetration in the hairless dog, weanling pig and grafted athymic nude mouse: evaluation of models for predicting skin penetration in man, *Brit. J. Dermatol.,* 111(Suppl. 27), 123, 1984.

9. **Hawkins, G. S. and Reifenrath, W. G.,** Influence of skin source, penetration fluid and partition coefficient on *in vitro* skin penetration, *J. Pharm.. Sci.,* 75, 378, 1986.

10. **Monteiro-Riviere, N. A.,** Ultrastructural evaluation of the porcine integument, in *Swine in Biomedical Research,* Vol. 1, Tumbleson, M. E., Ed., Plenum Press, New York, 641, 1986.

11. **Monteiro-Riviere, N. A. and Stromberg, M. W.,** Ultrastructure of the integument of the domestic pig *(Sus scrofa)* from one through fourteen weeks of age, *Anat. Histol. Embryol.,* 14, 97, 1985.

12. **Bronaugh, R. L., Stewart, R. F., and Congdon, E. R.,** Methods for *in vitro* percutaneous absorption studies. II. Animal models for human skin, *Toxicol. Appl. Pharmacol.,* 62, 481, 1982.

13. **Forbes, P. D.,** Vascular supply of the skin and hair in swine, in *Hair Growth, Advances in the Biology of Skin,* Vol. 9, Montagna, W. and Dobson, R. L., Eds., Pergamon Press, New York, 1969, 419.

14. **Ingram, D. L. and Weaver, M. E.,** A quantitative study of the blood vessels of the pig's skin and the influence of environmental temperature, *Anat. Rec.,* 163, 517, 1969.

15. **Meyer, W., Neurand, K., and Radke, B.,** Elastic fiber arrangement in the skin of pigs, *Arch. Dermatol. Res.,* 270, 391, 1981.

16. **Meyer, W., Neurand, K., and Radke, B.,** Collegen fiber arrangements in the skin of pigs, *J. Anat.,* 134, 139, 1982.

17. **Meyer, W., Schwarz, R., and Neurand, K.,** The skin of domestic mammals as a model for the human skin, with special reference to the domestic pig, *Curr. Probl. Dermatol.,* 7, 39, 1978.

18. **Ingram, D. L. and Weaver, M. E.,** A quantitative study of the blood vessels of the pig's skin and influence of environmental temperature, *Anat. Rec.,* 163, 517, 1969.

19. **Meyer, W. and Neurand, J.,** The distribution of enzymes in the skin of the domestic pig, *Lab Anim.,* 10, 237, 1976.

20. **Klain, G. J., Bonner, S. J., and Bell, W. G.,** The distribution of selected metabolic processes in the pig and human, in *Swine in Biomedical Research,* Vol. 1, Tumbleson, M. E., Ed., Plenum Press, New York, 1986, 667.

21. **Meyer, W., Schwarz, R., and Neurand, K.,** The skin of domestic mammals as a model for human skin, with special reference to the domestic pig, *Curr. Probl. Dermatol.,* 7, 39, 1978.

22. **Meyer, W. and Neurand, K.,** The distribution of enzymes in the skin of the domestic pig, *Lab. Anim.,* 10, 237, 1976.

23. **Gray, G. M. and Yardley, H. J.,** Lipid compositions of cells isolated from pig, human and rat epidermis, *J. Lipid Res.,* 16, 434, 1975.

24. **Nicolaides, N., Fu, H. C., and Rice, G. R.,** The skin surface lipids of man compared with those of eighteen species of animals, *J. Invest. Dermatol.,* 51, 83, 1968.

25. **Hedbert, C. L., Wertz, P. W., and Downing, D. T.,** The nonpolar lipids of pig epidermis, *J. Invest. Dermatol.,* 90, 225, 1988.

26. **Weinstein, G. D.,** Comparison of turnover time and of keratinous protein fractions in swine and human epidermis, in *Swine in Biomedical Research,* Bustad, L. K., McClellan, R. O., and Burns, M. P., Eds., Pacific Northwest Institute, Richland, WA, 287, 1966.

27. **Riviere, J. E., Bowman, K. F., Monteiro-Riviere, N. A., Carver, M. P., and Dix, L. P.,** The isolated perfused porcine skin flap (IPPSF). I. A novel *in vitro* model for percutaneous absorption and cutaneous toxicology studies, *Fund. Appl. Toxicol.,* 7, 444, 1986.

28. **Monteiro-Riviere, N. A., Bowman, K. F., Scheidt, V. J., and Riviere, J. E.,** The isolated perfused porcine skin flap (IPPSF): II. Ultrastructural and histological characterization of epidermal viability, *In Vitro Toxicol.*, 1, 241, 1987.

29. **Carver, M. P., Williams, P. L., and Riviere, J. E.,** The isolated perfused porcine skin flap (IPPSF). III. Percutaneous absorption pharmacokinetics of organophosphates, steroids, benzoic acid and caffeine, *Toxicol. Appl. Pharmacol.*, 97, 324, 1989.

30. **Bowman, K. F., Monteiro-Riviere, N. A., and Riviere, J. E,** Development of surgical techniques for preparation of *in vitro* isolated perfused porcine skin flaps for percutaneous absorption studies, *Am. J. Vet. Res.*, 52, 75, 1991.

31. **Monteiro-Riviere, N. A.,** Specialized technique: Isolated perfused porcine skin flap, in *Methods for Skin Absorption,* Kemppainen, B. W. and Reifenrath, W. G., Eds., CRC Press, Inc., 1990, 175.

32. **Riviere, J. E., Bowman, K. F., and Monteiro-Riviere, N. A.,** On the definition of viability in isolated perfused skin preparation, *Brit. J. Dermatol.*, 116, 739, 1987.

33. **Williams, P. L. and Riviere, J. E.,** Estimation of physiological volumes in the isolated perfused porcine skin flap, *Res. Commun. Chem. Pathol. Pharmacol.*, 66, 145, 1989.

34. **Monteiro-Riviere, N. A.,** Altered epidermal morphology secondary to lidocaine iontophoresis: *in vitro* and *in vivo* studies in porcine skin, *Fund. Appl. Toxicol.*, 15, 174, 1990.

35. **King, J. R., and Monteiro-Riviere, N. A.,** Cutaneous toxicity of 2-chloroethyl methyl sulfide in isolated perfused porcine skin, *Toxicol. Appl. Pharm.*, 104, 167, 1990.

36. **Williams, P. L. and Riviere, J. E.,** Definition of a physiologic pharmacokinetic model of cutaneous drug distribution using the isolated perfused porcine skin flap (IPPSF), *J. Pharm. Sci.*, 78, 550, 1989.

37. **Williams, P. L., Carver, M. P., and Riviere, J. E.,** A physiologically relevant pharmacokinetic model of xenobiotic percutaneous absorption utilizing the isolated perfused porcine skin flap (IPPSF), *J. Pharm. Sci.*, 79, 305, 1990.

38. **Riviere, J. E., Sage, B., and Monteiro-Riviere, N. A.,** Transdermal lidocaine iontophoresis in isolated perfused porcine skin, *Cutan. Ocular Toxicol.*, 9, 493, 1989/1990.

39. **Carver, M. P., Levi, P. E., and Riviere, J. E.,** Parathion metabolism during percutaneous absorption in perfused porcine skin, *Pest. Biochem. Physiol.*, 38, 245, 1990.

40. **Carver, M. P. and Riviere, J. E.,** Percutaneous absorption and excretion of xenobiotics after topical and intravenous administration to pigs, *Fund. Appl. Toxicol.*, 13, 714, 1989.

41. **Feldman, R. J. and Maibach, H. I.,** Percutaneous absorption of steroids in man, *J. Invest. Dermatol.*, 52, 89, 1969.

42. **Feldman, R. J. and Maibach, H. I.,** Absorption of some organic compounds through the skin of man, *J. Invest. Dermatol.*, 54, 399, 1970.

43. **Wester, R. C. and Maibach, H. I.,** Relationship of topical dose and percutaneous absorption in rhesus monkey and man, *J. Invest. Dermatol.*, 67, 518, 1976.

44. **Wester, R. C. and Maibach, H. I.,** *In vitro* percutaneous absorption and decontamination of pesticides in humans, *J. Toxicol. Environ. Health,* 16, 25, 1985.

Chapter 17

PERCUTANEOUS PENETRATION FROM SOIL: INHIBITION OF ENHANCEMENT?

Daniel A. W. Bucks, Ronald C. Wester, and Howard I. Maibach

TABLE OF CONTENTS

I. INTRODUCTION

A major dilemma in establishing regulatory limits for environmental pollutants is the establishment of environmental standards or limits for chemical concentrations in soil at industrial and residential sites. Factors and assumptions used to predict the bioavailability of a chemical from soil significantly affects the establishment of a virtually safe dose or acceptable daily exposure level of a compound in soil. The two major concerns in setting relevant contamination levels are public safety and cost/feasibility of clean-up.

The cost of remediation varies dramatically with the level to which soil must be decontaminated and excessive remediation means that limited resources will be spent without providing additional protection of public health. For example the cost of removing and disposing of soil containing more than 1 ppb 2,3,7,8-tetrachlorodibenzo-*p*-dioxin (TCDD) at the Castlewood site in Missouri would be $17,000,000. However, if the level for clean-up at that site were set at 10 ppb, the cost would drop to $6,000,000 and minimal action would be required at 100 ppb TCDD. The authors state other sites, such as Times Beach, Missouri, show an even more dramatic relationship between the cost of clean-up and the degree of remediation.

The following review addresses the current information on the methods and results obtained from investigations of chemical percutaneous penetration from a soil. Hopefully this exercise will stimulate consensus in (1) the significance of results obtained from previous studies in regards to the establishment of virtually safe chemical levels in soil and (2) the standardization of methodology employed in quantification of percutaneous penetration of chemicals from soil.

II. INVESTIGATIONS

Roels et al.[1] conducted an environmental study concerning blood lead (pb) levels among 11-year-old children attending schools situated less than 1 and 2.5 km from a primary lead smelter. Age-matched control children from a rural and an urban area were contemporaneously examined. Samples analyzed for lead were hand rinses, blood, air, and soil. The hand rinses were collected by slowly pouring 500 ml of $0.1N$ NHO_3 over the palm of one hand while the fingers were slightly spread.

As expected, lead exposure based upon blood lead levels decreased with distance from the smelter. Inhalation of airborne lead was the major source of slightly increased lead levels in adults; whereas, children demonstrated an increased blood lead level resulting from soiled hands. The authors conclude that in a lead smelter area, the enforcement of a permissible limit for airborne lead alone may not necessarily prevent an excessive exposure of children to environmental lead, since past emission of lead and possible transport of lead-containing dirt (e.g., through road transport) will maintain a high level of lead in soil dust and dirt irrespective of the current concentration of atmospheric lead.

Using the data published by Roels et al.[1] we estimate the mass of soil per area of skin adhering to the children as 2.3 ± 0.3 mg soil/cm^2 hand (assuming an area of 70 cm^2 for the child's palm and fingers and the rinsing procedure completely removes the dirt). See Table 1.

Poiger and Schlatter[2] studied the effect of solvents and adsorbents on the *in vivo* percutaneous penetration of TCDD using the hairless rat animal model (sex and site of application not published). The percutaneous penetration of TCDD from soil, methanol, activated carbon: water paste, petrolatum, polyethylene glycol 1500, or polyethylene glycol 1500: water was examined under aluminum foil occluded conditions. TCDD topical exposure levels ranged from 6.5 to 433 ng/cm^2. Soil was sieved to <160 μm, then ground in a mortar prior to formulation with TCDD. Each formulation was applied over 3 to 4 cm^2 for 24 h. Percutaneous

TABLE 1
Mass of Soil Per Unit of Area Adhering to Children

Location	Soil (pb μg/g)	Hand (μg pb)	mg Soil/cm² Hand[a]
Rural	114	17	2.1
Urban	112	20.4	2.6
2.5 km	466	62.2	<u>1.9</u>
< 1 km	2560	436	2.4

[a] Mean = 2.3; SD = 0.3.

Data taken from Roels, H. A., Buchet, J.-P., Lauwerys, R. R., Braux, P., Claeys-Thoreau, F., Lafontaine, A., and Verduyn, G., *Environment*, 22, 81, 1990.

penetration was assessed by measurement of radioactivity in the liver 24-h postapplication of compound.

The relative dermal bioavailability of TCDD from the above formulations was assessed by comparison of liver TCDD levels at 24 h from dosing. This type of analysis does not address the rate or total extent of TCDD percutaneous penetration. They reported that adsorption of TCDD to activated carbon completely prevented its percutaneous penetration. Our statistical analysis of their data indicated a significant different ($p < 0.01$, ANOVA) in TCDD penetration from the MeOH, petrolatum, and soil formulations. TCDD penetration from MeOH was significantly greater ($p < 0.05$, Newman-Keuls test) than from petrolatum or soil formulations. There was no significant difference ($p > 0.05$, Newman-Keuls test) between soil and petrolatum formulations. In addition, a significant difference ($p < 0.01$, ANOVA) in TCDD penetration from the polyethylene glycol 1500, polyethylene glycol: 15% water, and soil formulations was observed. Newman-Keuls test results indicated that TCDD penetration from soil was significantly less ($p < 0.05$) than from polyethylene glycol 1500 or polyethylene glycol—15% water formulations but no significant difference ($p > 0.05$) in TCDD penetration from the polyethylene glycol 1500 and polyethylene glycol: 15% water formulations. The authors conclude that TCDD penetration is highly dependent on the formulation in which it is applied and that mixing TCDD with soil or activated carbon results in reduced compound bioavailability.

Shu et al.[3] determined TCDD bioavailability from Times Beach, Missouri soil in the presence and absence of used crank case oil. The alluvial sandy loam soil was air-dried and sieved through 10-, 20-, and 40-mesh screens (<425 μm). Tritiated TCDD in hexane: methylene chloride (7:3, v/v) was added to the soil and mixed by passage through the 40-mesh screen three times. In studies employing crank case oil, the used oil was first added to the soil to achieve 0.5 or 2.0% (w/w) prior to [³H]TCDD addition. *In vivo* dermal bioavailability of soil-bound TCDD was estimated using male Sprague-Dawley and hairless (Naked ex Backcross and Holtzman strain) rats. A 12 cm² area of the back was the application site. Haired rats were lightly clipper shaven prior to dosing. The TCDD contaminated soil was applied by gentle rubbing. The site of application was covered with a nonocclusive perforated aluminum eye protector whose edges were covered with foam, backed with adhesive, and fixed in place with masking tape. Formulations of TCDD contaminated soil were left on the skin for 4 and 24 h.

As Poiger and Schlatter[2] did earlier, the relative dermal bioavailability of TCDD from the above formulations was assessed by comparison of liver TCDD levels at 24 h from dosing, therefore, the rate or total extent of TCDD percutaneous penetration is not known. However, relative differences in percutaneous penetration between the various treatment modalities should not be affected. The data indicated that variations in TCDD dose (10 compared to 100 ppb TCDD in soil) and oil concentration (0, 0.5, and 2.0%) did not affect the relative percent applied dose penetrating skin. A 4-h exposure resulted in about 60% of an applied dose penetrating from a 24-h exposure. There was no significant difference in TCDD percutaneous penetration from soil between lightly clipper shaven rats and hairless rats. Unfortunately the authors did not do the control study involving TCDD penetration in the absence of soil. As a side note, the holes perforating the protective cover are not small enough (~2 mm) to prevent surface loss of chemical adsorbed to soil or exfoliated keratinocytes from the application site. Using the data from Poiger and Schlatter,[2] Kimbrough et al.[4] of the Centers for Disease Control (CDC) have stated: "one ppb of 2,3,7,8-TCDD in soil is a reasonable level at which to begin consideration of action to limit human exposure for contaminated soil."

Brewster et al.[5] have studied the *in vivo* dermal absorption of ^{3}H-TCDD using male F344 rats. Radiolabeled TCDD was applied at 6 dose levels (0.00015, 0.001, 0.01, 0.1, 0.5, and 1.0 μmol/kg) from an acetone solution to the intracapular region of the back (total surface area of application not reported). The site of application was covered with a stainless steel perforated cap. The percentage of applied dose absorbed (defined as the difference between the administered dose and the amount in the application site) declined with increasing dose while the absolute total mass (μg/kg) absorbed increased nonlinearly with dose. Absorption of TCDD ranged from approximately 40% of the applied dose when applied at 0.001 and 0.00015 μg/kg to approximately 18% at the 0.1, 0.5, and 1.0 μg/kg dose levels. Major tissue depots included liver, adipose, skin, and muscle. The investigators conclude: "(Our) results indicate that the dermal absorption of these compounds is incomplete and that systemic toxicity following acute dermal exposure to levels found in the environment is unlikely." "Although the potential for systemic toxicity after acute environmental exposure to these chemicals is low, chronic low-level dermal exposure to these compounds . . . could result in bioaccumulation of [TCDD] body burdens sufficient to induce toxicity."

Evident from the statements from the several laboratories cited above is the lack of consensus on: (1) the veracity of the dermal penetration and (2) the target organ(s) of potential toxicity resulting from topical exposure. Age group (infants, toddlers, children, and adults) will affect the predominant route of exposure (oral vs. dermal absorption). This review addresses point 1 only since 2 is a separate issue.

Skowronski et al.[6,7] studied the *in vivo* percutaneous penetration of benzene and toluene when applied as pure chemical and when incorporated with sandy or clay soils. Male Sprague-Dawley rats were dosed on 13 cm^2 of lightly shaved costoabdominal skin. The application site was occluded with a glass cap. Chemicals were applied for 48 h. Percutaneous penetration was assessed by urinary excretion of rabiolabel.

The authors showed that the percutaneous penetration of benzene was significantly higher ($p < 0.05$) when applied as the pure chemical (86%) compared to application with either soils and penetration was significantly higher ($p < 0.05$) from sandy soil (64%) compared to clay soil (45%). However, no significant difference ($p > 0.05$) in toluene percutaneous penetration was observed between pure chemical (76%), sandy soil (85%), and clay soil (80%). The authors conclude that percutaneous penetration of a chemical from soil can approach levels observed when applied as pure compound.

Of importance are: (1) How did the investigators measure their evaporative losses used to correct the data? (2) What method was used to correct all data for volatile losses of 53, 51, and 55% of radioactivity during administration of toluene alone or in the presence of sandy or clay soil, respectively; and 67, 59, and 39% during benzene administration as pure chemical, with sandy soil, and with clay soil, respectively? (3) How did such large percentages of the chemical evaporate from the skin surface under "occluded" conditions? (4) The large volumes of compound per area of skin (24 μl/cm^2 benzene and 1.3 μl/cm^2 toluene) employed may not be applicable to the environmental situation.

Yang et al.[8] studied the *in vivo* and *in vitro* percutaneous penetration of benzo[a]pyrene (B[a]P) using female Sprague-Dawley rats dosed on lightly shaven dorsal skin. The application site was covered with a nonocclusive glass cell *in vivo* and simply left uncovered *in vitro*. B[a]P was applied *in vivo* and *in vitro* in either crude oil or crude oil contaminated air-dried loam soil. B[a]P concentration was 100 ppb in crude oil. B[a]P "contaminated" crude oil was applied as an acetone: carbon disulfide (1:1, v/v) solution with the solvent evaporated off with either air or $N_{2(g)}$. Soil was sieved and particle sizes <150 μm were used. The 1% crude oil contaminated soil was prepared by addition of the crude oil in 10 ml dichloromethane per gram soil and evaporating off the solvent. In the *in vitro* studies, skin was dermatomed to ~350 μm and a receptor fluid consisting of an aqueous solution of 6% Volpo 20 and 0.01% thimerosal employed. Formulations were applied for up to 96 h.

These authors reported no significant difference in B[a]P penetration between their *in vivo* and *in vitro* methodologies. The soil-sorbed crude showed a significant reduction in penetration (8.4%) compared with crude-oil-only experiments (38.1%) when B[a]P was applied at the same chemical dose. Increasing the mass of soil applied in excess of the minimal amount required to cover the skin surface (9.0 mg/cm^2 "monolayer" amount) did not increase the mass of B[a]P penetrating *in vivo*. The apparent conclusion from these studies is that B[a]P percutaneous penetration from crude oil contaminated soil will be significantly less than from crude oil alone.

Wester et al.[9] investigated the *in vivo* and *in vitro* percutaneous penetration of ^{14}C-labeled DDT from soil and acetone. The soil (Yolo Co. 65-Calif-57-8; 26% sand, 26% clay, 48% silt) was sieved and particles sizes of 180 to 300 μm were used. Radiolabeled DDT was added to the soil in hexane:methylene chloride (7.3, v/v). Soil was mixed well by hand, open to the air to allow dissipation of solvent. DDT concentration in soil and acetone was 10 ppm. The same masses and volumes were employed in the *in vitro* and *in vivo* studies.

In the *in vitro* studies, flow-through design cells, human cadaver skin dermatomed to 500 μm and human plasma as the receptor fluid were employed. After a 24-h exposure, the system was stopped and the skin surface washed with soap and water. The plasma receptor phase, skin washes, glass apparatus washes, and the skin itself were assayed for radiolabel.

In the *in vivo* studies, rhesus monkeys were dosed over 12 cm^2 of abdominal skin for 24 h. The site of application was protected by use of a nonocclusive device (two perforated aluminum eye protectors with a water permeable membrane sandwiched in between). Percutaneous penetration was determined by the method of Feldmann and Maibach[10,11] as modified for use with rhesus monkeys by Wester and Maibach.[12]

In vitro percutaneous absorption of DDT from acetone was significantly higher than from soil. DDT would not readily partition from human skin into the plasma receptor phase. Washing with soap and water removed most of the remaining surface material. The *in vivo* percutaneous penetration of DDT from acetone solution using the monkey (19% ± 9%) was not significantly different from that reported for man (10% ± 4%).[11] Significantly less DDT penetrated *in vivo* from soil in the rhesus monkey (3.3% ± 0.5%). For the acetone vehicle, the *in vitro* skin content of DDT (after soap and water washing) was the same as that penetrating the monkey skin *in vivo*. For DDT in soil, *in vitro* skin amounts were slightly lower than that penetrating *in vivo*. In conclusion, *in vivo* DDT absorption from soil will occur in man, but that amount will be less than a "surface deposit" of chemical from a volatile solvent (Figure 1).

III. DISCUSSION

Exposure assessment of an area with chemical contaminated soil is linked to considerations of excess risks of developing specific adverse health effects as a result of the total

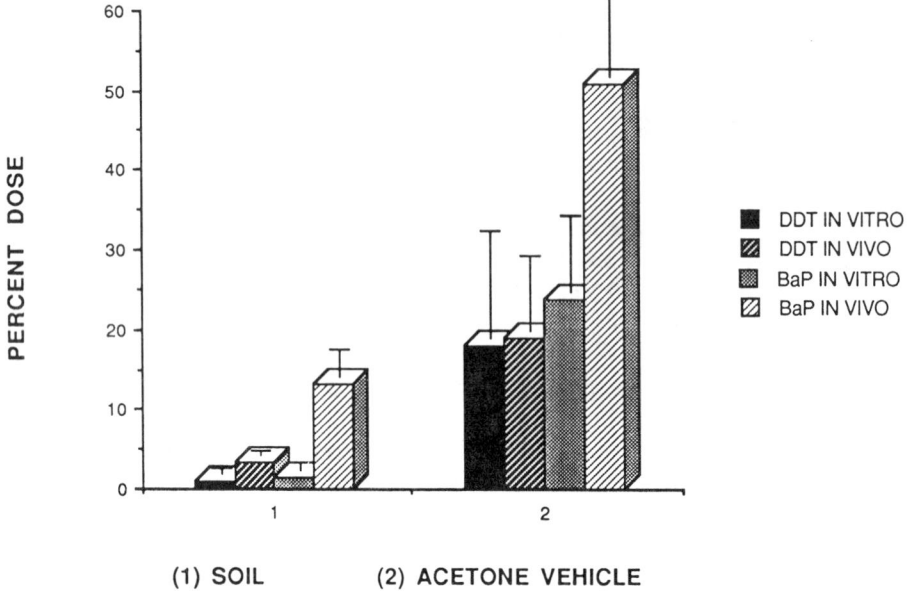

FIGURE 1.   Skin absorption of DDT and benzo[a]pyrene (B[a]P) from soil and acetone vehicles. *In vitro* compared to *in vivo* results.

cumulative dose an individual receives. The total cumulative dose is a function of several factors including:

1. Concentration(s) of contaminant(s) in the soil
2. Location of and access to contaminated areas
3. Type(s) of activity(ies) in contaminated areas
4. Duration of exposure (acute and chronic)
5. Specific exposure mechanisms
6. Soil type and moisture content
7. Amount of soil adhering to skin
8. Skin site and total area of exposure

Accumulation of compound in plants and the half-life of the compound in the soil are additional important factors in the determination of relevant clean-up levels and methodology to be employed. Dose rate may be an important factor in acute exposure assessment. The potential of increased risk from receiving high doses at susceptible life stages may be offset or exceeded by repair mechanisms operative at times of lesser dose. Clearly, the above mandates further study.

In general, the above investigations have revolved around a specific site with investigators employing many different techniques, methodologies *(in vitro* or *in vivo),* surface concentrations, and volumes of application to animals and/or humans. Table 2 lists the extensive ranges in mass of soil and volume of solution applied per square centimeter of skin utilized in the above studies. Of concern to all is the extrapolation of results obtained using animal models to estimate human body burden following topical exposure. With currently available limited data, animal skin is usually more permeable than human. When rank ordered, rat skin is more permeable than monkey or man and monkey skin closely approximates the permeability of man. Therefore, in studies employing rodents one expects the results to over estimate percutaneous penetration in man. Furthermore, we believe that results obtained

TABLE 2
Masses of Soil and Volumes of Solutions
Applied Per Square Centimeter of Skin

Investigator	Mass of oil (mg/cm^2)	Volume (μl/cm^2)
Roels et al. (calculated)	2.3	N.A.
Skowronski et al.	58—77	17—23
Yang et al.	9.0—56	47—93
Shu et al.	21	N.A.
Poiger and Schlatter	6.5—433	13—33
Wester et al.	40	5

Note: N.A., not applicable.

using *in vitro* systems must be substantiated *in vivo* before use in establishing regulatory guide lines because of the wide range in results one can obtain employing different *in vitro* methodologies.

In many cases one must realize that experimental conditions employed in some of the above investigations do not mimic actual environmental exposure situations and, therefore, the results reported may not be relevant to estimate potential human body burden following topical exposure. Results to date indicate that soil does not enhance chemical percutaneous penetration; however, under certain experimental conditions, levels of chemical absorption from soil may approach those following topical application of pure chemical.

Hopefully this review will stimulate discussion between the regulatory committees, industrial personnel, and academicians and generate an overall consensus as to what guidelines should be followed in conducting relevant studies to answer the questions raised.

REFERENCES

1. **Roels, H. A., Buchet, J.-P., Lauwerys, R. R., Braux, P., Claeys-Thoreau, F., Lafontaine, A., and Verduyn, G.,** Exposure to lead by the oral and the pulmonary routes of children living in the vicinity of a primary lead smelter, *Environment,* 22, 81, 1980.
2. **Poiger, H. and Schlatter, C. H.,** Influence of solvents and adsorbents on dermal and intestinal absorption of TCDD, *Food Cosmet. Toxicol.,* 18, 477, 1980.
3. **Shu, H., Teitelbaum, P., Webb, A. S., Marple, L., Brunck, B. Dei Rossi, D., Murray, F. J., and Paustenbach, D.,** Bioavailability of soil-bound TCDD: dermal bioavailability in the rat, *Fundam. Appl. Toxicol.,* 10, 335, 1988.
4. **Kimbrough, R. D., Falk, H., Stehr, P., and Fries, G.,** Health implications of 2,3,7,8-tetrachlorodibenzodioxin (TCDD) contamination of residential soil, *J. Toxicol. Environ. Health,* 14, 47, 1984.
5. **Brewster, D. W., Banks, Y. B., Clark, A.-M., and Birnbaum, L.,** Comparative dermal absorption of 2,3,7,8-tetrachlorobibenzo-*p*-dioxin and three polychlorinated dibenzofurans, *Toxicol. Appl. Pharmacol.,* 97, 156, 1989.
6. **Skrowronski, G. A., Turkall, R. M., and Abdel-Rahman, M. S.,** Soil adsorption alters bioavailability of benzene in dermally exposed rats, *Am. Ind. Hyg. Assoc. J.,* 49, 506, 1988.
7. **Skrowronski, G. A., Turkall, R. M., and Abdel-Rahman, M. S.,** Effects of soil on percutaneous absorption of toluene in male rats, *J. Toxicol. Environ. Health,* 26, 373, 1989.
8. **Yang, J. J., Roy, T. A., Krueger, A. J., Neil, W., and Mackerer, C. R.,** *In vitro* and *in vivo* percutaneous absorption of benzo[a]pyrene from petroleum crude-fortified soil in the rat, *Bull. Envir. Cont. and Tox.,* 43, 207, 1989.
9. **Wester, R. C., Maibach, H. I., Bucks, D. A. W., Sedik, L., Melendres, J., Liao, C., and DiZio, S.,** Percutaneous absorption of [^{14}C] DDT and [^{14}C]-benzo(a)pyrene from soil, *Fundam. Appl. Toxicol.,* in press.

10. **Feldmann, R. and Maibach, H.,** Percutaneous absorption of steroids in man, *J. Invest. Dermatol.*, 52, 89, 1969.
11. **Feldmann, R. and Maibach, H.,** Percutaneous penetration of some pesticides and herbicides in man, *Toxicol. Appl. Pharmacol.*, 28, 126, 1974.
12. **Wester, R. C. and Maibach, H. I.,** Percutaneous absorption in the rhesus monkey compared to man, *Toxicol. Appl. Pharmacol.*, 32, 392, 1975.

Chapter 18

SHORT-TERM *IN VITRO* SKIN EXPOSURE

Ronald C. Wester and Howard I. Maibach

TABLE OF CONTENTS

I. INTRODUCTION

The first step in percutaneous absorption is the partitioning of the chemical into the stratum corneum. Several factors will influence this partitioning, including the nature of the vehicle on the surface of the skin (assuming there is a vehicle), solubility of the chemical in the stratum corneum lipids, binding to stratum corneum proteins, etc. Rougier and co-workers[1-3] suggested that the amount of chemical absorbed within the stratum corneum after 30 min of application could reflect its stratum corneum/vehicle partitioning, but that it could also reflect its rate of entry into the skin. These authors went on to clearly show that the amount of various compounds which penetrated *in vivo* was strictly proportional to the time of application, thus providing indirect evidence that a constant flux of penetration really does exist *in vivo*.

This chapter will offer some insight into *in vitro* interactions between chemicals and human skin. It will reflect binding and absorption within the first 30 min of skin application which reflect circumstances shown *in vivo*. There is also a practical determinant in 30-min exposure because this can reflect real situation exposure. An example is exposure to chemical contaminants in water while bathing or swimming — an average exposure reflected in a 30-min time period.[4]

II. PARTITIONING INTO POWDERED HUMAN STRATUM CORNEUM

This is an *in vitro* model that utilizes the partition coefficient of the chemical contaminant in water with that of powdered human stratum corneum.[5] Adult foot calluses were ground with dry ice and freeze-dried to form a powder. That portion of powder that passed through a 40-mesh but not an 80-mesh sieve was used. The ^{14}C-labeled chemical as a solution in 1.5 ml water was mixed with 1.5 mg powdered human stratum corneum and the mixture was allowed to set for 30 min. The mixture was centrifuged and the proportions of chemical bound to human stratum corneum and that remaining in water were determined by scintillation counting.

Figure 1 shows the partition of benzene, PCBs, and *p*-nitroaniline from water into powdered human stratum corneum.[4] In 30 min (one assay time) all of the PCBs are bound to skin. Benzene partitioned between the atmosphere (volatilized) and skin, but 16.6 ± 1.4 percent bound to skin. *p*-Nitroaniline, a more water soluble compound, tended to stay on the water vehicle. That this was a true partitioning is reflected in Figure 2, where binding of *p*-nitroaniline was independent of dose.

III. *IN VITRO* PERCUTANEOUS ABSORPTION

In vitro percutaneous absorption utilized a section of fresh human skin from surgical reduction, firmly held above a continuous flowing water reservoir. The ^{14}C-labeled chemical as a solution in 1.5 ml water was placed on the outer surface of skin. Previous tests showed this volume of water just covered the surface of the skin and would represent the volume of water immediate to the surface of skin. The ^{14}C-labeled chemical in the surface water was in contact with skin for 30 min; then the surface of skin was rinsed with water to remove nonabsorbed chemical. That chemical that remained on the skin then absorbed into and through skin into the reservoir (maintained at 37°C).

That chemical that diffused through the human skin into the underlying reservoir would be considered the portion that was percutaneously absorbed and would become systematically available in humans *in vivo* (labeled systemic in Figures 3 to 6).

The skin was then tape-stripped twice to determine how much material was surface-

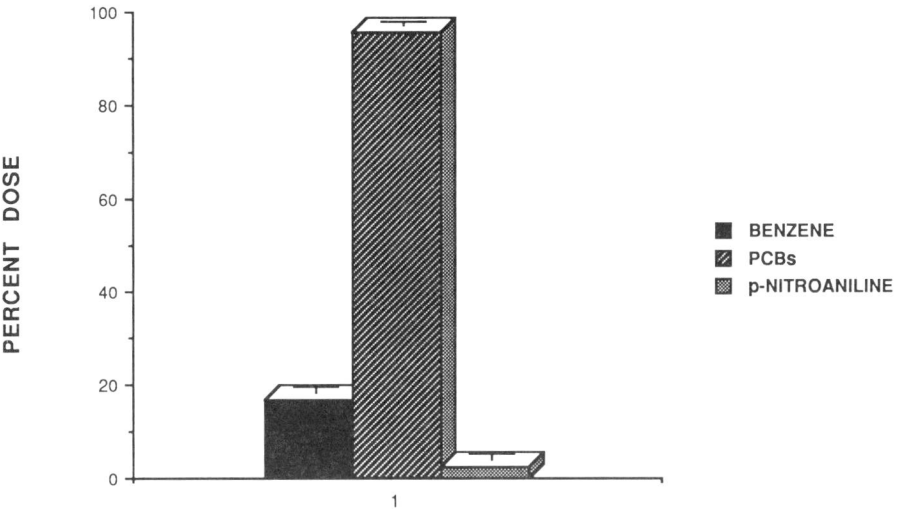

FIGURE 1. Partitioning of benzene, PCBs, and p-nitroaniline from water into powdered human stratum corneum within a 30-min time period.

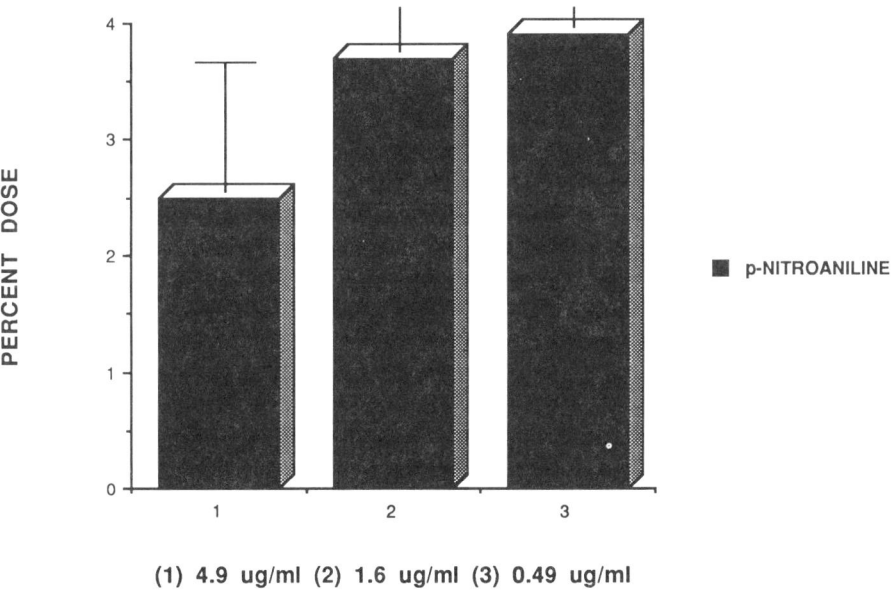

(1) 4.9 ug/ml (2) 1.6 ug/ml (3) 0.49 ug/ml

FIGURE 2. Partitioning of 10-fold dose levels of p-nitroaniline to powdered human stratum corneum.

bound to the stratum corneum (outer layer of skin). The skin is fragile after the *in vitro* procedure, and two tape strippings are able to pull the stratum corneum away from the rest of the skin (labeled stratum corneum in Figures 3 to 6). The remaining portion of the inner skin (epidermis and dermis) was digested (Soluene 350 for 5 h at 45°C) and its portion of ^{14}C-labeled chemical also determined (labeled skin in Figures 3 to 6).

Figure 3 shows that within 30 min considerable p-nitroaniline had been absorbed through human skin. Figure 4 shows that PCBs also readily penetrated human skin in 30 min. With benzene the percent dose systemic, and in the stratum corneum and lower portions of skin are small after 30 min. This reflects evaporation of benzene. However, if the single dose

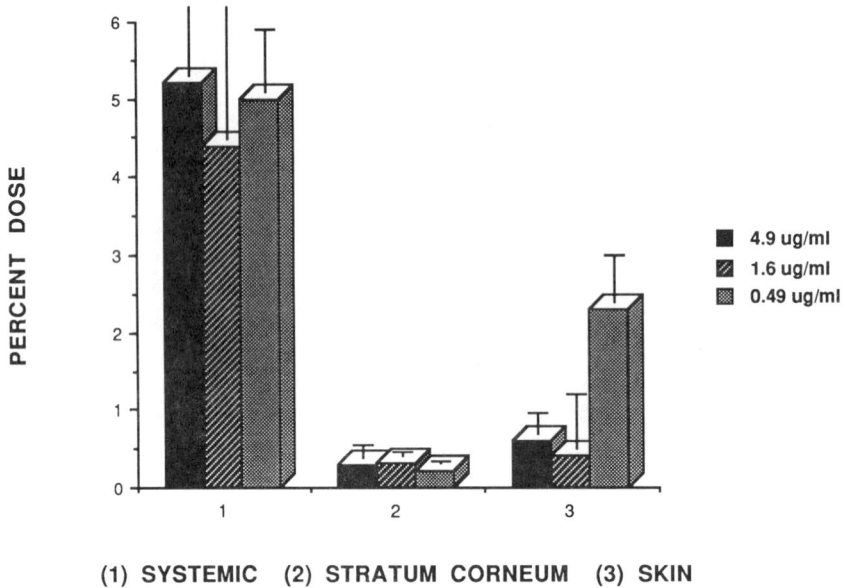

FIGURE 3.   *In vitro* absorption and skin content of *p*-nitroaniline following 30-min skin exposure.

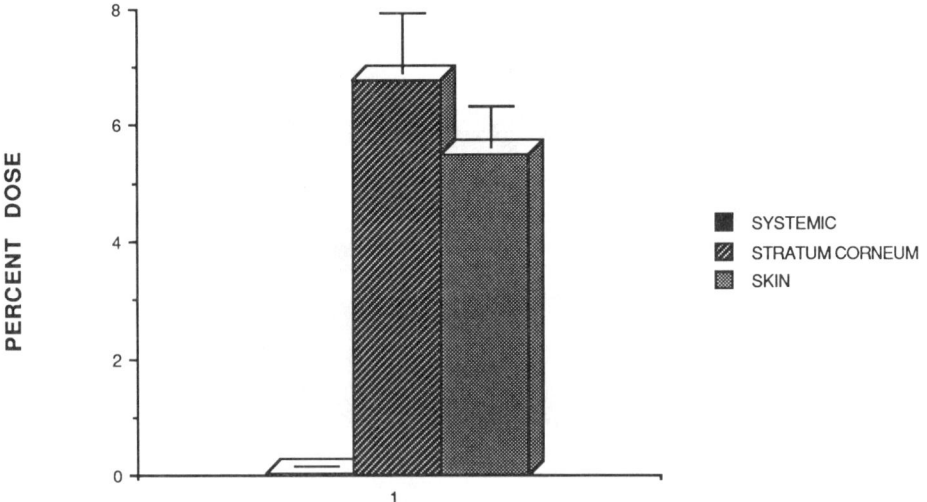

FIGURE 4.   *In vitro* absorption and skin content of PCBs following 30-min skin exposure.

of benzene is divided in 3, then applied to skin 3× in successive or spaced multiple doses, then the amounts of benzene absorbed and in skin are increased (Figure 6). This probably reflects the longer presence of benzene on skin due to multiple applications.

Table 1 summarizes the 30-min exposures for *p*-nitroaniline and suggest that for this compound, the various models compliment each other.

Table 2 gives the *in vitro* percutaneous absorption of DDT and B[a]P in human skin. The percent dose as skin content are given for 24-h exposure and the shorter exposure time of 25 min (6). With DDT the results are uncanny in that the 25-min exposure amount equals that for 24-h exposure *in vivo*. B[a]P skin contact is significant for 25-min exposure, but falls far short of that accumulated in 24-h *in vitro* or absorbed *in vivo* in 24 h.

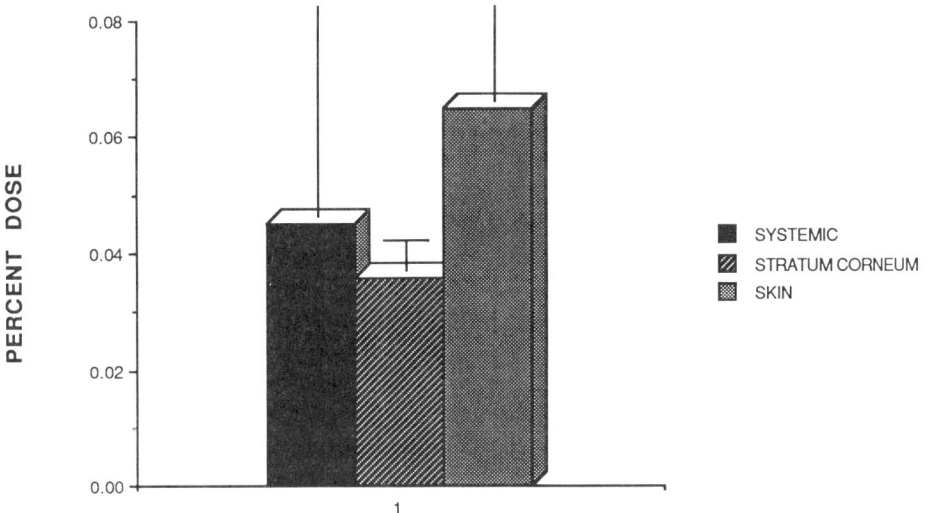

FIGURE 5. *In vitro* absorption and skin content of benzene following 30-min skin exposure.

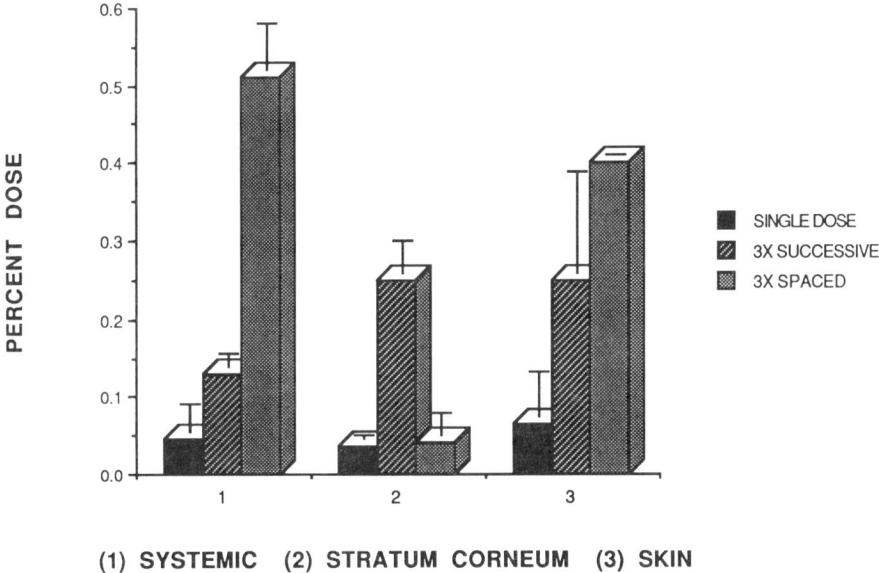

(1) SYSTEMIC (2) STRATUM CORNEUM (3) SKIN

FIGURE 6. *In vitro* absorption and skin content of benzene follow 30-min skin exposure to single and multiple doses.

IV. SUMMARY

Skin exposure from drug therapy and from environmental chemical contaminants is usually of an extended period of a day or longer. However, some exposures will be of a considerably shorter duration in terms of minutes. In either case, the first steps of percutaneous absorption happen within the first 30 min. Rougier and co-workers[1,2,3] have shown that *in vivo* absorption for the longer term can be predicted from this first 30-min exposure. With binding to powdered human stratum corneum, and with *in vitro* percutaneous absorption. Our data reflect these early happenings. Our *in vivo* data confirm some of the predicting potential for longer exposure.

TABLE 1
In Vivo Percutaneous Absorption of *p*-Nitroaniline in the Rhesus Monkey Following 30-Minute Exposure to Surface Water: Comparisons to *In Vitro* Binding and Absorption

Phenomenon	Percent dose absorbed/bound
In vivo percutaneous absorption, rhesus monkey	4.1 ± 2.3
In vitro percutaneous absorption, human skin	5.2 ± 1.6
In vitro binding, powdered human stratum corneum	2.5 ± 1.1

TABLE 2
Skin Exposure Time and Skin Content During *In Vitro* Percutaneous Absorption

Chemical	Formulation	Percent dose (24-h exposure)	Skin content (25-min exposure)
DDT	Acetone	18.1 ± 13.4	16.7 ± 13.2
	Soil	1.0 ± 0.7	1.8 ± 1.4
Benzo[a]pyrene	Acetone	23.7 ± 9.7	5.1 ± 2.1
	Soil	1.4 ± 0.8	0.14 ± 0.13

Note: In vivo percutaneous absorption in the rhesus monkey for 24-h application time was 18.9 ± 9.4% (acetone vehicle) and 3.3 ± 0.5% (soil) for DDT. For benzo[a]pyrene 5.10 ± 2.0 (acetone vehicle) and 13.2 ± 3.4% (soil) was absorbed.

The message is that even short-term chemical skin exposure can result in significant skin chemical content. This may be good for some drug therapy. This may not be good for some environmental exposure.

REFERENCES

1. **Rougier, A., Dupuis, D., Lotte, C., Roguet, R., and Schaefer, H.,** Correlation between stratum corneum reservoir function and percutaneous absorption, *J. Invest. Dermatol.,* 81, 275, 1983.
2. **Dupuis, D., Rougier, A., Roguet, R., Lotte, C., and Kalopissis, G.,** *In vivo* relationship between horny layer reservoir effect and percutaneous absorption in human and rat, *J. Invest Dermatol.,* 82, 353, 1984.
3. **Rougier, A., Dupuis, D., Lotte, C., Roguet, R., Wester, R. C., and Maibach, H. I.,** Regional variation in percutaneous absorption in man: measurement by the stripping method, *Arch. Dermatol. Res.,* 278, 465, 1986.
4. **Wester, R. C., Mobayen, M., and Maibach, H. I.,** *In vivo* and *in vitro* absorption and binding to powdered human stratum corneum as methods to evaluate skin absorption of environmental chemical contaminants from ground and surface water, *J. Toxicol. Environ. Health.,* 21, 367, 1987.
5. **Wester, R. C. and Maibach, H. I.,** Dermatopharmocokinetics in clinical dermatology, *Sem. Dermatol.,* 2, 81, 1983.
6. **Wester, R. C., Maibach, H. I., Bucks, D. A. W., Sedik, L., Melendres, J., Liao, C., and DiZio, S.,** Percutaneous absorption of [^{14}C]-DDT and [^{14}C]-benzo(a)pyrene from soil, *Fund. Appl. Toxicol.,* 15, 510, 1990.

Chapter 19

PROTOCOL FOR *IN VITRO* PERCUTANEOUS ABSORPTION STUDIES

Robert L. Bronaugh and Steven W. Collier

TABLE OF CONTENTS

I. INTRODUCTION

It would be unwise to attempt to write a rigid protocol suitable for every *in vitro* absorption study. It is important to recognize that although one can present general guidelines for conducting skin absorption and metabolism experiments, the specific conditions used will be determined by the properties of the chemical examined and also the goals of the study. The following protocol will briefly cover each step in conducting a percutaneous absorption study. Most of these steps have been discussed in detail in other chapters in this book.

II. CHOICE OF MEMBRANE

A. HUMAN OR ANIMAL SKIN

Absorption data for regulatory concerns related to human health should obviously be obtained with human skin in order to have the most relevant information. The difficulty in obtaining large quantities of human skin often requires that some studies be performed with animal skin. Experiments with animal skin can be conducted with more confidence if some human skin studies are performed so as to "calibrate" the skin of the animal model with the test compound. Animal skin is generally more permeable to chemicals than human skin.[1] The choice of an animal model may depend in part on the ease of preparation of skin for the diffusion cells. A hairless animal such as the fuzzy rat or hairless guinea pig allows for very thin sectioning of the skin with a dermatome so that most of the dermis is removed.[2] This is particularly important for water insoluble compounds.

B. NUMBER OF SUBJECTS

Data from at least three animal or human subjects should be obtained and averaged to allow for biological variation between subjects. The variability in absorption is often small (less than twofold) for inbred strains of laboratory animals but can be fivefold in different samples of normal human skin.

C. REGIONAL VARIABILITY

Variability in skin permeation is well known to occur in different regions of human skin. The trunk (back and abdomen) and the extremities (arms and legs) have reasonably similar barrier properties (< twofold differences). Enhanced absorption can be observed in regions of the face (fourfold) and especially the scrotum (twentyfold). In dealing with absorption of a facial product, one should therefore expect faster absorption than through abdominal skin. Small differences in regional absorption may be unimportant when one considers the large variability in skin permeation from subject to subject.

D. VALIDATION OF HUMAN SKIN BARRIER

Because human skin is obtained following trauma such as surgery, illness or accident it can be damaged prior to preparation for a study and this damage can occur without the knowledge of the investigator. Scrubbing skin with detergent prior to surgery or autopsy is a frequent cause of barrier alteration. It is highly recommended, therefore, that the barrier properties of the skin be pretested with a standard compound such as tritiated water[3] prior to conducting an experiment with the test compound.

III. PREPARATION OF MEMBRANE

Full thickness skin should generally not be used. Since absorbed chemicals are taken up by blood vessels directly beneath the epidermis *in vivo*, an *in vitro* study should use a membrane with most of the dermis removed. This is particularly important for water insoluble compounds that would diffuse slowly through the dermis. A suitable membrane can be prepared from fresh skin with a dermatome at a thickness of 200 to 350 μm and the skin viability maintained for use in metabolism studies. With nonhairy skin, heat or enzyme separation at the epidermal-dermal junction can be achieved, but loss of viability of skin preparations may result.

IV. DIFFUSION CELL

The choice of diffusion cell depends on the type of study to be conducted. The advantages and disadvantages of each are described.

A. FLOW CELL

Flow cells are necessary for maintaining viability of skin in diffusion cells since nutrient media must be continually replaced. Also, these cells are preferable for studies requiring round-the-clock sampling since samples can be collected automatically in a fraction collector. Flow cells of adequate design, however, will have only small exposed areas of skin for applying test compounds.[4] This is because the receptor volume must be small so that the cell contents can be rapidly exchanged. A large transdermal device cannot be applied to skin in its entirety in a flow cell. Both finite and infinite studies can be performed.

B. STATIC CELL

The static cell is ideal for studies requiring large surface areas of skin. Some designs incorporate stirring devices for mixing of solutions on both sides of the membrane when this is desired.

V. RECEPTOR FLUID

A. VIABLE OR NONVIABLE SKIN

Skin viability can be maintained through the appropriate use of a balanced salt solution containing glucose or tissue culture media.[5] If nonviable skin is satisfactory, a phosphate buffered saline solution (pH 7.4) could be used. The use of viable skin in a diffusion cell study is warranted even if metabolism is not of interest. The credibility of *in vitro* data is enhanced by maintaining the skin in a viable state.

B. HYDROPHOBIC TEST COMPOUND

The receptor fluids can be altered for water insoluble compounds to facilitate partitioning of these compounds into the receptor fluid. Absorbed test compound remaining in the skin at the end of the experiment should still be determined (after washing off the unabsorbed material). Addition of bovine serum albumin (4%) to HHBSS and adequate mixing of receptor contents improves the partitioning of compound into the receptor fluid.[6] For nonviable skin experiments, use of a lipophilic receptor fluid, such as 6% PEG 20 oleyl ether has been successful.[7]

VI. TEMPERATURE

Skin should be maintained at a physiological temperature. We have used 32°C as an average skin surface temperature based on measurements taken in our laboratory. To achieve this skin temperature, water is heated to 35°C and pumped through the holding blocks in our flow cells. Required water temperature will vary with the design of the diffusion cell.

VII. VEHICLE

The vehicle of choice in a permeation study is determined by a number of factors. For absorption measurements of a chemical contained in a specific product, the most valid results are obtained if the identical vehicle is used. For absorption results that must be predictive of expected permeation from a wide range of vehicles, then several vehicles with differing solubility properties representative of commercial products should be used (i.e., emulsion, cream, volatile vehicle, polar vehicle, nonpolar vehicle).

If one vehicle only is to be used, the choice of a volatile vehicle (such as acetone or ethanol) often gives the highest absorption values obtainable for the test compound and eliminates vehicle effects on absorption. A small amount of volatile vehicle ($10 \ \mu l/cm^2$) should be applied so that it will evaporate rapidly (<1 min) and therefore not damage the barrier properties of the membrane.

VIII. DOSE

The dose of chemical applied to skin will depend on the type of study being conducted. For determination of absorption of a chemical in a commercial product it is important to apply a similar dose to skin. The percentage of the applied dose that is absorbed will often decrease as the dose applied to the skin is increased. The dose applied to skin must be in solution otherwise it is not available for absorption. If radiolabeled compounds are used, only a few μg of material will often contain enough radioactivity for measurement of absorption.

IX. DAMAGED SKIN

The extent of absorption of chemicals through damaged skin can be useful information for situations in which the compound can come in contact with this type of skin or when the worst case absorption scenario is of interest. The barrier layer of skin can be compromised reproducibly by cellophane tape stripping the stratum corneum or by abrading the surface of skin with a needle or other sharp instrument.[8] This topic has been covered in detail in Chapter 10.

X. VOLATILE COMPOUNDS

Since the absorption of volatile compounds is influenced by the evaporation of compound from the surface of the skin, it may be desirable to do at least some of the measurements under occluded skin conditions. The top of a one-chambered diffusion cell can be covered with a stopper or with Parafilm to prevent evaporation. However, occlusion of the skin for longer than a few hours can alter barrier properties by affecting the hydration of the membrane.

XI. DURATION OF STUDY

Percutaneous absorption studies are frequently run for 24 h unless conditions of use of the test compound make this an unrealistic time. Other considerations, such as the duration of viability of skin in the diffusion cell may limit the length of a study. A 24-h duration is useful since toxicity data from other routes is often expressed on a daily basis. The exposed skin of people is generally washed at least once in 24 h so that absorption into the skin does not occur after that time. Finally, the turnover of cells in the stratum corneum can result in the loss of unabsorbed material on the surface of the skin after about 24 h.

XII. EXPRESSION OF DATA

The steady-state rate of absorption is very useful information, if it can be obtained from the data; it can be divided by the applied concentration to calculate a permeability constant. This method of expression is only possible when an infinite dose is applied in solution to the skin so that the concentration remains unchanged during the course of the experiment.

In many instances only a finite dose comes in contact with skin during exposure. Under these conditions, the percent of the applied dose appearing in the skin and receptor fluid is frequently determined. A measurement of absorption rate and total absorption can be obtained.

REFERENCES

1. **Bronaugh, R. L., Stewart, R. F., and Congdon, E. R.,** Methods for *in vitro* percutaneous absorption studies. II. Animal models for human skin, *Toxicol. Appl. Pharmacol.,* 62, 481, 1982.
2. **Bronaugh, R. L. and Stewart, R. F.,** Methods for *in vitro* percutaneous absorption studies; VI: Preparation of the barrier layer, *J. Pharm. Sci.,* 75, 487, 1986.
3. **Bronaugh, R. L., Stewart, R. F., and Simon, M.,** Methods for *in vitro* percutaneous absorption. VII: Use of excised human skin, *J. Pharm. Sci.,* 75, 1094, 1986.
4. **Bronaugh, R. L. and Stewart, R. F.,** Methods for *in vitro* percutaneous absorption studies. IV: the flow-through diffusion cell, *J. Pharm. Sci.,* 74, 64, 1985.
5. **Collier, S. W., Sheikh, N. M., Sakr, A., Lichtin, J. L., Stewart, R. F., and Bronaugh, R. L.,** Maintenance of skin viability during *in vitro* percutaneous absorption/metabolism studies, *Toxicol. Appl. Pharmacol.,* 99, 522, 1989.
6. **Bronaugh, R. L., Stewart, R. F., and Storm, J. E.,** Extent of cutaneous metabolism during percutaneous absorption of xenobiotics, *Toxicol. Appl. Pharmacol.,* 99, 534, 1989.
7. **Bronaugh, R. L. and Stewart, R. F.,** Methods for *in vitro* percutaneous absorption studies. III: Hydrophobic compounds, *J. Pharm. Sci.,* 73, 1255, 1984.
8. **Bronaugh, R. L. and Stewart, R. F.,** Methods for *in vitro* percutaneous absorption studies. V: Permeation through damaged skin, *J. Pharm. Sci.,* 74, 1062, 1985.

Chapter 20

FINITE DOSE *IN VITRO* PERCUTANEOUS PERMEATION BASED ON THE DIFFUSION MODEL: ESTIMATION OF PHARMACOKINETIC PARAMETERS AND PREDICTION OF THE PHARMACOKINETIC PROFILES

Kiyoshi Kubota and Howard I. Maibach

TABLE OF CONTENTS

I. INTRODUCTION

Finite dose percutaneous absorption processes are often studied in *in vitro* experiments. For instance, the FDA and AAPS report[1] of the workshop on principles and practices of *in vitro* percutaneous penetration studies indicates that "Lag times and steady state fluxes should be set forth in those instances in which the data lend themselves to such treatment. Whenever possible, the drug content in both the tissue and receptor at the end of the experiment should be noted and the total mass balance, which includes measurement of the residual drug in skin, should be determined." However, this report does not describe analysis of the observed kinetic profiles when steady state fluxes are not attained and lag times and steady state fluxes cannot be delineated; there seems no standard method to obtain the kinetic parameters from the data of *in vitro* finite dose permeation experiments. With the exception of transdermal therapeutic systems[2] in which steady state is attained and maintained, finite dosing is usually administered in clinical practice. Therefore, *in vitro* studies applying a finite dose resembling that utilized clinically should offer important information. When a finite dose is administered to the donor site, the drug flux, J, increases and reaches the maximum value, J_{max}, and then decreases. However, even when drug diffusivity through vehicle is considered large J_{max} is usually a complex function of the thickness (or volume) of donor cell solution, partition coefficient between donor cell and skin, and diffusivity through skin. In Table 1, the relationship between J_{max} and those parameters is listed assuming 5×10^{-3} cm/h for the permeability coefficient, K_p, and 0.1% or 1 mg/cm³ for the initial donor cell concentration. In this case, if the vehicle concentration is maintained at 0.1%, Jss = 5 µg/cm²/h. As shown in Table 1, J_{max} represents a complicated function of three factors as above (vehicle thickness, skin/vehicle partition, and diffusivity through skin).

The reliable parameter values are obtainable from the *in vitro* finite dose permeation studies. In the first section, the relationship between the "terminal elimination rate constant" and the pharmacokinetic parameter is elucidated. The analytical method (a mathematical method to treat the diffusion equation) is discussed to introduce the concept, "terminal elimination rate constant" in the diffusion process. In the second section, "model-independent" methods to estimate the pharmacokinetic parameters from the *in vitro* finite dose percutaneous permeation experiments are described. The relation between the statistical moment theory[3] and diffusion model as a theoretical basis of the parameter estimation method is discussed. In the third section, the methods to predict the percutaneous permeation pharmacokinetic profiles from the given parameter values based upon the diffusion model are discussed. We stress that most of the proposed methods remain theoretical; only some have been used for few compounds.[4] The methods are not considered established but rather as models to be tested for the *in vitro* studies utilizing various compounds.

II. TERMINAL ELIMINATION RATE CONSTANT IN DIFFUSION MODEL — INTRODUCTION BY THE ANALYTICAL METHOD

The analytical method is a mathematical method to handle the diffusion equation. The procedure may be divided into three steps: [Step 1] where the general solution satisfying Fick's Second Law of Diffusion is given; [Step 2] where the eigenvalue, α_i in Equation 1 below, is specified for the given boundary conditions; and [Step 3] where the coefficients A_i and/or B_i which satisfy the initial condition are specified. Generally, Step 1 and Step 2 are not complicated, while Step 3 is sometimes complicated. Fortunately, the terminal elimination rate constant can be estimated independently of Step 3. Thus, the discussion in this section focuses on Step 1 and Step 2.

[Step 1]: The general solution, C_i, (i = 1, 2, ---) given in Equation 1 below satisfies Fick's Second Law of Diffusion (Equation 2).

TABLE 1
Values for J_{max} ($\mu g/cm^2/h$) as a Function of Kinetic Parameter Values[a]

k_d (h^{-1})	l_d (cm)	l_v^b (cm)				
		0.1	0.2	0.3	0.4	0.5
0.125	0.04	3.91	4.34	4.52	4.62	4.69
		(3.21)	(3.74)	(4.06)	(4.29)	(4.45)
0.05	0.1	3.06	3.73	4.04	4.22	4.34
		(6.42)	(7.61)	(8.36)	8.91)	(9.35)
0.02	0.25	2.03	2.82	3.25	3.53	3.73
		(12.8)	(15.2)	(16.8)	(18.5)	(19.0)
0.005	1	0.77	1.31	1.71	2.03	2.28
		(39.3)	(44.0)	(47.9)	(51.1)	(54.0)

Note: $K_p = 0.005$ cm/h is assumed for all cases; the initial concentration of a drug in vehicle is 1 mg/ml; the diffusivity of a drug through vehicle is assumed to be great; J_{max} = the maximum flux from skin to donor cell; t_{max} = time to J_{max}; K_p = permeability coefficient which is given as $k_d \cdot l_d$; l_v = thickness of vehicle; k_d = diffusion parameter defined as D/l_s^2, where D and l_s are the effective diffusion coefficient through skin and effective thickness of skin, respectively; l_d = partition parameter or apparent length of diffusion defined as $Km \cdot l_s$, where Km is the skin/vehicle partition coefficient.

[a] Calculated using Program 1 in Appendix 1.
[b] Values in parentheses are t_{max} (h).

$$C_i = [A_i \sin(\alpha_i \cdot x/l) + B_i \cos(\alpha_i \cdot x/l)] \exp[-(\alpha_i^2 D/l^2)t] \qquad (1)$$

$$\partial C/\partial t = D \partial^2 C/\partial x^2 \qquad \text{[Fick's Second Law]} \qquad (2)$$

where C is the drug concentration in skin, which is regarded as a single simple membrane with thickness l, x is distance [x = 0 for the vehicle-skin boundary and x = l for the skin-sink boundary], t is time, A_i and B_i are the coefficients not specified in Step 1, α_i is the eigenvalue also not specified in Step 1 and D is the effective diffusion coefficient through skin. It is easy to show that the general solution, C_i in Equation 1 satisfies Fick's Second Law of Diffusion given in Equation 2 if C_i is differentiated with respect to t and x.

[Step 2]: In Step 2, each eigenvalue, α_i, is specified for the given boundary conditions. Some relationship between coefficients A_i and B_i is also determined in this step. The "boundary conditions" are unique to the partial differential equations including the parabolic partial differential equation (Equation 2). In Equation 2, the drug concentration in the skin, C, is a function of two variables, t (time), and x (distance). As in the ordinary differential equations frequently employed for the compartment model in standard pharmacokinetics, C in Equation 2 should be specified for the initial condition, C(t = 0) in respect of the variable, t. Similarly, C should be specified in respect of the variable, x, for the boundary conditions. Provided that the unidirectional diffusion is considered, two boundary conditions should be given, one for the (left) donor site-skin boundary and one for the (right) skin-sink (receptor cell) boundary. Herein, two cases for *in vitro* studies are considered; (A) the processes after the removal of a drug from the donor site and (B) the processes of finite dose permeation.

A. PROCESSES AFTER THE REMOVAL OF DRUG IN DONOR SITE

Two boundary conditions are given for x = 0 (left boundary) and x = *l* (right boundary) as

Left boundary: $\partial C(x = 0)/\partial x = 0$ (3)

Right boundary: $C(x = l) = 0$ (4)

The left boundary condition (Equation 3) designates that there is no drug flux at this boundary. The drug flux per unit area, J, is given as

$$J = -D \, \partial C/\partial x \qquad \text{[Fick's First Law]} \qquad (5)$$

When $\partial C(x = 0)/\partial x$ is not zero, the inward (when the value is negative) or outward (when positive) flux should exist at the left boundary. Because the left boundary is the dead end, $\partial C(x = 0)/\partial x$ should be zero. Two boundary conditions should be given for every C_i in Equation 1. Thus, $A_i = 0$ is derived from Equations 1 and 3. From Equation 1 where $A_i = 0$ and Equation 4, the following relation is derived

$$\cos(\alpha_i = 0 \quad \text{or} \quad \alpha_i = \pi/2, \, 3\pi/2, \, 5\pi/2, \, ---, \, (2i - 1)\pi/2, \, --- \qquad (i = 1, 2, ---) \quad (6)$$

Every C_i in Equation 1, where $A_i = 0$ and α_i is specified in Equation 6, satisfies both Fick's Second Law of Diffusion (Equation 2) and the boundary conditions (Equations 3 and 4). Thus their sum, ΣC_i, also satisfies those equations. If some combination of the values for B_i in C_i (Equation 1) (i = 1, 2, ---) can be specified so that the initial condition or the concentration-distance curve at t = 0 is modeled by the sum $\Sigma C_i(t = 0)$, this sum satisfies all of Fick's Second Law of Diffusion (Equation 2), boundary conditions (Equations 3 and 4) and initial condition. Thus, this ΣC_i is C to be eventually obtained. Details for the procedure to specify B_i in C_i in Equation 1 for the initial condition are not given here. However, in Figure 1, how $\Sigma C_i(t = 0)$ can model the initial condition is shown. In Figure 1a, the concentration-distance curve at t = 0 modeled by sum of the first three cosine curves is shown for the case where the vehicle is removed after the steady state is attained. In this case, B_i in Equation 1 is specified as

$$B_i = Km \cdot Cv \cdot 8/[(2i - 1)^2 \pi^2] \qquad (i = 1, 2, —) \qquad (7)$$

where Km and Cv are the skin/vehicle partition coefficient and the drug concentration in the vehicle which is considered constant before vehicle removal, respectively. Figure 1a shows that even when the first three cosine curves with different "wave lengths" are added to each other, this sum or ΣC_i (t = 0) can model the initial linear concentration-distance curve well (the steady state concentration-distance curve which had been established in skin before the vehicle was removed) except for the part near the left boundary.

Figure 1b shows the case for drug release from the vehicle (solution case) directly into sink. This process is also described by the boundary conditions Equations 3 and 4 where C is read as the drug concentration in vehicle because the left boundary (the side of the backing membrane) is the "dead end" and the right boundary is the sink. In this case, B_i in Equation 1 is given as

$$B_i = Cv \cdot 4/[(2i - 1)\pi] \cdot (-1)^{i-1} \qquad (i = 1, 2, —) \qquad (8)$$

As shown in Figure 1b, to model the initial condition (even horizontal concentration in

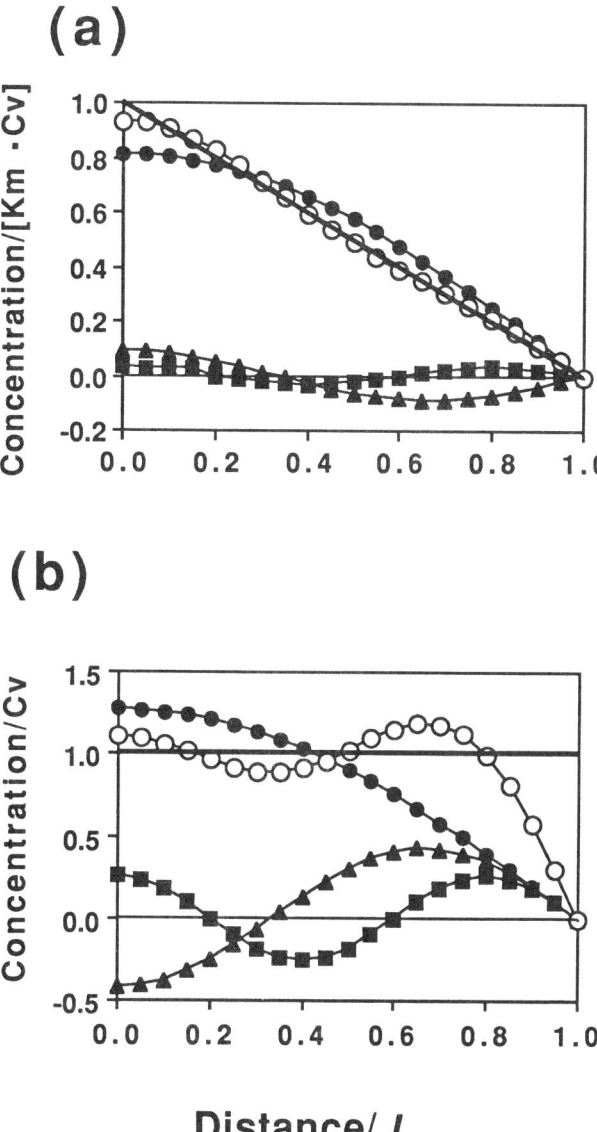

(a)

(b)

Distance/ *l*

FIGURE 1. Initial drug concentration-distance curves in membrane (bold solid lines), expressed as the sum of the first three cosine curves in Equation 10. The thickness of membrane is given as *l*. The symbols, ●, ▲ and ■ represent the first, second and third cosine curves in Equation 10, respectively, and ○ represents the sum of them. (a) Case where vehicle is removed at t = 0 when the steady state has already been established. The drug concentration at the uppermost epidermis is given as Km · Cv, where Km and Cv are the skin/vehicle partition coefficient and drug concentration in vehicle, respectively. The value for B_i in Equation 10 is given in Equation 7. (b) Case where drug is released directly into sink. The B_i value is given in Equation 8.

vehicle) by a sum of cosine curves is difficult as compared to the case in Figure 1a. However, one can imagine that if many terms are employed, ΣC_i (t = 0) will approach the horizontal linear line in Figure 1b as n approaches infinity.

When λ_i is defined as

$$\lambda_i = \alpha_i^2 D/l^2 = (\pi^2/4)D/l^2, \ (9\pi^2/4)D/l^2, \ (25\pi^2/4)D/l^2, \ —, \ [(2i-1)^2\pi^2/4]D/l^2, \quad (9)$$

C_i in Equation 1 can be rewritten as

$$C_i = B_i \cos(\alpha_i \cdot x/l) \exp(-\lambda_i \cdot t) \tag{10}$$

Because $\lambda_1 < \lambda_2 < \lambda_3 < \lambda_4$ —, when sufficient time, t, elapses, C or ΣC_i approaches C_1 which is given as

$$C_1 = B_1 \cos[(\pi/2)(x/l)] \exp(-\lambda_1 \cdot t) \tag{11}$$

When D/l^2 is defined as the diffusion parameters,[5] k_d,[6] λ_1 is given as

$$\lambda_1 = (\pi^2/4)k_d \simeq 2.74 \, k_d \tag{12}$$

Note that the eigenvalues α_i and λ_i, including λ_1, in Equations 6, 9 and 12 were specified only for the boundary conditions [Step 2] and are independent of the initial condition. Therefore, the terminal elimination rate constant, λ_1, and the terminal half-life, $0.693/\lambda_1$, are independent of the mode of the application of a drug to the donor site before removal [e.g., whether the vehicle concentration was maintained to be constant or significantly changed with time before vehicle removal.][7,8]

When skin is "frozen" at a specific time t = T, from Equation 11, the concentration-distance curve in skin approaches a cosine curve irrespective of the initial condition. Similarly, when skin is "cut" at a specific distance x = X, from Equation 11, the concentration-time curve at any point in the skin becomes log-linear with slope—λ_1 given in Equation 12.

Drug flux is given in Equation 5 and when sufficient time elapses, J approaches

$$J = -D \partial C_1(x = 0)/\partial x = (B_1 \cdot D \cdot \alpha_1/l) \exp(-\lambda_1 \cdot t)$$

$$= [B_1 \cdot D \cdot \pi/(2l)] \exp(-\lambda_1 \cdot t) \tag{13}$$

where λ_1 is given in Equation 12. Thus, the terminal elimination rate constant of J is the same as that in concentration of drug at every point within the skin. From Equation 12, we obtain

$$k_d = \lambda_1/(\pi^2/4) \simeq \lambda_1/2.74 \tag{14}$$

Equation 14 indicates that if the terminal elimination rate constant after removal of the vehicle, λ_1, is estimated in *in vitro* or *in vivo* experiment, a pharmacokinetic parameter, k_d, can be estimated. For instance, in *in vivo* experiment, the terminal half-life of hydrocortisone is 44 (40 to 48) h, independent of the initial treatment (with or without occlusion and/or stripping).[9] The value for λ_1 is, thus, estimated to be 0.693/44 = 0.016 h^{-1}. From Equation 12, k_d is 0.016/2.74 = 0.006 h^{-1}. This value for k_d is one half of that obtained from *in vitro* experiment[10] where the value of $1/(2k_d)$ is estimated to be 44.7 h or k_d = 1/(2 × 44.7) = 0.011 h^{-1}.

B. PROCESSES OF FINITE DOSE PERMEATION

In this section, we consider the case where a vehicle (solution matrix) or donor cell solution with thickness, l_v, is applied to skin with thickness, l_s. The distance, x, designates points from the outermost portion of the donor site [x = $-l_v$] through the donor cell (or vehicle) and skin *via* the donor site-skin boundary [x = 0] to the skin-sink boundary [x = l_s] ($-l_v \leq x \leq l_s$). When the diffusion coefficient through the donor cell solution (or vehicle) is considered large (the donor cell or vehicle can be considered as a compartment and drug concentration in the donor site is even at every time, t), the following two relations at the vehicle (donor cell)-skin boundary hold between the drug concentration in vehicle (donor cell), U, and drug concentration in skin, C.

$$\text{\textbf{Left boundary:}} \quad \text{Km} \cdot \text{U} = \text{C}(x = 0) \tag{15}$$

(i.e., the ratio of the drug concentration in skin at the left boundary to that in donor cell solution is the partition coefficient) and

$$-(\text{dU/dt})l_v = -D\,\partial\text{C}(x = 0)/\partial x \tag{16}$$

(i.e., the rate of change in drug amount per unit area in donor site [left hand side] is equal to the drug flux at the donor site-skin boundary in skin [right hand side]). The right boundary is the same as in (A) and is given as

$$\text{\textbf{Right boundary:}} \quad \text{C}(x = l_s) = 0 \tag{17}$$

In addition to C_i in Equation 1, U_i for U (which is given as $U = \Sigma\ u_i$) given below is used as the general solution

$$U_i = w_i \exp(-\alpha_i^2 \cdot k_d \cdot t) \tag{18}$$

where w_i is a constant.

The boundary conditions, Equations 15 to 17 should be given for every C_i and corresponding U_i. From Equations 1, 15, 16, and 18, Equations 19 and 20 below are derived, respectively, as

$$B_i = \text{Km} \cdot w_i \tag{19}$$

$$A_i = -w_i \cdot l_v \cdot \alpha_i/l_s \tag{20}$$

Because $C_i(x = l_s) = 0$ (Equation 17), and using Equations 19 and 20, the following is derived from Equation 1.

$$\tan \alpha_i = \text{Km } l_s/(lv \cdot \alpha_i) = (l_d/l_v)/\alpha_i \tag{21}$$

where l_d is the partition parameter[4,5] or ''apparent length of diffusion''[11] and defined as $l_d = \text{Km} \cdot l_s$. This parameter is one of the three parameters employed in the *in vitro* pharmacokinetics, k_d, l_d, and l_v (vehicle thickness). When the diffusion coefficient of drug through vehicle cannot be considered large, a fourth parameter, diffusion coefficient through vehicle, D_v, should also be employed.[11] The value for α_i in Equation 21 can be estimated as each x value of the points of intersection of $y = \tan x$ and $y = (l_d/l_v)/x$ (Figure 2). The terminal elimination rate constant, λ_1, is given by

$$\lambda_1 = \alpha_1^2 \cdot k_d \tag{22}$$

where α_1 is given as the first positive value of α_i in Equation 21. From Figure 2, one can see that $0 \le \alpha_1 \le \pi/2$. Thus, $\lambda_1 \le (\pi^2/4)k_d$. Equation 22 indicates that the terminal elimination rate constant, λ_1, is smaller (or the terminal half-life $0.693/\lambda_1$ is longer) than that in the process after vehicle removal in Equation 12. Because α_1 is determined by the

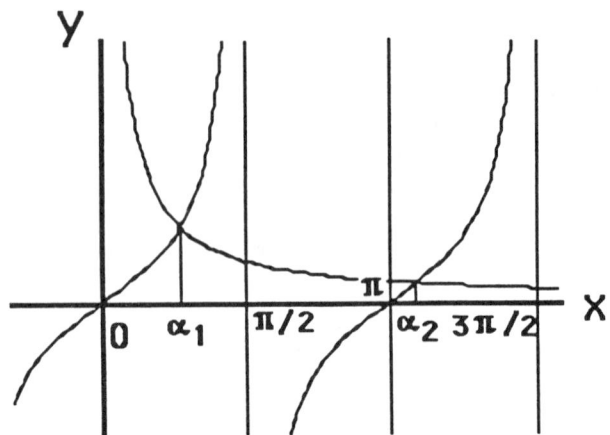

FIGURE 2.   A method to estimate the α_i value in Equation 21. Each value for α_i is between $[(i - 1)\pi, (2i - 1)\pi/2]$ (i = 1, 2, ---) and given as x value of the intersect of $y = \tan x$ and $y = (l_d/l_v)/x$.

ration, l_d to l_v, in Equation 21, the value of λ_1 in Equation 22 is determined by all of the three parameter values, k_d, l_d and l_v. Equation 14 should not be used to derive the k_d value from the λ_1 value estimated from the finite dose permeation study. If this λ_1 in Equation 22 is used instead of that in Equation 14, k_d will be underestimated.

III. PARAMETER ESTIMATION—STATISTICAL MOMENT THEORY AND DIFFUSION MODEL

A. GENERAL CONSIDERATION

In standard pharmacokinetics, the "model-independent" parameters such as the steady state volume of distribution, Vd_{ss}, are estimated by statistical moment theory.[3] The statistical moment theory can cover several pharmacokinetic processes including the linear diffusion process. One of the major conceptual parameters in the statistical moment theory is the mean residence time, MRT. The MRT has a close relationship with the steady-state pharmacokinetics because, at the steady state, one may consider that the residence time of every drug molecule as being equal to the MRT. The relation between the MRT and flux at the steady state, Jss, and amount at the steady state, Ass, is formulated as

$$Ass = Jss \cdot MRT \qquad (23)$$

Equation 23 may be realized by considering the relation between the number of students in a college and their mean residence time in the college. If the "mean residence time", MRT, of students in the college is 5.1 years and when the "steady state" is attained and the "steady state flux" (Jss) is 1000 students per year (the number of students who enter and leave the college per year), then, the college has the total students at the "steady state" (Ass), 1000 students/year × 5.1 year = 5100 students as in Equation 23.

In a one-compartment model, Jss is given as $Jss = Css \cdot CL = Css \cdot V \cdot k = Ass \cdot k$ because $CL = V \cdot k$ and $Ass = Css \cdot V$ where Css, CL, V, and k are the steady state drug concentration in blood, clearance, volume of distribution, and apparent first-order elimination rate constant, respectively. Thus, from Equation 23, MRT = 1/k. Generally in a compartment model, the steady state volume of distribution can be defined as $Ass = Vd_{ss} \cdot Css$. Because $Jss = Css \cdot CL$, from Equation 23, Vd_{ss} is given as $Vd_{ss} = CL \cdot MRT$.[12]

B. MRT FOR THE DIFFUSION MODEL

In the diffusion model where the skin receptor site boundary can be regarded as sink, the steady-state flux per unit area, Jss, and amount per unit area, Ass, at the steady state are given, respectively, as

$$Jss = D \cdot Km \cdot Cv/l_s \tag{24}$$

$$Ass = Km \cdot Cv \cdot l_s/2 \tag{25}$$

From Equations 23, 24, and 25, the MRT through skin, MRTs, is given as[13]

$$MRTs = Ass/Jss = l_s^2/(2D) = 1/(2k_d) \tag{26}$$

When the diffusivity through the vehicle is great and the vehicle can be regarded as a compartment, in the fictitious experiment where the constant flux is supplied into the vehicle from the outside and the steady state is attained, the amount in the vehicle per unit area at the steady state, Avss, is given as

$$Avss = Cv \cdot l_s \tag{27}$$

Jss is given in Equation 24. Thus, MRT through vehicle, MRTv, is given as[11]

$$MRTv = Avss/Jss = l_v \cdot l_s/(D \cdot Km) = l_v/[(D/l_s^2) \cdot Km \cdot l_s] = l_v/(k_d \cdot l_d) \tag{28}$$

Because the overall MRT is a simple sum of each MRT,[3] the overall MRT is given as[11]

$$MRT = MRTv + MRTs = l_v/(k_d \cdot l_d) + 1/(2k_d) \tag{29}$$

C. MASS BALANCE AND STATISTICAL MOMENTS

In this section, the definition of the mean residence time is based on the probability density function (p.d.f.) of the event [a drug molecule is irreversibly removed from a system at time t (or resides in the system until t)]. When this p.d.f. is described as f, the mean residence time, MRT, is defined as

$$MRT = \int_0^\infty t \cdot f \, dt \tag{30}$$

Because the drug amount irreversibly removed during [t, t + Δt] is J(t) · Δt, the probability of the event [a drug molecule is irreversibly removed from a system during [t, t + Δt]] is also proportional to J(t) · Δt. Therefore, f is given as $J(t)/AUC_J$ where AUC_J is the area under the curve of the flux, J, because the integral of f from 0 to infinity is AUC_J/AUC_J = 1. AUC_J or Ae(∞) (cumulative amount eventually excreted) is equal to the initial amount in the system, A_0 if all the initial amount is eventually released. Thus, MRT is defined as

$$MRT = \int_0^\infty t \cdot f \, dt = \frac{\int_0^\infty t \cdot J \, dt}{AUC_J} = \frac{AUMC_J}{Ae(\infty)} \tag{31}$$

where $AUMC_J$ is the area under the moment curve[12] of the flux, J, and defined as $AUMC_J = \int_0^\infty t \cdot J \, dt$.

Generally, the flux, J, from a system can be defined as $J = -dA/dt$ where A is the amount of drug in the system. Because of the principle of mass balance,

$$A(t) + \int_0^t J d\tau = A_0 \tag{32}$$

When f in the product rule for integration $\int f \, dt = t \cdot f - \int t \cdot f' \, dt$ ($f' = df/dt$), is specified as $f = A$, and noting that $J = -f'$, the following equation is derived

$$\int A \, dt = t \cdot A + \int t \cdot J \, dt \tag{33}$$

If A in Equation 32 is substituted into the right hand side of Equation 33, we obtain

$$\int A \, dt = t \left(A_0 - \int J \, dt \right) + \int t \cdot J \, dt \tag{34}$$

Herein, the value of the definite integral of Equation 34 from $t = 0$ to ∞ is considered. We define $\int_0^\infty A \, dt$ as AUC_A. As in Equation 31, $\int_0^\infty t \cdot J \, dt$ is $AUMC_J$. In addition, $\int_0^\infty J \, dt$ should be equal to A_0 if the whole initial amount in the system is eventually released. Therefore, from Equation 34, we develop the following equation[14]

$$AUC_A = AUMC_J$$
$$= MRT \cdot Ae(\infty) \tag{35}$$

When Equation 35 is applied to the relation between amount in the vehicle, Av, and the flux from the vehicle, Jv, we get

$$AUC_{Av} = AUMC_{Jv} = MRTv \cdot Ae(\infty) \tag{36}$$

where MRTv is given in Equation 28 and $Ae(\infty)$ is the cumulative amount eventually released from the vehicle and considered to be equal to A_0.

When vehicle and skin are considered to be one system, the amount in this system, A, is given as $A = Av + As$. Because AUC of amount of this system, AUC_A is given as $\int_0^\infty A \, dt = \int_0^\infty Av \, dt + \int_0^\infty As \, dt = AUC_{Av} + AUC_{As}$, it follows that

$$AUC_{Av} + AUC_{As} = AUMC_{Js} = MRT \cdot Ae(\infty) \tag{37}$$

where MRT is given in Equation 29 and Ae (∞) is the cumulative amount eventually excreted from skin and should be also equal to A_0. When MRT in Equation 29 is substituted into Equation 37, we get

$$AUC_{Av} + AUC_{As} = (MRTv + MRTs) \cdot Ae(\infty) \tag{38}$$

and it follows from Equations 36 and 38 that

$$AUC_{As} = MRTs \cdot Ae(\infty) \tag{39}$$

Because MRTs is given in Equation 26, from Equation 39, the following equation is derived[4]

$$k_d = 2Ae(\infty)/AUC_{As} \tag{40}$$

Similarly, from Equations 28, 36, and 40 the following equation is derived[4]

$$l_d = Ae(\infty)/AUC_{Av} \cdot l_v/k_d = AUC_{As}/(2AUC_{Av}) \cdot l_v \qquad (41)$$

Noting that $Av = Cv \cdot l_v$, it follows that $AUC_{Av} = \int_0^\infty Av\ dt = \int_0^\infty Cv\ dt \cdot l_v = AUC_{Cv} \cdot l_v$ where AUC_{Cv} is the AUC of the drug concentration in the solution of donor site, and from Equations 40 and 41, the following important equation is derived to estimate the permeability constant, Kp, from the *in vitro* finite dose percutaneous permeation study[15]

$$Kp = Km \cdot D/l_s = l_d \cdot k_d = Ae(\infty)/AUC_{Cv} \qquad (42)$$

IV. EXPERIMENTS WHERE THE PROPOSED METHODS ARE EMPLOYED

To use Equations 40 to 42, several conditions should be satisfied. First, the amount in the vehicle and in the skin should be measured precisely. Soap and water are sometimes adopted to remove drug in the vehicle.[16] However, this procedure may extract drug absorbed in the skin or may not remove ointment on the skin surface completely. It is preferable to use a polymer adhesive[4] where drug is incorporated or to use a liquid easily removed from the skin surface. Second, to use Equation 41, a vehicle composed of solution matrix[17] with exact thickness should be employed. Thus, the use of a reliable adhesive with a precise thickness is most preferable. Being different from Equation 41, Equation 40 can be applied to any case where $Ae(\infty)$ and AUC_{As} are finite values, or, flux from skin eventually becomes zero. For instance, when the complex transdermal therapeutic system is applied for 24 h and then removed, $Ae(\infty)$ and AUC_{As} have finite values, thus, Equation 40 may be employed to estimate the k_d value, though Equation 41 cannot be used in this case.

To use Equations 40 and 41, skin obtained from the same subject is cut into several pieces.[4] On each piece, the same vehicle is applied and the vehicle-skin complex is placed between the donor and receptor cells. Using several vehicle-skin complexes, the experiment is conducted until, for example, 6, 12, 24, 48, and 72 h have passed.[4] Then, vehicle and skin are separated and stored until analyzed. From this study, Av and As at 6, 12, 24, 48, and 72 h are obtained the AUC_{Av} and AUC_{As} in Equations 40 and 41 can be estimated.

To use Equation 42, liquid (usually water) is used in the donor cell. Several small samples of donor cell solution are taken to measure the drug concentrations in donor cell solution, Cv, at various times, t. For instance, 10 samples of 10 to 30 μl/cm^2 donor cell solution are taken from the initial volume of donor cell solution, 500 μl/cm^2. In Table 2, the estimated Kp values are listed from the *in vitro* experiments simulated supposing $Kp = k_d \cdot l_d = 0.005$ cm/h. The initial concentration in the vehicle, Cv(0) is 1 mg/ml. Thus, if the vehicle concentration were maintained to be the initial concentration, $Jss = 5$ μg/cm^2/ h. Ten donor cell samples, 10 to 30 μl/cm^2 each, are assumed taken from the initial donor cell volume, 500 μl/cm^2 at 3, 6, 12, 24, 30, 36, 48, 54, 60, and 72 h. The receptor fluid is assumed to be collected every 4 h until 72 h. As shown in Table 2, Equation 42 can be used even when a considerable change in the volume of the donor cell occurs (e.g., when 10 samples of 30 μl/cm^2 are taken from the initial donor cell volume, 500 μl/cm^2, and the final donor cell volume at 72 h is 200 μl/cm^2). However, to estimate the value for Kp by Equation 42, the experiment should be conducted long enough so that the terminal elimination rate constant estimated from the last several Cv-t plots is actually the same as that from the last several J-t plots. In addition, the Kp value is estimated precisely when $Ae(\infty)$, which is obtained by the extrapolation using the terminal elimination rate constant of J-t profiles, is actually the same as the initial amount subtracted by the amount removed as the donor cell samples.[15] Otherwise, the Kp value should be overestimated irrespective of the sample

TABLE 2
Values for Permeability Coefficient Kp Estimated by Equation 42[a]

Sample vol. (μl/cm²)	k_d (h⁻¹)	l_d (cm)	J_{max}[b] (μg/cm²/h)	t_{max} (h)	$t^{1/2}_{cv}$ (h)	$t^{1/2}_j$ (h)	ΔA (μg/cm²)	As(72) (μg/cm²)	Ae(72) (μg/cm²)	Av(72) (μg/cm²)	Ae(∞) (μg/cm²)	AUC(72) (μg/cm²·h)	AUC(∞) (μg/cm²)	Kp (cm/h × 10⁻³)	$J_{max}/C_v(0)$[b] (cm/h × 10⁻³)
10	0.125	0.04	4.59	4.62	60.8	60.2	70.2	9.1	240.0	180.8	435.8	49.65	89.31	4.88	4.59
10	0.05	0.1	4.25	9.61	63.6	63.4	68.4	22.9	229.3	179.5	438.5	48.45	89.65	4.89	4.25
10	0.02	0.25	3.64	19.4	70.9	71.8	64.5	57.1	203.6	174.8	445.7	45.69	90.38	4.93	3.64
10	0.005	1	2.12	54.3	97.5	251.4	51.8	197.8	107.9	142.5	876.3	36.44	86.55	10.12	2.20
20	0.125	0.04	4.59	4.60	48.9	47.8	136.7	8.2	233.6	121.5	373.5	48.30	76.90	4.86	4.59
20	0.05	0.1	4.25	9.54	51.8	51.3	133.5	20.8	223.9	121.8	378.9	47.22	77.56	4.88	4.25
20	0.02	0.25	3.63	19.2	59.0	60.7	126.3	53.2	199.9	120.7	392.3	44.66	78.93	4.97	3.63
20	0.005	1	2.17	53.1	86.4	208.0	101.7	190.4	106.6	101.3	726.4	35.71	77.80	9.34	2.17
30	0.125	0.04	4.59	4.59	37.0	35.4	198.5	7.0	225.8	68.7	314.7	46.65	65.00	4.84	4.59
30	0.05	0.1	4.24	9.47	39.9	39.2	194.3	18.1	217.4	70.2	321.8	45.73	65.92	4.88	4.24
30	0.02	0.25	3.61	18.9	47.0	49.5	184.5	48.2	195.5	71.8	340.0	43.45	67.83	5.02	3.61
30	0.005	1	2.13	51.9	75.1	172.8	149.4	182.2	105.1	63.3	605.4	34.92	69.23	8.74	2.13

Note: Sample vol. = Volume of a sample taken from the donor cell; t_{max} = the maximum flux from skin to donor cell; t_{max} = time to J_{max}; $t^{1}/_{2_{Cv}}$ = terminal half-life of Cv-time profiles calculated by the linear regression method for the last 4 log Cv-time plots; $t^{1}/_{2_3}$ = terminal half-life of J-time profiles calculated by the linear regression method for the last 4 log J-time plots; ΔA = total drug amount removed from donor cell as samples; As(72) = drug amount in skin at 72 h; Ae(72) = cumulative drug amount excreted into receptor cell until t = 72 h; Av(72) = drug amount in vehicle at 72 h; Ae(∞) = cumulative drug amount eventually excreted into receptor cell estimated by extrapolation using $t^{1}/_{2_3}$; AUC(72) = AUC of Cv until t = 72 h estimated by trapezoidal method; AUC(∞) = AUC extrapolating to t = ∞ using $t^{1}/_{2_{Cv}}$; Cv(0) = the initial drug concentration in donor cell solution.

a   Calculated using Program 2 in Appendix 2 where n = 50 was adopted for the number of sublayers for skin; Kp = 0.005 cm/ h, l_v = 0.5 cm, Cv(0) = 1 mg/ml (i.e., the initial amount in donor cell = 500 μg/cm^2) are assumed for all the cases.

b   If Cv is not changed with time, J approaches 5 μg/cm^2/h and J/Cv approaches Kp.

volume as in the case of $k_d = 0..005 \, h^{-1}$ and $l_d = 1$ cm in Table 2 where $t_{max} > 50$ h and the terminal log-linear phase is not attained at 72 h.

V. PREDICTION OF PHARMACOKINETIC PROFILES

When C_i in Equation 1 is specified for the boundary conditions and the initial condition, pharmacokinetic profiles can be predicted if the parameter values are known or postulated. Though the basic parameter values should be estimated from the experimental *in vitro* data for each compound in each vehicle, once those values are known, we can predict the expected profiles supposing that one or more parameter values are varied based upon those parameter values. For instance, whether the pharmacokinetics will be significantly changed by the use of other types of vehicle with different thickness or skin/vehicle partition coefficient can be predicted. Those modifications will sometimes yield a dramatic change in the pharmacokinetic profiles; in other cases, however, only indiscernible changes will occur by the modification of those factors. Thus, the prediction may help the process in the development of the appropriate vehicle. The pharmacokinetic profiles after the complex dosing form, where the repeated topical application at the same site is involved, are also predicted using the parameter values and may give the theoretical basis for the topical application of a drug in clinical practice.

Two methods for the prediction are discussed in this last section. In Appendices 1, 2, and 3, simple programs (Programs 1, 2, and 3, respectively) written in BASIC to predict the pharmacokinetic profiles of percutaneous drug absorption which can work in an inexpensive pocket computer are listed. The first method is the analytical method used for the prediction. The second method is the "random walk method",[18] one of the numerical methods. The latter can be used to predict the complicated processes including the repeated dose. Instead of this advantage, the numerical method has two disadvantages. (1) The pharmacokinetic profiles at specific time, $T (= n\Delta t)$, is predicted based on those at $(n - 1)\Delta t$. Thus, the information is available only after the whole profiles of n time steps are calculated by iteration. In the analytical method, the profiles at T are calculated without knowing the values at times preceding T. (2) Some pharmacokinetic parameters such as the terminal elimination rate constant and half-life are obtained secondarily from the calculation results. In the analytical method, those parameters are clearly incorporated in Equations 12 and 22.

A. PREDICTION BY THE ANALYTICAL METHOD

When vehicle is removed, the drug concentration in the skin is given in Equation 10. However, the absolute values of the concentration in skin cannot be predicted unless the skin/vehicle partition coefficient, Km, is itself estimated by another experiment (e.g., the measurement of drug amounts in horny layer and vehicle solution when equilibrium is attained[19]). This is because the partition parameter,[4,5] l_d, (Km $\cdot$ l_s) and diffusion parameter[4-6] k_d, (D/l_s^2) are estimated from the pharmacokinetic experiment where AUC of Av and that of As and the flux from skin are analyzed (Equation 41 and 42) whereas the values for Km, l_s or D itself are not obtainable. However, the amount in the skin is estimated using the pharmacokinetic parameters l_d and k_d. At time t, the amount in skin As, is given by

$$As = \int_0^{l_s} \Sigma \, B_i \cos(\alpha_i \cdot x/l_s) \exp(-k_d \cdot \alpha_i^2 \cdot t) dt$$

$$= \Sigma \, B_i \cdot l_s/\alpha_i \sin\alpha_i \exp(-k_d \cdot \alpha_i^2 \cdot t) \qquad (43)$$

When the vehicle is removed after the steady state is reached, α_i is given in Equation 6 and B_i is given in Equation 7. Thus,

$$As = l_d \cdot Cv \sum 16/[(2i - 1)^3\pi^3] \cdot (-1)^{i-1}\exp[-k_d \cdot (2i - 1)^2\pi^2t/4] \qquad (44)$$

Flux from skin, J, and cumulative amount excreted from skin, Ae, are given by

$$J = -D\partial C(l_s)/\partial x$$

$$= D \sum B_i \cdot \alpha_i/l_s\sin\alpha_i\exp(-k_d \cdot \alpha_i^2 \cdot t)$$

$$= k_d \cdot l_d \cdot Cv \sum 4/[(2i - 1)\pi] \cdot (-1)^{i-1}\exp[-k_d \cdot (2i - 1)^2\pi^2t/4] \qquad (45)$$

$$Ae = l_d \cdot Cv \sum B_i \cdot l_s/\alpha_i\sin\alpha_i[1 - \exp(-k_d \cdot \alpha_i^2 \cdot t)]$$

$$= l_d \cdot Cv \sum 16/[(2i - 1)^3\pi^3] \cdot (-1)^{i-1}\{1 - \exp[-k_d \cdot (2i - 1)^2 \pi^2t/4]\} \qquad (46)$$

Thus, J and Ae are also estimated using the pharmacokinetic parameters.

For the case of release from vehicle, α_i is also given in Equation 6 while B_i in Equation 8 is used in Equations 43, 45, and 46.

When a finite dose in a solution matrix with a great diffusivity is applied on the skin, α_i is obtainable iteratively from Equation 21 using several methods such as the fixed-point iteration method[20] employing a one-point iteration function, $\alpha_i = \arctan[(l_d/l_v)/\alpha_i] + (i - 1)\pi$ (i = 1, 2, —) when a calculator with a function, arctangent, is available. In Program 1 in Appendix 1, however, another method, the interval bisection method is used. A_i and B_i in Equation 1 and w_i in Equation 18 are given as

$$A_i = -2KmCv\alpha_i/(\beta + \beta^2 + \alpha_i^2) \qquad (47)$$

$$B_i = 2KmCv\beta/(\beta + \beta^2 + \alpha_i^2) \qquad (48)$$

$$w_i = 2Cv\beta/(\beta + \beta^2 + \alpha_i^2) \qquad (49)$$

where Cv is the initial donor cell concentration (or Cv = U(t = 0)) and β is given as[21]

$$\beta = Km \cdot l_s/l_v = l_d/l_v \qquad (50)$$

Thus, the vehicle concentration, U, at time t is given as

$$U = 2Cv \cdot \beta\sum 1/(\beta + \beta^2 + \alpha_i^2)\exp(-k_d \cdot \alpha_i^2 \cdot t) \qquad (51)$$

The flux from vehicle to skin, J(0, t), and cumulative amount excreted from the vehicle, Ae(0, t) are given by

$$J(0,t) = 2l_d \cdot k_d \cdot Cv\sum \alpha_i^2/(\beta + \beta^2 + \alpha_i^2)\exp(-k_d \cdot \alpha_i^2 \cdot t) \qquad (52)$$

$$Ae(0,t) = 2l_d \cdot Cv\sum 1/(\beta + \beta^2 + \alpha_i^2)[1 - \exp(-k_d \cdot \alpha_i^2 \cdot t)] \qquad (53)$$

The drug concentration in the skin is given in Equation 1 where A_i and B_i are given in Equations 47 and 48, respectively. However, the absolute concentration in skin cannot be estimated from the parameters obtainable from the pharmacokinetic experiment as in the process after the vehicle is removed. The amount in skin, As, is given as

$$As = 2l_d \cdot Cv\sum 1/(\beta + \beta^2 + \alpha_i^2) \cdot [(1 - \cos\alpha_i)/\cos\alpha_i] \cdot \exp(-k_d \cdot \alpha_i^2 \cdot t) \qquad (54)$$

Thus, As can be estimated using the pharmacokinetic parameters k_d, l_d and l_v. Flux from skin to receptor cell, $J(l, t)$, and cumulative amount excreted into receptor cell, $Ae(l, t)$, are given, respectively, by[21]

$$J(l,t) = 2l_d \cdot k_d \cdot Cv \Sigma \alpha_i^2/[(\beta + \beta^2 + \alpha_i^2)\cos\alpha_i] \exp(-k_d \cdot \alpha_i^2 \cdot t)] \quad (55)$$

$$Ae(l,t) = 2l_d \cdot Cv \Sigma 1/[(\beta + \beta^2 + \alpha_i^2)\cos\alpha_i] \cdot [1 - \exp(-k_d \cdot \alpha_i^2 \cdot t)] \quad (56)$$

Because to estimate α_i in the finite dose case (Equation 21, Figure 2) is a task suitable for a computer, a short program (Program 1) written in BASIC to estimate the values for α_i and to predict the pharmacokinetic profiles is listed in Appendix 1. Practically, the first several terms are employed to estimate the values in Equations 51 to 56. The question "how many terms should be employed?" is not an easy one to answer; it is related to time to be estimated (when time is sufficiently large, all the terms except for the first can be ignored) and the values to be estimated (generally, $J(l, t)$ in Equation 55 requires more terms as compared to $J(0, t)$ in Equation 52, to attain the similar magnitude of the precision). However, in Program 1 in Appendix 1, as an indicator to select the number of terms employed and also to show that the calculation is conducted correctly, the $Ae(\infty)$ and MRT calculated as $Ae(\infty) = 2l_d \cdot Cv \Sigma 1/[(\beta + \beta^2 + \alpha_i^2) \cos \alpha_i]$ and $AUMC_{J(l, t)}/Ae(\infty)$, where $AUMC_{J(l, t)}$ is calculated as $2l_d \cdot Cv/k_d \Sigma 1/[\alpha_i^2(\beta + \beta^2 + \alpha_i^2) \cos \alpha_i]$, are compared with the initial volume, $A_0 (= Cv \cdot l_v)$, and the MRT given in Equation 29, respectively.

B. RANDOM WALK METHOD

The random walk method[18] is suitable for the prediction of the pharmacokinetic processes under the complicated dosing form such as the repeated topical application at the same site. This method is based on the following relationship[22] where a drug molecule is assumed to diffuse according to the principle of the random walk

$$2D\Delta t = (\overline{\Delta x})^2 \quad (57)$$

where D is the diffusion coefficient and $(\overline{\Delta x})^2$ is the mean squared distance a molecule will move in a time Δt and D is the diffusion coefficient. Conversely, when skin is subdivided into n sublayers, and Δx is fixed as $\Delta x = l_s/n$, the corresponding Δt is estimated at $2D\Delta t = (l_s/n)^2$ or $\Delta t = (l_s^2/D)/(2n^2)$. Because D/l_s^2 is the diffusion parameter, k_d,

$$\Delta t = 1/(2n^2 k_d) \quad (58)$$

At every Δt given in Equation 58, a molecule in the i th sublayer "jumps" into the $(i - 1)$ and $(i + 1)$ th sublayers with a probability of 0.5, respectively. Conversely, half of the molecules which were in the $(i - 1)$ th layer and half of them in the $(i + 1)$ th layer will "jump" into the i th layer or

$$A_i(t + \Delta t) = 0.5A_{i-1}(t) + 0.5A_{i+1}(t) \qquad (i = 2 \text{ to } n-1) \quad (59)$$

where $A_i(t)$ is the amount of drug in the i th layer at t. If the sink is located at the right boundary, half of the molecules which were in the n th layer are removed into sink as flux, J, and the following equations are used

$$A_n(t + \Delta t) = 0.5A_{i-1}(t) \quad (60)$$

$$J(t + \Delta t/2)\Delta t = 0.5A_n(t) \quad (61)$$

where J(t) is the flux from skin to sink at time t. When the left boundary is the "dead end", half of the molecules in the first layer remain in the same layer and half of the molecules, which were in the second layer, "jump" into the first layer so that

$$A_1(t + \Delta t) = 0.5A_1(t) + 0.5A_2(t) \tag{62}$$

When a finite dose is applied and the diffusivity of a drug through the vehicle is large, the whole vehicle is considered to be one layer with diffusion coefficient Dv^*. Because Δt in Equation 1 should be the same for both vehicle and skin,[18] Dv^* is defined as in the following equation

$$l_v^2/(2Dv^*) = 1/(2n^2 k_d) \tag{63}$$

Conversely, if the diffusivity of the vehicle is incorporated into the prediction precisely, n should be chosen so that Dv^* in Equation 63 is near the actual value Dv. [For a discussion of the general case where vehicle is also subdivided, see reference 18]. At the vehicle-skin boundary, $C(x = 0) = Km \cdot U$, where U is the drug concentration in vehicle at t. Thus, the ratio of amount in the first layer of skin to that in the vehicle, K, is given as

$$K = [Km \cdot U \cdot l_s/n]/[U \cdot l_v] = l_d/(n \cdot l_v) \tag{64}$$

where l_d and l_v are the partition parameter[4,5] and effective thickness of vehicle,[11] respectively. When $K < 1$, half of $A_1(t)$ will "jump" into vehicle but only $0.5KAv(t)$ will "jump" into the first layer of skin where $Av(t)$ is the amount in vehicle at t and the other molecules remain in vehicle or

$$Av(t + \Delta t) = 0.5A_1(t) + (1 - 0.5K)Av(t) \tag{65}$$

$$A_1(t + \Delta t) = 0.5A_2(t) + 0.5KAv(t) \tag{66}$$

[For the case, $K > 1$, see reference 18]. In Appendix 2, Program 2 is listed for a case where a finite dose is applied. In Appendix 3, Program 3 is listed for a case for skin after the removal of the vehicle where the left boundary is the "dead-end" as in Equation 62. Program 3 can be also used to predict the processes of drug release directly into sink from the solution matrix with the initial conditions $A_i(0) = Cv \cdot l_v/n$. When the vehicle is applied to skin containing no drug, $Av(0) = Cv \cdot l_v$ and $A_i(0) = 0$ (i = 1, 2, —, n). When the vehicle is removed, $A_i(T)$, where T is the time until when the vehicle was applied in the previous process (calculated by Program 2), are the initial conditions, $A_i(0)$ (i = 1, 2, —, n), for the new process after the vehicle is removed (calculated by Program 3). When the new vehicle is applied on skin where drug molecules are already distributed, $A_i(T)$, where T is the time when the previous process is interrupted and obtained by Program 2 or 3, are used as the initial conditions, $A_i(0)$, for the new process. In this case, the initial condition for the vehicle is given as $Av(0) = Cv \cdot l_v$. The number of sublayers, n, should be selected specifically. For instance, the cumulative amount excreted is usually well estimated even for n = 10, while the percentage error is great when the amount remaining in the skin after the vehicle removal is estimated using small n. However, when n is large, Δt in Equation 58 is small so that the calculation speed will slow down.

Finally, though the diffusivity through liquid is large (e.g., the diffusion coefficient of most substances through water is $\simeq 10^{-5}$ cm²/s),[23] because the l_v employed in Equation 64 is sometimes much larger than l_d, Dv^*, calculated in Equation 63 may be greater than the actual diffusivity through vehicle solution [e.g., the value for Dv^* calculated in LINE 60 in Program 2 for aqueous donor cell solution is greater than 1 to 5×10^{-5} cm²/s]. Thus, we recommend stirring the vehicle solution frequently in such a case. If a polymer vehicle is used, sublayers should be employed also for the vehicle to predict the pharmacokinetic processes.[18]

ACKNOWLEDGMENTS

We thank Dr. E. H. Twizell, Reader in Computational Mathematics, Department of Mathematics and Statistics, Brunel University, Uxbridge, Middlesex, England UB8 3PH, for his valuable comments on the manuscript.

APPENDICES: PROGRAMS WRITTEN IN BASIC FOR THE PREDICTION OF THE PHARMACOKINETIC PROFILES

APPENDIX 1: PROGRAM 1
Analytical Method for the Finite Dose Application

```
10 REM PROGRAM 1 ANALYTICAL METHOD  FINITE DOSE
20 P=3.14159:W=10^(-9)
30 INPUT "KD (H^-1)=";K:INPUT "LD (MICROM)=";D
40 INPUT "LV (MICROM)=";V:INPUT "C0 (%, W/V)=";Z
50 INPUT "N FOR SERIES =";N:DIM A(N):B=D/V
60 FOR I=1 TO N:L=0:X=(4*I-3)*P/4:G=P/8
70 T=TAN(X)-B/X:IF (L>32) OR (ABS(T)<W) THEN GOTO 100
80 R=1:IF T>0 THEN LET R=-1
90 X=X+R*G:G=G/2:L=L+1:GOTO 70
100 A(I)=X:NEXT I
110 REM MRT
120 C=0:M=0:FOR I=1 TO N
130 Q=1/(B+B*B+A(I)*A(I))/COS(A(I))
140 C=C+Q:M=M+Q/K/A(I)/A(I)
150 NEXT I:PRINT "AMOUNT CALCULATED (TRUE VALUE)"
160 PRINT 2*D*Z*C;"(";V*Z;" MICROG/CM2)"
170 PRINT "MRT CALCULATED (TRUE VALUE)"
180 PRINT M/C;"(";V/K/D+1/2/K;" H)"
190 PRINT "TERMINAL ELIMINATION CONSTANT =";A(1)*A(1)*K;"(H^-1)"
200 PRINT "TERMINAL T1/2 =";.693/A(1)/A(1)/K;"(H)"
210 REM CALCULATION OF U, J, AS & AE
220 DIM E(2):DIM J(2)
230 INPUT "T (HR)=";T:E(1)=0:E(2)=0:J(1)=0:J(2)=0
240 U=0:S=0:FOR I=1 TO N:C=COS(A(I))
250 Q=1/(B+B*B+A(I)*A(I)):Y=EXP(-K*A(I)*A(I)*T)
260 U=U+B*Q*Y:J(1)=J(1)+A(I)*A(I)*Q*Y
270 E(1)=E(1)+Q*(1-Y):S=S+Q*(1-C)/C*Y
280 J(2)=J(2)+A(I)*A(I)*Q/C*Y:E(2)=E(2)+Q/C*(1-Y):NEXT I
290 U=2*Z*U:J(1)=2*D*K*Z*J(1):E(1)=2*D*Z*E(1)
300 S=2*D*Z*S:J(2)=2*D*K*Z*J(2):E(2)=2*D*Z*E(2)
310 PRINT "CONC IN VEHICLE=";U;"(%, W/V)"
320 PRINT "AMOUNT IN VEHICLE=";U*V;"(MICROG/CM2)"
330 PRINT "J FROM VEHICLE=";J(1);"(MICROG/CM2/H)"
340 PRINT "AE FROM VEHICLE=";E(1);"(MICROG/CM2)"
350 PRINT "A IN SKIN=";S;"(MICROG/CM2)"
360 PRINT "J FROM SKIN=";J(2);"(MICROG/CM2/H)"
370 PRINT "AE FROM SKIN=";E(2);"(MICROG/CM2/H)"
380 REM CONC IN SKIN LS=20 MICROM IS SUPPOSED
390 INPUT "X (MICROM FROM SURFACE)=";X
400 U=0:FOR I=1 TO N:C=COS(A(I)*X/20)
410 S=SIN(A(I)*X/20):Q=1/(B+B*B+A(I)*A(I))
420 Y=EXP(-K*A(I)*A(I)*T):U=U+(-A(I)*Q*S+B*Q*C)*Y:NEXT I
430 U=2*Z*U*10
440 PRINT "CONC IN SKIN =";U;"(*KM (MILIG/CM3))
450 INPUT "CHANGE DISTANCE-1    CHANGE TIME-2";H
460 ON H GOTO 390,230:END
```

APPENDIX 2: PROGRAM 2
Random Walk Method to Estimate the Processes Where Solution Matrix Vehicle with Large Diffusion Coefficient is Applied

```
10 REM PROGRAM 2  RANDOM WALK METHOD VEHICLE IS APPLIED
20 INPUT "KD (H^-1)=";K:INPUT "LD (MICROM)=";D
30 INPUT "LV (MICROM)=";V:INPUT "C0 (%,W/V)=";Z
40 INPUT "N OF SUBLAYERS=";N
50 T=1/(2*N*N*K):DIM A(N):DIM B(N):DIM E(2):DIM J(2)
60 P=D/V/N:A(0)=Z*V:IF P>1 THEN PRINT "USE LARGER N":END
70 PRINT "DV SUPPOSED=";N*N*V*V*K*10^(-6)/3.6;"(X10^-5 CM2/S)"
80 INPUT "FINAL TIME (H)=";F:F=-INT(-F/T)
90 PRINT "* INPUT INITIAL CONDITION IN SKIN"
100 FOR I=1 TO N
110 PRINT "AMOUNT(T=0) (MICROG/CM2) IN";I;"TH LAYER="
120 INPUT A(I):NEXT I:E(1)=0:E(2)=0:L=0
130 J(2)=A(N)/2/T:J(1)=(P*A(0)-A(1))/2/T
140 PRINT "T=";L*T;"(HR)"
150 PRINT "  AMOUNT IN VEHICLE=";A(0);"(MICROG/CM2)"
160 PRINT "  CONC IN VEHICLE=";A(0)/V;"(%, W/V)"
170 PRINT "  AMOUNT EXCRETED FROM VEHICLE=";E(1);"(MICROG/CM2)"
180 PRINT "  AMOUNT IN SKIN=";S;"(MICROG/CM2)"
190 PRINT "  AMOUNT EXCRETED FROM SKIN=";E(2);"(MICROG/CM2)"
200 IF L=F THEN GOTO 300
210 PRINT "T=";(L+.5)*T;"(HR)"
220 PRINT "  FLUX FROM VEHICLE=";J(1);"(MICROG/CM2/H)"
230 PRINT "  FLUX FROM SKIN=";J(2);"(MICROG/CM2/H)"
240 B(0)=A(1)/2+(1-P/2)*A(0):B(1)=P*A(0)/2+A(2)/2
250 B(N)=A(N-1)/2:E(2)=E(2)+A(N)/2
260 FOR I=2 TO N-1:B(I)=A(I-1)/2+A(I+1)/2:NEXT I
270 FOR I=0 TO N:A(I)=B(I):NEXT I
280 S=0:FOR I=1 TO N:S=S+A(I):NEXT I:E(1)=S+E(2)
290 L=L+1:GOTO 130
300 FOR I=1 TO N
310 PRINT "AMOUNT IN";I;"TH LAYER=";A(I);"(MICROG/CM2)"
320 NEXT I:END
```

APPENDIX 3: PROGRAM 3
Random Walk Method to Estimate the Processes after Vehicle is Removed

```
10 REM PROGRAM 3 RANDOM WALK METHOD AFTER VEHICLE REMOVAL
20 INPUT "KD (H^-1)=";K:INPUT "N OF SUBLAYERS=";N
30 T=1/(2*N*N*K):DIM A(N):DIM B(N)
40 INPUT "FINAL TIME (H)=";F:F=-INT(-F/T)
50 PRINT "* INPUT INITIAL CONDITION IN MEMBRANE"
60 FOR I=1 TO N
70 PRINT "AMOUNT (MICROG/CM2) IN";I;"TH LAYER="
80 INPUT A(I):NEXT I:E=0:L=0
90 J=A(N)/2/T
100 PRINT "TIME=";L*T;"(H)"
110  S=0:FOR I=1 TO N:S=S+A(I):NEXT I
120 PRINT "   AMOUNT IN SKIN=";S;"(MICROG/CM2)"
130 PRINT "   AMOUNT EXCRETED FROM SKIN=";E;"(MICROG/CM2)"
140 IF L=F THEN GOTO 200
150 PRINT "TIME=";(L+.5)*T;"(H)"
160 PRINT "   FLUX FROM SKIN=";J;"(MICROG/CM2/H)"
170 B(1)=A(1)/2+A(2)/2:B(N)=A(N-1)/2:E=E+A(N)/2
180 FOR I=2 TO N-1:B(I)=A(I-1)/2+A(I+1)/2:NEXT I
190 FOR I=1 TO N:A(I)=B(I):NEXT I:L=L+1:GOTO 90
200 FOR I=1 TO N
210 PRINT "AMOUNT IN";I;"TH LAYER=";A(I);"(MICROG/CM2)"
220 NEXT I:END
```

REFERENCES

1. **Skelly, J. P., Shah, V. P., Maibach, H. I., Guy, R. H., Wester, R. C., Flynn, G., and Yacobi, A.,** FDA and AAPS report of the workshop on principles and practices of *in vitro* penetration studies: relevance to bioavailability and bioequivalence, *Pharmaceut. Res.,* 4, 265, 1987.

2. **Yum, S. I.,** Transdermal therapeutic systems and rate controlled drug delivery, *Med. Prog. Tech.,* 15, 47, 1989.

3. **Yamaoka, K., Nakagawa, T., and Uno, T.,** Statistical moments in pharmacokinetics, *J. Pharmacokinet. Biopharmaceut.,* 6, 547, 1978.

4. **Kubota, K. and Yamada, T.,** Finite dose percutaneous drug absorption: theory and its application to *in vitro* timolol permeation, *J. Pharm. Sci.,* 79, 1015, 1990.

5. **Okamoto, H., Hashida, M., and Sezaki, H.,** Structure-activity relationship of 1-alkyl- or 1-alkenylaza-cycloalkanone derivatives as percutaneous penetration enhancers, *J. Pharm. Sci.,* 77, 418, 1988.

6. **Guy, R. H. and Hadgraft, J.,** Physicochemical interpretation of the pharmacokinetics of percutaneous absorption, *J. Pharmacokinet. Bipharmaceut.,* 11, 189, 1983.

7. **Kubota, K. and Ishizaki, T.,** A calculation of percutaneous drug absorption. I. Theoretical, *Comput. Biol. Med.,* 16, 7, 1986.

8. **Kubota, K. and Ishizaki, T.,** A calculation of percutaneous drug absorption. II. Computation results, *Comput. Biol. Med.,* 16, 21, 1986.

9. **Feldmann, R. J. and Maibach, H. I.,** Penetration of 14C hydrocortisone through normal skin, *Arch. Dermatol.,* 91, 661, 1965.

10. **Siddiqui, O., Roberts, M. S., and Polack, A. E.,** Percutaneous absorption of steroids: relative contribution of epidermal penetration and dermal clearance, *J. Pharmacokinet. Biopharmaceut.,* 17, 405, 1989.

11. **Kubota, K. and Ishizaki, T.,** A diffusion-diffusion model for percutaneous drug absorption, *J. Pharmacokinet. Biopharmaceut.,* 14, 409, 1986.

12. **Benet, L. Z. and Galeazzi, R. L.,** Noncompartmental determination of the steady state volume of distribution, *J. Pharm. Sci.,* 72, 857, 1979.

13. **Kubota, K. and Ishizaki, T.,** A theoretical consideration of percutaneous drug absorption, *J. Pharmacokinet. Biopharmaceut.,* 13, 55, 1985.

14. **Nakashima, E. and Benet, L. Z.,** General treatment of mean residence time, clearance, and volume parameters in linear mammillary models with elimination from any compartment, *J. Pharmacokinet. Biopharmaceut.,* 16, 475, 1988.

15. **Kubota, K. and Maibach, H. I.,** An estimation of the permeability coefficient from finite dose *in vitro* percutaneous drug permeation study, *J. Pharm. Sci.,* in press.

16. **Bucks, D. A., Wester, R. C., Mobayen, M. M., Yang, D., Maibach, H. I., and Coleman, D. L.,** *In vitro* percutaneous absorption and stratum corneum binding of alachlor: effect of formulation dilution with water, *Toxicol. Appl. Pharmacol.,* 100, 417, 1989.

17. **Kubota, K., Yamada, T., Ogura, A., and Ishizaki, T.,** A novel differentiation method of vehicle models for topically applied drugs: application to a therapeutic timolol patch, *J. Pharm. Sci.,* 79, 179, 1990.

18. **Kubota, K., Koyama, E., and Yasuda, K.,** A random walk method for percutaneous drug absorption pharmacokinetics: an application to repeated administration of therapeutic timolol patch, *J. Pharm Sci.,* in press.

19. **Scheuplein, R. J.,** Mechanism of percutaneous adsorption: I. Routes of penetration and the influence of solubility, *J. Invest. Dermatol.,* 45, 334, 1965.

20. **Twizell, E. H.,** Nonlinear equations and nonlinear system, in *Numerical Methods, with Applications in the Biomedical Sciences,* Ellis Horwood Ltd., New York, 1988, 75.

21. **Cooper, E. R. and Berner, B.,** Finite dose pharmacokinetics of skin penetration. *J. Pharm. Sci.,* 74, 1100, 1985.

22. **Burnette, R. R.,** A Monte-Carlo model for the passive diffusion of drugs through the stratum corneum, *Int. J. Pharm.,* 22, 89, 1984.

23. **Cussler, E. L.,** Values of diffusion coefficients, in *Diffusion: Mass Transfer in Fluid Systems,* Cambridge University Press, Cambridge, 1984, 107.

Chapter 21

EFFECT OF SODIUM LAURYL SULFATE-INDUCED SKIN IRRITATION ON *IN VITRO* PERCUTANEOUS ABSORPTION

Klaus-P. Wilhelm, Christian Surber, and Howard I. Maibach

TABLE OF CONTENTS

*   Modified from Wilhelm, K-.P. et. al., *J. Invest. Derm.*, in press. With permission.

I. INTRODUCTION

Penetration of drugs into and through skin is important for the efficacy and toxicity of topically applied compounds. Permeation of compounds is usually determined by *in vitro* or *in vivo* application of a compound under investigation to healthy skin. However, many drugs are used to treat various skin diseases where the cutaneous barrier is in doubt[1-3] and transdermally delivered drugs may themselves cause irritant or allergic dermatitis. This may cause extensive variation in systemic absorption and/or local skin concentration.

An increase in permeability to water and some drugs has been demonstrated in atopic dermatitis and psoriasis.[1-3] Similar pathological conditions can be induced in animal models by EFA-deficient diet, UV-B or UV-C irradiation, or application of chemicals.[5-9] Sodium lauryl sulfate (SLS) reproducibly induces irritant dermatitis reactions with increased blood flow, transepidermal water loss (TEWL), and mitotic activity[10,11] and has frequently been used as a model substance to study irritant dermatitis.[10,12-15] We employed SLS irritation as a model for diseased skin to study percutaneous penetration and skin drug concentration after topical application of compounds with different physicochemical properties.

II. MATERIAL AND METHODS

Chemicals — Specific activities were (1) 4.12 mCi/mmol ibuprofen (IB); (2) 56 mCi/mmol acitretin (AC); (3) 39.9 mCi/mmol indomethacin (IM); and (4) 55.0 mCi/mmol hydrocortisone (HC). All radiochemicals has a purity greater than 95%. To yield maximum thermodynamic activity for each compound, penetration studies were performed with saturated drug solutions (Table 1). Isopropylmyristate (IPM) was chosen as the vehicle because it is a frequently used vehicle constituent in various dermatological formulations. Because of the photolability of acitretin, the preparation and use of the acitretin formulation was performed under minimal light exposure.

Animals — Female hairless guinea pigs (430 to 550 g) were housed at constant ambient conditions (20 ± 2°C; 45 to 65% r.h.) with a controlled diurnal cycle and free access to food and water. Starting 24 h before exposure until the end of exposure, animals were housed in single cages.

Pretreatment — SLS (0.5 ml of 0.5%) in aqueous solution was applied to the upper back of the animals with an occlusive plastic chamber (2.5 cm²; Hilltop®, Cincinnati, OH). The plastic chamber was removed after 24 h of application. Distilled water served as control pretreatment.

Transepidermal water loss measurements — TEWL was measured *in vivo* with an Evaporimeter (EP 1, Servo Med, Stockholm, Sweden)[16] 24 h after removal of the SLS application (prior to sacrifice of the animals).

***In vitro* percutaneous penetration** — Animals were sacrificed 24 h after termination of pretreatment, and full-thickness skin was excised from the dorsum of the guinea pig and washed in phosphate buffered saline, pH 7.2 (PBS). Subcutaneous fat and other extraneous tissue were carefully trimmed, if necessary, from the dermal surface with a scalpel blade. The pretreated aspect of the skin was mounted in conventional flow-through diffusion cells (Laboratory Glass Apparatus, Inc., Berkeley, CA).[17] The cells allowed 1 cm² of skin to be exposed without occlusion to the chemical, to ambient temperature (20 to 22°C), and relative humidity (40 to 65%), while the dermis was bathed in 4-ml receptor solution (PBS, 37°C). The receptor solution was stirred by a teflon-coated magnetic bar at 600 rpm and was continuously replaced at a rate of 5 ml/h with a peristaltic pump (Casette® Pump, Manostat®, New York, NY). The receptor solution was collected hourly for 22 h.

Sample handling and analysis — After a 22-h exposure, the remaining drug formulation was removed from the skin by a standardized procedure in order to apply a mass balance

TABLE 1
Investigated Compounds[a]

	HC	IM	IB	AC
Mol. wt.	363	358	206	326
Melt point (C°)	218	76	159	226
Log [P][b]	1.6	3.1	3.5	6.0
IPM (g/l) solub.	0.35	2.0	140.0	0.33
mg Appl.	0.15	0.9	63.0	0.15
μmol Appl.	0.41	2.51	305.4	0.46

[a] Physicochemical data of compounds used in penetration experiments.

[b] Log [P]: octanol water partition coefficient adapted from Ref. 23 or determined experimentally (AC).

technique.[18] Exposed skin was separated from unexposed skin and both parts were solubilized in a solubilizer.

Calculation and statistics — The cumulative amount of drug per cm^2 was plotted agaist time. Steady-state flux and lag time were calculated by linear regression of the linear portion of the graph (pseudo-steady state; reached after 5 to 10 h for investigated compounds).

Permeability coefficients (K_p) were determined as:

$$K_p = \frac{J_{ss}}{C}$$

where C is the concentration in the donor solution and J_{ss} being steady-state flux. Enhancement factors $Ef(Q)$ and $Ef(C_{skin})$ were calculated as:

$$Ef(Q) = \frac{Q(SLS\ group)}{Q(control\ group)} \qquad Ef(C_{skin}) = \frac{C_{skin}(SLS\ group)}{C_{skin}(control\ group)}$$

and relative drug dose in the skin (D_{skin}) as:

$$D_{skin} = \frac{C_{skin} \times 100}{Q}$$

with Q being the cumulative amount of compound recovered in the receptor fluid after a 22-h penetration through 1 cm^2 skin surface and C_{skin} being the skin drug concentration per 1 cm^2.

III. RESULTS

Skin irritation was evaluated 24 h after the end of the pretreatment to minimize the effects of occlusion on TEWL readings. SLS pretreatment resulted in uniform, moderate-to-intense erythema; control (water) pretreatment did not induce visible reactions (Table 2). Water permeability *in vivo* (TEWL) was increased 4.5-fold in the SLS-pretreated animals as compared with controls (Table 2).

Percutaneous penetration was increased in skin pretreated with SLS for all compounds studied (Figure 1). The flux of all drugs with the exception of indomethacin demonstrated a "steady state" as evidenced by the flux over time. Enhancement of the cumulative dose was highest for HC (5.9) and lowest for AC (3.4) (Figure 2). The mass balance was 81.8% or higher for all experiments (Table 3).

TABLE 2
TEWL and Visual Assessment

	TEWL (gm^{-2} h^{-1})	Visual[a]
Control	3.2 ± 1.0	0.0 ± 0.0
SLS	14.5 ± 2.8*	2.2 ± 0.3*

Note: TEWL measurements *in vivo* and visual scores 24 h after 24-h pretreatment with 0.5% SLS or water (control); Means ± SD; N = 10.

[a] Arbitrary units (5 point scale from 0 to 4)
* Statistically significantly different from controls; p ≤ 0.05 (*t*-test for unpaired data)

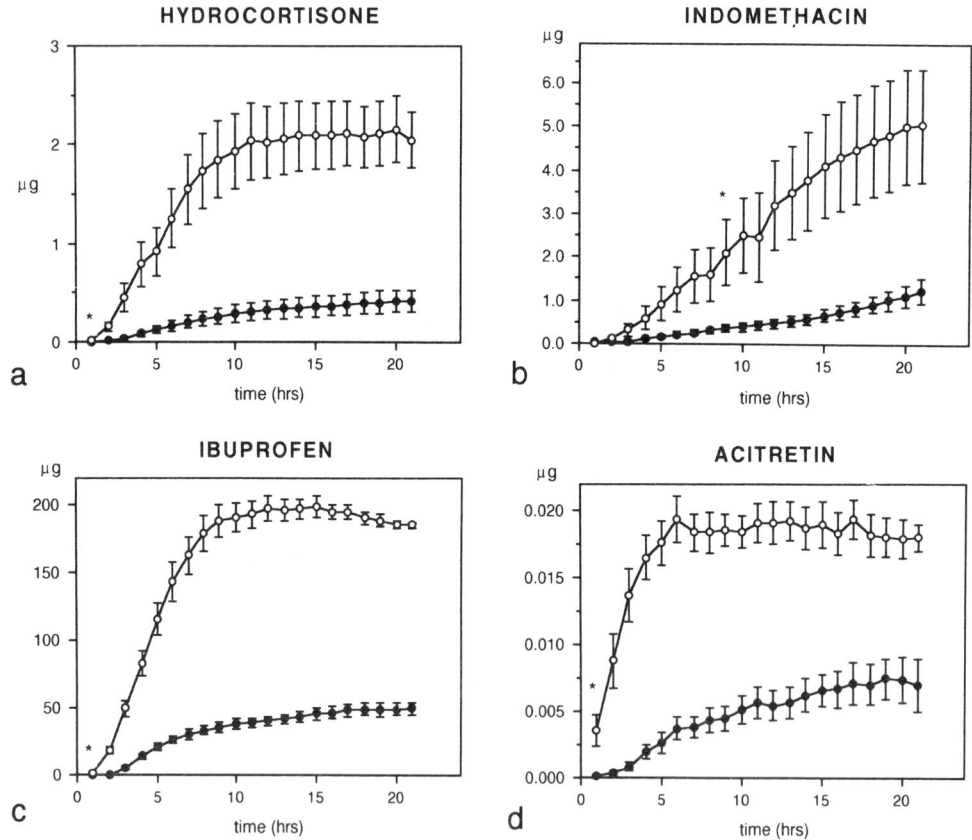

FIGURE 1.   Shown are flux rates over time for (a) hydrocortisone, (b) indomethacin, (c) ibuprofen, and (d) acitretin. Animals had been pretreated for 24 h with 0.5% SLS (open circles) or water (control, closed circles); Mean ± 1 SEM; N = 5 to 6.* Flux rates are statistically significantly different from controls for all time point beyond star (*p* ≤0.05; *t*-test for unpaired data).

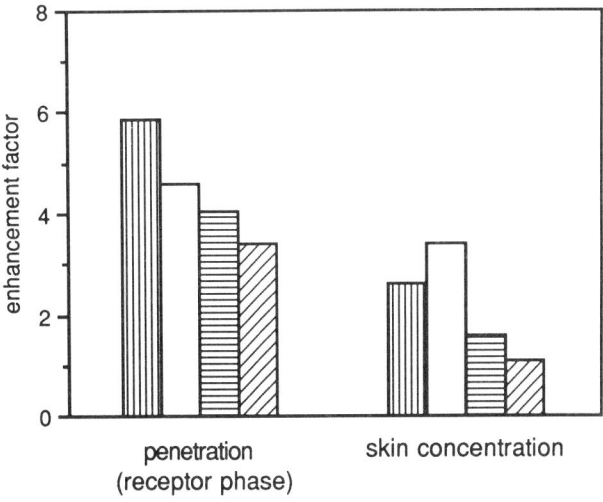

FIGURE 2.   Illustrated are enhancement factors for cumulative amount compound penetrated (per cm²) and recovered in the receptor phase and for skin concentrations after decontamination (per cm² surface). Enhancement factors were calculated as described in the materials and methods section. Compounds are ⦀ hydrocortisone; ☐ indomethacin; ☰ ibuprofen; and ▨ acitretin.

TABLE 3
Mass Balance

Compound	Formulation	Skin	Receptor fluid	Total
HC (contr.)	81.4 ± 3.88	2.3 ± 0.70	4.1 ± 2.89	87.9 ± 1.58
HC (SLS)	57.7 ± 13.06	6.1 ± 2.13	24.2 ± 10.64	88.1 ± 2.58
IM (contr.)	90.0 ± 0.99	3.1 ± 1.34	1.4 ± 0.72	94.5 ± 2.01
IM (SLS)	77.9 ± 5.19	10.5 ± 2.09	7.0 ± 5.05	95.4 ± 1.06
IB (contr.)	103.4 ± 2.41	0.7 ± 0.11	1.5 ± 0.33	105.7 ± 2.47
IB (SLS)	95.7 ± 2.53	1.2 ± 0.44	5.8 ± 0.48	102.7 ± 2.08
AC (contr.)	89.7 ± 1.86	0.4 ± 0.15	0.1 ± 0.04	90.2 ± 1.91
AC (SLS)	81.1 ± 2.68	0.4 ± 0.07	0.3 ± 0.05	81.8 ± 2.74

Note: Summarized is the recovery of applied substance in percent of applied dose. Values listed under formulation include compound removed by the wash procedure; Means ± SD; N = 5 to 6.

Drug concentrations in the skin were not different between the irritated and control group for acitretin but were increased for the other compounds between factor 1.6 (IB) and 3.4 (IM) (Table 4). Since drug concentrations in the skin were less increased in SLS-treated animals than the amount of drug recovered in the receptor phase, the relative drug dose in the skin (in percent of the cumulative dose recovered in the receptor fluid) was smaller in SLS-treated groups than in controls for all drugs with the exception of indomethacin (Figure 3).

Lag time and permeability coefficients (K_p) are summarized in Table 5. K_p was significantly increased in the SLS-treated groups between a factor of 3.2 (AC) and 5.5 (HC). The increases in flux in the SLS-treated groups were paralleled by shortened lag times for AC and HC, but not for IM and IB.

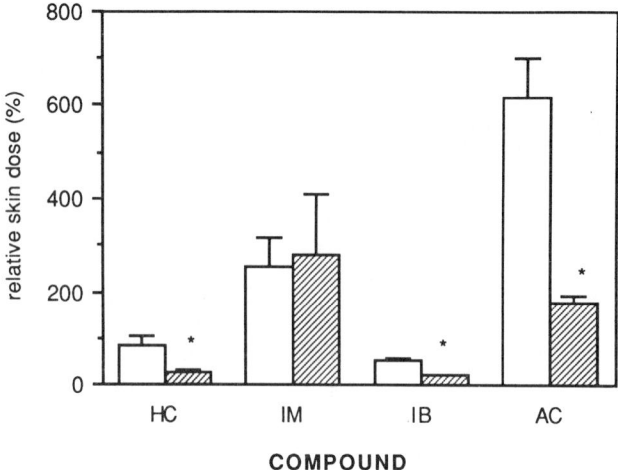

FIGURE 3. Shown are relative drug doses in the skin (as a percentage of the cumulative dose penetrated and recovered in the receptor phase) of control (open bars) and SLS-pretreated skin (hatched bars); Mean ± 1 SEM; N = 5 to 6.* Statistically significantly different from controls ($p \leq 0.05$; *t*-test for unpaired data).

TABLE 4
Skin Drug Concentration

Drug	Control	Irritant dermatitis	Ratio
HC	3.52 ± 1.05	9.13 ± 3.20*	2.6
IM	27.50 ± 12.09	94.26 ± 18.83*	3.4
IB	469.17 ± 69.04	735.31 ± 277.29*	1.6
AC	0.62 ± 0.23	0.66 ± 0.11[a]	1.1

Note: Shown are skin drug concentrations (μg per cm^2 surface) after 22-h penetration experiment *in vitro* and after skin surface decontamination with soap and water. Animals had been pretreated for 24 h with 0.5% SLS (irritant dermatitis) or water (controls); Mean ± SD; N = 5 to 6.

[a] Not significant (p >0.05)
* Statistically significantly different from controls (p ≤ 0.05; two-tailed *t*-test for unpaired data).

IV. DISCUSSION

To study the influence of irritant dermatitis on percutaneous penetration of diverse drugs, hairless guinea pigs were exposed to SLS *in vivo*. This pretreatment was chosen because SLS is a widely used model substance to study irritant dermatitis.[10,12-15] SLS pretreatment reproducibly induced irritant skin reactions in all animals. Water permeability (TEWL) *in vivo* was increased in SLS-treated animals 4.5-fold as compared with controls. The TEWL increase may partially be due to increased blood flow and rise in skin temperature in irritated skin.[20,21] However, that SLS induces an alteration of stratum corneum (SC) barrier function to water, independent of blood flow, was demonstrated *in vitro* for tritiated water by Scheuplein and Ross.[22] Guinea pigs are more susceptible to irritation by many chemicals than humans. In the present study, SLS-induced irritation was morphologically and functionally

TABLE 5
Penetration Parameters

	Lag time (h)		Permeability coefficient (K_p) (cm h^{-1} × 10^3)	
	Control	SLS	Control	SLS
HC	6.7 ± 0.9	4.9 ± 0.8*	1.05 ± 0.72	5.80 ± 2.44*
IM	8.4 ± 0.8	8.5 ± 0.8[a]	0.38 ± 0.20	2.17 ± 1.49*
IB	4.6 ± 0.6	4.1 ± 0.8[a]	0.27 ± 0.07	1.38 ± 0.11*
AC	5.3 ± 0.5	2.0 ± 0.8*	0.02 ± 0.01	0.06 ± 0.01*

Note: Shown are parameters for percutaneous absorption through irritated skin (SLS) and controls; Means ± SD.

[a] Not significant (p >0.05).
* Statistically significantly different from controls (p ≤0.05; two-tailed *t*-test for unpaired data).

(TEWL increase) similar to reactions in humans but slightly more intense than in humans. Treatment of adult human volunteers with 0.5% SLS for 24 h resulted in 2.5-fold TEWL increase.[10]

After sacrificing the animals, percutaneous penetration from saturated drug solutions was studied *in vitro* using the removed (pretreated) skin. Comparing irritated vs. control skin, *in vitro* percutaneous penetration was increased for all compounds studied, with the greatest effect for the most hydrophilic compound (hydrocortisone). Similarly, Bronaugh and Stewart observed the greatest difference between penetration in healthy and in damaged skin for those compounds most poorly absorbed in healthy skin.[6] Interestingly, in the present investigation, the enhancement of the cumulative amount of compound penetrated was paralleled by a shorter lag time for HC and AC, but not for IM and IB.

In a simplified, diagrammatic concept, penetration of xenobiotics takes place in different consecutive steps. First, a drug has to partition from the vehicle into SC, then from SC into the viable epidermis, and from there into the blood. In a heterogeneous structural model of the SC proposed by Michaels et al.,[23] SC is represented in the form of "bricks and mortar", with the keratinocytes being the bricks and the intercellular lipid bilayers being the mortar.[23,24] It is now generally agreed that the intercellular lipid bilayers are implicated in the barrier function of SC.

Aqueous solutions of SLS swell and disrupt SC; both lipid and protein structure are affected:[22,28-30] It is hypothesized that SLS uncoils and extends α-keratin structure resulting in spatial expansion and in increase in SC surface and SC thickness.[22,28-30] Insertion of SLS into the lipid structure may reduce the ability of the lipids to pack together resulting in a fluidization of intercellular lipids and may also remove lipids.[28,30] These mechanisms would explain the observed increase in diffusivity. Since the diffusional resistance of SC is greater to polar substances than to nonpolar materials[25-27] a disruption of this barrier should enhance penetration of hydrophilic compounds to a greater extent than the penetration of lipophilic compounds. The present data support this assumption. Whether SLS pretreatment also increased drug partitioning into SC cannot be answered with the present data since SC was not isolated from epidermis and dermis.

For topical therapy, it is important to deliver effective drug concentrations into the skin, whereas systemic absorption can be regarded as an unintended side effect. Comparing the permeability characteristics in irritated vs. healthy skin, it was found that skin concentrations were increased in irritated skin by factor 1.6 (IB) to 3.4 (IM) but unchanged for AC. Enhancement of skin drug concentration was lower than enhancement of flux for all compounds. As a consequence, the relative skin dose was lower in the SLS group for all compounds except for indomethacin.

V. CONCLUSION

Percutaneous penetration of diverse drugs was equivocally influenced by SLS-induced irritation. SLS-irritated skin was more permeable for all drugs studied; however, enhancement of cumulative absorption was higher than enhancement of skin concentration and enhancement was higher for hydrophilic compounds than for lipophilic. Our results suggest that topical drug therapy in irritant dermatitis should be done with special caution and with the awareness of probably increased systemic absorption. In addition, SLS, which is considered as a penetration enhancer for transdermal drug delivery sytems, may not be an appropriate additive to topical vehicle formulations to increase dermatological therapy.

ACKNOWLEDGMENT

A postdoctoral fellowship for K-. P. Wilhelm from the Deutsche Forschungsgemeinschaft (DFG, Bonn, Germany) is gratefully acknowledged.

REFERENCES

1. **Schaefer, H., Farber, E. M., Goldberg, L., and Schalla, W.**, Limited application period for dithranol in psoriasis, *Br. J. Dermatol.*, 102, 571, 1980.
2. **Turpeinen, M., Mashkilleyson, N., Björksten, F., and Salo, O. P.**, Percutaneous absorption of hydrocortisone during exacerbation and remission of atopic dermatitis in adults, *Acta Derm. Venereol. (Stockholm)* 68, 331, 1988.
3. **Frithz, A.**, Cellular changes in the psoriatic epidermis. IX. Neutron activation analysis of mercury in patients topically treated with ammonium mercuric chloride, *Acta Derm. Venereol. (Stockholm)* 50, 345, 1970.
4. **Maibach, H. I.**, Clonidine: irritant and allergic contact dermatitis assays, *Contact Dermatitis*, 12, 192, 1985.
5. **Lamaud, E. and Schalla, W.**, Influence of UV irradiation on penetration of hydrocortisone. *In vivo* study in hairless rat skin, *Br. J. Dermatol.*, 111 (Suppl 127), 152, 1984.
6. **Bronaugh, R. L. and Stewart, R. F.**, Methods for *in vitro* percutaneous absorption studies V: permeation through damaged skin, *J. Pharm. Sci.*, 74, 1062, 1985.
7. **Solomon, A. E. and Lowe, N. J.**, Percutaneous absorption in experimental epidermal disease, *Br. J. Dermatol.*, 100, 717, 1979.
8. **Osamura, H., Jimbo, Y., and Ishihara, M. J. D.**, Skin penetration of nicotinic acid, methyl nicotinate, and butyl nicotinate in the guinea pig, *J. Dermatol.*, 11, 471, 1984.
9. **Lambrey, B., Schalla, W., Kail, N., Lahmy, J. P., and Schaefer, H.**, Influence of an essential fatty acid deficient diet on absorption of topical hydrocortisone in the rat, *Br. J. Dermatol.*, 117, 607, 1987.
10. **Wilhelm, K. P., Surber, C., and Maibach, H. I.**, Quantification of sodium lauryl sulfate irritant dermatitis in man; comparison of four techniques: skin color reflectance, transepidermal water loss, laser Doppler flow measurement and visual scores, *Arch. Dermatol. Res.*, 281, 293, 1989.
11. **Wilhelm, K. P., Saunders, J. C., and Maibach, H. I.**, Increased stratum corneum turnover induced by subclinical irritant dermatitis, *Br. J. Dermatol.*, 122, 793, 1990.
12. **Agner, T. and Serup, J.**, Sodium lauryl sulphate in irritant patch testing (Abstr.), in Int. Symposium on Contact Dermatitis, Stockholm, Sweden, 1990.
13. **Bettley, F. R.**, The irritant effect of detergents, *Trans. St. Johns Hosp. Dermatol. Soc.*, 58, 65, 1972.
14. **Rietschel, R. L.**, Irritant contact dermatitis, *Dermatol. Clinics*, 2, 545, 1984.
15. **Van der Valk, P. J. M., Nater, J. P., and Bleumink, E.**, Skin irritancy of surfactants as assessed by water vapor loss measurements, *J. Invest. Dermatol.*, 82, 291, 1984.
16. **Nilsson, G. E.**, Measurement of water exchange through skin, *Med. Biol. Eng. Comput.*, 15, 209, 1977.
17. **Gummer, C. L., Hinz, R. S., and Maibach, H. I.**, The skin penetration cell: a design update, *Int. J. Pharm.*, 40, 101, 1987.
18. **Bucks, D. A. W., McMaster, J. R., Maibach, H. I., and Guy, R. H.**, Bioavailability of topically administered steroids by a "mass balance" technique, *J. Invest. Dermatol*, 90, 29, 1988.
19. **Zar, J. H.**, *Biostatistical Analysis*, Prentice-Hall, Englewood Cliffs, NJ, 1974, 101.
20. **Grice, K., Sattar, H., Sharratt, M., and Baker, H.**, Skin temperature and transepidermal water loss, *J. Invest. Dermatol.*, 57, 108, 1978.
21. **Mathias, C. G. T., Wilson, D. M., and Maibach, H. I.**, Transepidermal water loss as a function of skin surface temperature, *J. Invest. Dermatol.*, 77, 219, 1981.
22. **Scheuplein, R. and Ross, L.**, Effects of surfactants and solvents on the permeability of epidermis. *J. Soc. Cosm. Chem.*, 21, 853, 1970.
23. **Michaels, A. S., Chandrasekaran, S. K., and Shaw, J. E.**, Drug permeation through human skin, theory and *in vitro* experimental measurement, *A. I. Ch. E. J.*, 21, 985, 1975.
24. **Elias, P. M.**, Epidermal lipids, membranes, and keratinization, *Int. J. Dermatol.*, 20, 1, 1981.
25. **Scheuplein, R. J. and Blank, I. H.**, Permeability of the skin, *Physiol. Rev.*, 51, 702, 1971.
26. **Guy, R. H. and Hadgraft, J.**, Physicochemical aspects of percutaneous penetration and its enhancement, *Pharmaceut. Res.*, 5, 753, 1988.
27. **Kasting, G. B., Smith, R. L., and Cooper, E. R.**, Effect of lipid solubility and molecular size on percutaneous absorption, in *Skin Pharmacokinetics*, Shroot, B. and Schaefer, H., Eds., Karger, Basel, 1987, 138.
28. **Goodman, M. and Barry, B. W.**, Action of penetration enhancers on human stratum corneum as assessed by differential scanning calorimetry, in *Percutaneous Absorption, Mechanisms-Methodology-Drug Delivery*, 2nd ed., Bronaugh, R. L. and Maibach, H. I., Eds., Marcel Dekker, New York, 1989, 567.
29. **Rhein, L. D., Robbins, C. R., Ferne, K., and Cantore, R.**, Surfactant structure effects on swelling of isolated human stratum corneum, *J. Soc. Cosm. Chem.*, 37, 125, 1986.
30. **Imokawa, G., Sumura, K., Katsumi, M.**, Study on skin roughness caused by surfactants II. Correlation between protein denaturation and skin roughness, *J. Am. Oil Chem. Soc.*, 52, 484, 1975.

INDEX